MODERNISM'S INHUMAN WORLDS

MODERNISM'S INHUMAN WORLDS

RASHEED TAZUDEEN

CORNELL UNIVERSITY PRESS
Ithaca and London

First published 2024 by Cornell University Press

Library of Congress Cataloging-in-Publication Data

Names: Tazudeen, Rasheed, 1983– author.
Title: Modernism's inhuman worlds / Rasheed Tazudeen.
Description: Ithaca [New York] : Cornell University Press, 2024. | Includes bibliographical references and index.
Identifiers: LCCN 2023047130 (print) | LCCN 2023047131 (ebook) | ISBN 9781501776496 (hardcover) | ISBN 9781501776502 (pdf) | ISBN 9781501776519 (epub)
Subjects: LCSH: English literature —19th century—History and criticism. | Animals in literature. | Modernism (Literature)—English-speaking countries. | Metaphor in literature.
Classification: LCC PR468.A56 T39 2024 (print) | LCC PR468.A56 (ebook) | DDC 820.909034—dc23/eng/20231116
LC record available at https://lccn.loc.gov/2023047130
LC ebook record available at https://lccn.loc.gov/2023047131

To my mother, Farida Tazudeen, and to my grandmother,
Mehrunnisa Ariff (1936–2020)

Contents

Acknowledgments

I have come to learn that the word *earth* is also a verb. You can be earthed—given earth to inhabit, physically, imaginatively, and spiritually—through relations of care and kinship both with your fellow beings and with the lands and waters that have cherished and sustained you and your forebears. I have been earthed in so many immeasurable ways during this writing, by so many beings, lands, and ancestors. And I have been earthed by so many voices: the voices of my family; loved ones; human and nonhuman kin; my academic and spiritual mentors; ancestral voices; and the voices of forests, rivers, and oceans that spoke their mysteries to me when my heart was open enough to hear.

I wrote the first words of this book across the street from University Press Books in Berkeley, California, on a scrap of paper that has stayed with me throughout all of the various iterations of this work, a little seedling growing waywardly amidst so many different soils. And I wrote the last words twelve years later, in Nelahozeves, Czech Republic, at a small outdoor café next to the birthplace and ancestral home of Antonín Dvořák, a spiritual and musical mentor whose works—alongside those of Johann Sebastian Bach—have guided and held me through most of the years of my life. I went to Nelahozeves and, a week earlier, to Leipzig and to Köthen, Germany, to immerse myself in the sacredness of these lands and of the creations that emerged from them. I wanted, too, to give these works back, to use my violin as a means of planting them in the lands where they first arose, as offerings to the spirits dwelling there still. The next time I go back, *insha'Allah*, I will offer the scrap of paper with my first words on it to the waters and the spirits of the Vltava, a river that I have fallen in love with more than once, and a river whose being and creation I hope to commune with long enough to become a river like her. For I think of this book as one modest offering, given first to the Almighty, Most Merciful, and Cherisher and Sustainer of Worlds; given to one river and to many rivers; given to many soils, forests, and all of the other enigmatic materials of the earth; given to my family, my ancestors, my mentors, and all of my kin that have supported me with unfathomable generosity throughout my life.

I'll start with kin and, specifically, with four special friendships that have made both my work and my academic life possible: Seulghee Lee, Juno Richards, Sunny Xiang, and Yiren Zheng. Seulghee, thank you for being a constant source of inspiration and wild and beautiful thinking since the day we met as first-year graduate students (by a hot dog stand on the Berkeley campus), for all the late-night conversations, pick-up basketball games, and mid-week *Sopranos* marathons that made graduate life magical as well as bearable. Juno, thank you for always being there for me, for helping me to feel seen and valued, for giving me the emotional and practical support to be able to navigate hierarchies and exclusion and other unpleasant aspects of academic life, and for modeling generosity and care in addition to fiery and uncompromising brilliance in your work and thought. Sunny, I have been privileged to be able to develop ideas alongside you and to learn with and from you from our first year of grad school to the present. Thank you for your support and for your care during some of the most difficult years of my life, for your ongoing friendship, and for your oceans of intellectual and spiritual depth from which I continue to draw insight and inspiration. Yiren, thank you for restoring my faith in academia—and for continuing to sustain it—just by your way of being and thinking, your devotion to producing some of the most innovative and truly interdisciplinary scholarship I've ever read, and your willingness to create the best and most needed kinds of dialogue across disciplines and fields.

I owe profound gratitude to Elizabeth Abel, who has been there with this project from its first planting and has helped to cultivate and shape it in so many innumerable ways. Elizabeth, your Virginia Woolf graduate seminar was the ground from which this book sprouted, and all of our conversations since then have been its nourishing rain, and you its sky. Thank you for believing in this work and for believing in me from the beginning. So much of my thinking, too, has been inspired by conversations with Anne-Lise François and with fellow classmates/colleagues in Anne-Lise's biopolitics and ecocriticism graduate seminar, particularly Seulghee Lee, Adam Ahmed, and Juliana Chow. Anne-Lise, I feel like I had to relearn how to think ecologically—and how to think at all—after every seminar meeting you held or talk you gave and after every conversation we had. It is my deepest wish that this work reflects at least some of what your wondrous, enigmatic thought and being have breathed into it.

I have benefitted from the guidance of so many excellent mentors and colleagues at UC Berkeley, among them John Bishop, C. D. Blanton, Mitchell Breitwieser, Pheng Cheah, Spencer Coldren, Catherine Gallagher, Steven Goldsmith, Lyn Hejinian, Donna Jones, Manya Lempert, Laura Kolb, Rosa Martinez, Jocelyn Rodal, Michael Rubenstein, Charis Thompson, and Emily Thornbury. I also want to give thanks to the brilliant community of scholars that I had the good fortune to work with at the Jackman Humanities Institute at the University of Toronto. My sense of the interdisciplinary was (re)shaped in the most profound ways through conversations and other forms of collaboration with Alan Bewell, Alyson Brickey, Matthew Cohn, Rebecca Comay, Christopher Dingwall, Robert Gibbs, Elizabeth Harvey, Peter Jones, Atreyee Majumdar, Yan Liu, Michael Nicholson, John Paul Ricco, Noa Reich, and Erin Soros, among others. I also want to thank the students in my posthumanism seminar (and later, our posthumanism reading group) at Toronto, especially Miranda Alksnis, Kalila Badali, Anna Deregowski, Lana Maleva, Eric Schwenger, and James Watkins-Strand.

This work has also been sustained by the generosity and grace of many of my wonderful colleagues in the English department at Yale University. I would like to thank Jonathan Howard and Cajetan Iheka both for being amazing and inspiring scholars and for organizing the Yale Black Environmentalisms Symposium that completely altered my sense of what was intellectually, politically, and socially possible in both environmental studies and in the academy as a whole. I also want to thank all of the symposium participants, especially Allegra Ayida, Hannah Cole, Elizabeth DeLoughrey, Kassidi Jones, Elleza Kelley, Ernest Mitchell, J. T. Roane, Marisa Solomon, and Kathryn Yusoff for their brilliant work and for the conversations ongoingly inspired by this work. Thank you to Jill Campbell and Margaret Homans for their pedagogical and intellectual mentorship and for their commitments to inclusivity and antiracist teaching and thought. Thank you to Ben Glaser for reaching out to a fellow modernist even before I set foot in New Haven and for the sharing of ideas and pedagogies during and beyond our semester of coteaching. I could not have asked for more caring, gracious, and attentive department chairs than Jessica Brantley and Langdon Hammer, to whom I also give my deepest gratitude. And I am grateful to have known, worked with, and been amidst the brilliance of so many other colleagues in English as well as comparative literature; philosophy; and women's, gender, and sexuality studies at Yale, among them Serena Bassi, Felisa Baynes-Ross, Ardis Butterfield, Joe Cleary, Robin Dembroff, Anastasia Eccles, Marcel Elias, Greg Ellermann, Marta Figlerowicz, Alanna Hickey, Jill Jarvis, Heather Klemann, Greta LaFleur, Naomi Levine, Maia Lwin, Priyasha Mukhopadhyay, Catherine Nicholson, Stephanie Newell, Joseph North, John Durham Peters, Evren Savci, Arthur Wang, and Ruth Yeazell. And special thanks to all of the wonderful students that I have been privileged to teach in my seminars, Black and Indigenous Ecologies, Hearing the African Diaspora, and Global Sound, among others. You have enriched my understanding of ecology, sound, music, empire, and the possibility of a decolonial future, a future that you are all well on your way toward creating.

Almost every chapter in this book began its life as a conference paper, most often at the conferences held by the American Comparative Literature Association (ACLA). I am grateful for the ideas, conversations, collaborations, and friendships that emerged through the ACLA and other conferences over the years. And while I would need a far more capacious memory than the one I possess in order to mention all of the brilliant people with whom I have exchanged and refined ideas since the inception of this project, a provisional list would include Peter Adkins, Christina Alt, Valérie Bénéjam, Shane Butler, Katherine Ebury, James Ford III, Darrell Moore, John Mowitt, Anahid Nersessian, Sarah Nooter, Paul Saint-Amour, Derek Ryan, Fritz Senn, Kelly Sultzbach, Keston Sutherland, Michelle Witen, and Dora Zhang.

Since she always likes to be in the middle of everything, here is probably the best place to give gratitude and love to Fuga, my amazing dog (and true dog-philosopher, in the lineage of the philosophizing dogs of Kafka's and Woolf's imaginations), who has been a constant, loving companion to me during almost all of my writing sessions for this book. Thank you for filling my life with your warmth and joyfulness and calm energy and love of the earth; for accompanying me almost everywhere I went to write; for always insisting on sitting together in the grass at Yale cross-campus; and for the countless hours we spent (and continue to spend) thinking, learning, and imagining alongside one another.

Blessings and all of the love I have to give go to my caring, supportive, and endlessly generous family: my parents, Farida and Vathiar Tazudeen, my brothers, Noor and Wahid Tazudeen, my sister-in-law, Stephanie Wang, and the newest members of our kin—my niece, Sara Tazudeen (a little over two years old now), and nephew, Noah Tazudeen (one month old). Sara, I don't know when you will be old enough to read this book, but I hope that whenever the time comes, you find it at least half as interesting as the cat and owl books that you currently devour on a daily basis. I think you were only a year old when you fell in love with the night sky, and I've been inspired by the infinitude of your being and feeling ever since. I want to love like you love the moon, the way your world stops when the moon appears to you, the way you glorify its every appearance with song, with rapture, your being arched toward it, its light your light. Noah, I pray every night for a beautiful future for you and your sister, and I will never stop believing in the world that your hands and heart will help to shape.

To Sandra Huber, thank you for helping me to feel and to be amidst enchantment, for the ritual and magic that we shared, for your spiritual and intellectual dedication to the unseen forces of this world, and for your worship of the night and its beings. You are, now and always, a field of flowers in communion with the stars, the way your feet absorb light from the cosmos while your hair sews stars into the soil.

I could not have been earthed—in body, imagination, or spirit—during this writing without the support of Alexandra Lambertson. Thank you, Alexa, for the joy, love, and music we shared amidst what felt like an ending world during the pandemic. I couldn't have written a single word without you, couldn't have found my own breath or felt the ground I walked on without your generosity, care, and empathy and your beautiful music resonating in every room of all the places that we lived in as this work took shape. The breath you give to your instruments and to the world is sacred to me, and may it always bring joy and nourishment to the beings around you; may it always make fertile the earth that hears you.

I dedicate this work—along with my entire being—to my mother, Farida Tazudeen, to the memory of her mother, Mehrunnisa Ariff, and to the spirits of all of my ancestors. Mom, I often meditate on your name and its syllables made from the same substance as the corals of the oceans you crossed to give me life. Everything I do and everything I create is maybe just my attempt to find my way back to the rhythms and tonalities—the infinitudes—of the first Tamil *juju* songs you sang to me. My first memory was your voice, the first measure of your song where my world began and where it remains still.

MODERNISM'S INHUMAN WORLDS

Introduction

What Rhymes with Odradek?

What does metaphor gather, and what gathers in metaphor? We might ask the same knotted questions of the term "species," which is also a means of giving or transferring meaning to the diverse beings that it holds in relation across time. In its classical origins, metaphor (*metaphora*, μετάφορά) is a "carrying over"—from the combination of the Greek *meta* (μετά, "between") and *phero* (φέρω, "to carry")—of one framework of conceptual meaning to another, or what I. A. Richards calls a "transaction between contexts."[1] In the *Poetics*, Aristotle writes, "Metaphor [μεταφορά, *metaphora*] consists in giving [επιφορά, *epiphora*] the name [όνομα, *onoma*] that belongs to something else; the transference being either from genus to species, or from species to genus, or from species to species, or on grounds of analogy."[2] Sedimented within Aristotle's definition, however, murmur other logics. Metaphor gives name, meaning, place. Yet metaphor's giving is also a gathering, or an *epiphora*, which is the word that Aristotle uses for "giving a name." *Epiphora*, as John T. Kirby notes in "Aristotle on Metaphor," shares a similar root with the *phora* ("carry") of *metaphora*. But instead of referring to a transfer or carrying across, "*epi*- as a prefix may designate movement over or beyond boundaries" as well as "a sense of accumulation of one thing *over* or *besides* another."[3] Thus, "*epipherein* may mean to put, or pile, something on top of something else."[4] And the noun *epiphora* may mean both "an additional payment" and "the new or additional designation of a(n) (unusual or unaccustomed) name to something that already has a(n) (ordinary) name."[5] At what may be its (Western) philosophical origins, metaphor is noncoincident with itself, marked by a split between the functions of giving names (*metaphora*) and gathering beings (*epiphora*), or between meaningful transference and agglutinative profusion. And yet this split, this dual striving, as we will explore throughout this book, *is* metaphor. Metaphor names the excesses gathered into itself, along with its own inability to give name or place to what gathers. So we return to our questions, now become an irresolute *ainigma*, a knottedness insisting through modernism's varied ecologies, worlds, and planetarities: what does metaphor gather, and what gathers in metaphor?

Gathering's (il)logics themselves are strange, and every gathering is potentially a riddle. In *Black Gathering*, Sarah Jane Cervenak opens with the thought of gathering as "either (or simultaneously) noun and verb, a collectivity or state" as well as "a fugitive movement in which togetherness is unmoored from the enclosures of category and purpose."[6] In this sense, metaphor's gathering offers another modality of the social in which beings gather (and are gathered) outside of the logics of the concept. This book proposes *multispecies metaphor* as an eco-trope, or an analytic of planetary gathering, through which to think the inhuman socialities made possible in and through metaphor.

Everything that I want to say—on the inhumanist and planetary dimensions of modernism, metaphor, and species—can be read as sought attunement to the *ainigma* of one peculiar, knotted being: Kafka's Odradek, from the miniature, "The Cares of a Family Man" (*Die Sorge des Hausvaters*, 1919).[7] Odradek—a thrown-together, yet sentient patchwork of thread, "star-shaped" spool and asymmetrical wooden rods—unspools in fugitivity from concepts. For Odradek is "*sinnlos*": sign-less and senseless, inhuman and enigmatic in his incommensurability with the laws of *Haus* and *Vater*, or the structures of the human and the family that he inhabits while remaining undeterminable by and unanswerable to.[8] Yet Odradek also weaves other worlds into existence through the creative expression of his being, un- and respooling his "knotted and tangled together" bits of motley thread (*aneinander geknotete, aber auch ineinander verfitzte Zwirnstücke*) through and beyond the time and space of the home, laughing with a sound like the "rustling of falling leaves," "lurking" (*sich hälten*) under stairs and in corners of the house that he makes into dwellings.[9] Odradek's "aim"-less (*Ziel*) and alogical weaving/unweaving, which is not subject to the same temporal constraints and gradual degeneration as human "activity" (*Tätigkeit*), is the primary anxiety or care (*Sorge*) of the *Hausvater*: "He does no apparent harm to anyone," the man muses, "but the idea that he is likely to survive me [*überleben*] I find almost painful."[10]

It is the ongoingness of what insists beyond—and what remains oblique to—humanistic determination toward which *Modernism's Inhuman Worlds* strives. How do we imagine (with) such knotted beings—beings gathered, along with everything that is gathered into beings—that resist epistemological capture, worlding indifferently to the conceptual structures of time, family, home, and lineage? In Kafka's *Zürau Aphorism* 57, "language can only be used very obliquely [*andeutungsweise*] of things outside the physical world [*der sinnlichen Welt*] . . . since all it knows how to do—according to the nature of the physical world—is to treat of ownership [*Besitz*] and its relations."[11]

Language, in one sense, is nothing more than a mechanism by which to possess (land, beings, time) and to enact relations of ownership. Yet from language's modality as an otherwise- or oblique (*andeutungsweise*) speaking, there emanates what Paul North calls a "magical language"—or a language of "hinting"—"that would not possess but abandon."[12] One needs a mode of thought, a manner of saying, weak enough to gather multitudes without possessing them. We might also call this language a language of riddle or *ainigma* in which what is named retains an obliqueness to concepts and insists beyond determination. As *ainigma*, Odradek is both a bounded, individualized being *and* an unpossessable living-on (*überleben*) of manifold histories, futures, expressivities (inhuman laughter among them) unbound by—and unspooling beyond—humanistic thought and time

Like metaphor itself, then, Odradek is simultaneously an identity and a multitude, a named entity as well as an epiphoric profusion of names. Even the name "Odradek," as the narrator notes in the miniature's first sentences, is shrouded amidst untraceable linguistic origins: Slavonic perhaps, or Slavonic-inflected German, or neither? "The uncertainty of both interpretations [*Deutungen*]," we are told, "allows one to assume with justice that neither is accurate, especially as neither of them provides an intelligible meaning [*Sinn*] of the word."[13] In the *Rhetoric*, Aristotle claims that "metaphors speak in riddles [*metaphorai . . . ainittontai*],"[14] and in the *Poetics*, "the very nature of the riddle [*ainigma*, αἴνιγμα] is this, to describe a fact in an impossible combination of names."[15] Metaphor, and Odradek as metaphor, enigmatizes not by concealing but rather by providing too many identifications, more names and more truths than can be contained (or possessed) within a given order of sense or meaning. The riddle (or ur-*ainigma* even) of the Sphinx is only solvable with the name of Man from within the terms of a particular humanism rigidly bound to the dogma of being's singularity and determinability. Maybe there exists a metamorphic being—or a being that is a gathering of many beings and many temporalities—or some other humanistically unimaginable mingling of time and form that has four legs at dawn, two legs at noon, and three at night.

No riddle, including the Sphinx's, is fully solvable, as Gilles Deleuze argues in *Difference and Repetition*. Every solution works only within the determinate conditions that the solution itself—as a response grounded in a particular set of spatial, temporal, and epistemological coordinates—sets in place. The enigmatic or "mad" core of the riddle, however, persists beyond each and every determination: "It may be that there is necessarily something mad in every question and every problem, as there is in their transcendence in relation to answers, in their insistence through solutions[,] and the manner in which they maintain their own openness."[16] In this sense, the nature of the riddle is also, perhaps, its ongoingness: its Odradekian insistence through all determinations, its capacity to transform the given reality through which it is both approached and given provisional solution.

In Darwinian thought, "metamorphic being" is a redundancy, since every being or species is reshaped in and through relations with other beings, environments, and worlds over time. In *On the Origin of Species* (hereafter *OS*), as we will explore in chapter 2, Darwin envisions the possibility of a "race of [swimming] bears" becoming so habitually aquatic that they transform into a "monstrous," "whale"-like species over the course of geo-evolutionary time (*OS* 210). A being metamorphoses from having four legs to no legs (and two flippers). (Interestingly, a set of paleontological discoveries in 2008 revealed that the first whales *did* have legs before they evolved to swim more effectively in water, after which their front legs gradually became flippers and their back legs and hips disappeared. Morphological *ainigma* of a being that, like all beings, is composed of both a historical series of shifting forms *and* agitations against the enclosures of form.)[17] The inter-implication of time and form is also what informs Deborah Bird Rose's philosophy of "multispecies knots of ethical time," which proceeds from Aboriginal eco-cosmologies[18] to arrive at a similar understanding of the relational character of species. "Every creature," writes Rose, "has a multispecies history—it came into being through its own forebears and through others. Each individual is both itself in the present, and the history of its forebears and mutualists."[19] Time is embodied in beings, and each being is a knot of embodied time, shrouded in the excesses of its own histories and futures.

The trope of multispecies metaphor, I want to argue, gives a means to imagine (alongside) the enigmatic knottedness of beings in which every being is simultaneously its bounded self and all of the relations that have co-composed a self: a named entity and an entity oversaturated with names, temporalities, threads. In its dual orientation toward both giving and gathering—the transfer of identity (*metaphora*) and the sedimenting or piling up of names (*epiphora*)—metaphor itself is a knot that helps us to think toward the knottedness of being, or what Rose calls "a wider world of multispecies knots."[20]

We might imagine modernism itself as such "a wider world"—or perhaps many worlds—"of multispecies knots." The ongoing indeterminacy, or *Sinnlosigkeit*, of Odradek in this context limns the inhumanism of both metaphor and of a modernism that is routed through relations between species and metaphor. In Susan Stanford Friedman's planetary modernist analytic (which we will revisit in more detail shortly), the term "planetary" is meant to gesture toward the ecological and multispecies valences of a vastly temporally and spatially rescaled modernism. "*Planetary*," writes Friedman, "suggests the Earth as a place of matter and climate, life and the passage of time, and an array of species of which the human is only one" and "opens up the possibility of thinking about nonhuman modernities or the interconnections of the human and nonhuman in rethinking modernity and modernism."[21] The planetarity of modernism(s), then, is animated by strivings toward modes of thought and expression through which to imagine the enigmatic relations of beings with selves, times, other beings, the earth, and finally, the cosmos.

For reasons that will become clearer in the next section, I want to shift Friedman's terms slightly to think toward an *inhuman* modernism, rather than a nonhuman modernism or what might also be called a posthuman modernism.[22] The category of the nonhuman, as Claire Colebrook and Jami Weinstein have argued, tends to identify beings in terms of an opposition to the still-centered figure of the human. The term "posthuman," similarly, remains bound to the humanist temporal and conceptual frames that it seeks to "surpass."[23] The inhuman, however, "orients us to all that is not human, not just that which comes after the human. It also pushes us to scales beyond the human—temporalities and spatialities both deep and astronomical."[24] This line of thinking is based on the idea that because most humanisms have articulated themselves *as* a kind of transcendence of the limits of the human, humanism itself may already be a posthumanism. As they write in the preface to the same volume, "The human (if there is such a thing) has, more often than not, imagined itself by way of what surpasses the human (whether universal rationality, life, or spirit)."[25]

Branching from these definitions, I also want to think the inhuman through its ongoing, Odradekian *indifference* to humanism and its determinants, as that which cannot be situated within traditional humanist organizations and distributions of being, yet also rolls, rustles, lurks, and weaves restlessly across these and all structures. Inhuman relations and world-makings are thus only expressible through otherwise speakings, or in our analysis, a language and a metaphorics of riddle. In this vein, Kathryn Yusoff theorizes the inhuman in *A Billion Black Anthropocenes or None* as "a call across categories, material and symbolic, corporeal and incorporeal, intimacies cut across life and nonlife in the indifferent register of matter."[26] The inhuman thus also gives a means to imagine *with* multispecies, other-than-human worlds—including the worlds still to be made on the other side of mass extinction—without recourse to humanistic conceptualities. As we will explore in chapter 6, Alexis Pauline Gumbs's hybrid ecopoetic / "speculative documentarian" work *M Archive: After the End of the World* theorizes through a strangely temporalized "after." Gumbs's after indicates both the aftermath of climate apocalypse and mass extinction as well as an after *to* the end of the world: an imagining of the futurities emerging from an ended world and a grasping toward a form fluid enough to think them.[27] In this sense, the after is also a means of writing with and alongside the end of what is called world, an ending that is not singular but multiplicate and endlessly recurrent across the vast, overlapping temporal and spatial scales of an inhuman modernist imaginary.

In six chapters, *Modernism's Inhuman Worlds* tracks the oblique speakings and expressivities of metaphor—as an inhumanist, enigmatic gathering of beings—across modernist and contemporary metamodernist literatures. We will think with—and in the early words of Virginia Woolf, "lie on the earth" with—a kinship of inhuman modernists stretching from Charles Darwin and Lewis Carroll to Alexis Pauline Gumbs and Cody-Rose Clevidence, among other contemporary poets. Throughout, we will stay attuned to questions of the unbounding of time, space, and world as given in critical accounts of both modernist / metamodernist and Anthropocenic periodicity.[28] After a theoretical foray into multispecies metaphor's (anti-)lineages in chapter 1, I wager in chapter 2 a set of provisional origins for an inhuman modernism in the ungrounding of humanist logos and the opening to the nonsensical growths and gatherings of species in Darwin's *Origin* and Carroll's *Alice* books. Proceeding from this unground and the nonsense worlds grown from it, chapter 3 explores what we will call a metaphorics of contagion, made amidst the resonances between Gustave Flaubert's *La tentation de Saint Antoine* (The Temptation of Saint Anthony, 1874) and the "Circe" episode of James Joyce's *Ulysses* (1922). Contagion in both works becomes a means of epiphoric gathering that undermines humanistic logics of individuated being and form by giving world to otherwise joinings of bodies, forms, earths, and the expressive materials of (aesthetic and earthly) creation itself.

In chapter 4, we turn to / with Kafka's oblique articulations of an inhumanist language and metaphorics through the earth-philosophies and earth-faiths of the animal narrators of *Der Bau* (*The Burrow*, 1924) and *Forschungen eines*

Hundes (*Investigations of a Dog*,1924). Both *Hundeschaft* (dog-philosophy) and what we might call *Bauschaft* (burrow-philosophy) are modes of devotion to earth rooted in an epistemology of abandon to—rather than possession of—what is enigmatically given by the ground: sounds, shelter, food, knowledge, meaning, and other forms of nourishment. As praxes, comportments, faiths, and philosophies, *Bauschaft* and *Hundeschaft* gesture toward the possibility of a language that begins from the ontological sociality of being in relation to earth ("*dem warmen Beisammensein*" in the dog-philosopher's words)[29] and insists beyond conceptual and other determinations. With Virginia Woolf's *Between the Acts* (1941) and her unfinished manuscript "Anon" (1941) in chapter 5, we reach a point where inhuman modernist and contemporary Anthropocenic imaginings of apocalypse commune. Woolf's late aesthetics of impersonality speak to the ongoing (self-)expressivity of the earth beyond the extinctions anticipated—the potential fascist annihilation of Europe in 1938–41; global climate catastrophe in the present—in comic indifference to humanistic time and thought. What comes after humanism, Woolf's last works ask, and how to imagine what emerges after oblivion through epistemologies that have only ever been humanist? In what I call the inhuman comedy of *Between the Acts*, Woolf modulates thought of human extinction into the ongoingness of the earth's noise- and color worlds, the enigmatic resonance still of what a young Virginia Stephen once called an earthly "sound like a vast breath, as though it were the very inspiration of earth itself, & all the living things on her."[30] We conclude amidst the metamodernist ecopoetics of the present age, which become for us a means of attunement toward what can only be given in the modality of riddle—earth sounds and earth breath and everything gathered into the time and world of the earth's inhuman breathing—in and through the apocalypses to come. Alexis Pauline Gumbs, Cody-Rose Clevidence, Anne Waldman, and Sawako Nakayasu, among other contemporary poets discussed in chapter 6, draw from the experimental and creative energies of modernisms past in order to cultivate inhumanist grammars, geographies, and metaphorics by which we might philosophize (with) extinction and post-extinction worlds.

I.

Finding a means to think the gatherings of beings outside of traditional modes of representation—primary among these the humanistic-conceptual—has long been a preoccupation of both multispecies and modernist studies.[31] The question has taken on greater ethical urgency as we find ourselves amidst an age of mass extinction theorized variously as Anthropocene or as one or more of its many reconfigurations.[32] As Rose reminds us, what is called Anthropocene is not only the potential erasure of beings and worldwide ecosystems, but also the elimination of a diverse multiplicity of inhumanly embodied and practiced temporalities and forms of time: an extinction of beings as well as times. W. B. Yeats's "The Song of Wandering Aengus" (1899; originally "A Mad Song"), a poem that speaks simultaneously in mythological and multispecies registers, is strangely prophetic on this theme. Aengus wanders into a hazel wood, where with a carved hazel wand he fishes in a stream and catches a silver trout that metamorphoses into "a glimmering girl / With apple blossom in her hair," a girl who is then etherealized into the "brightening air." In seeking this ungraspable *ainigma*—of a figure in whom is amalgamated sea, land, and air; a being that twice calls Aengus "by my name" yet cannot itself be named—Aengus also gives his being over to devotional strivings toward the earth and the cosmos themselves. He promises to "find out where she has gone,"

> And kiss her lips and take her hands;
> And walk among long dappled grass,
> And pluck till time and times are done,
> The silver apples of the moon,
> The golden apples of the sun.[33]

We might see in Aengus's devotions a hinting toward an apocalyptic erasure of both "time and times," or the end of human(ist) time and the end of the imaginable times of all beings. Unlike Odradek's, however, Aengus's strivings seemingly cannot persist beyond the limit point of human extinction. Yet we might also read the final two lines in terms of a cyclical renewal or reiterativity of cosmic time (from moon back to sun) and of the time of Aengus's plucking. Harvest that is also the making of ongoing relations between earth and sky, an enigmatic communion between earth's grass and the apples of the cosmos.

Something of the enigmatic, too, inheres within early critical formulations of multispecies studies, as in Thom van Dooren, Eben Kirksey, and Ursula Münster's in *Multispecies Studies: Cultivating Arts of Attentiveness*:

> A multispecies approach focuses on the multitudes of lively agents that bring one another into being through entangled relations that include, but always also exceed, dynamics of predator and prey, parasite and host, researcher and researched, symbiotic partner, or indifferent neighbor. But these larger contexts are not mere environments in the

> sense of a homogeneous, static background for a focal subject. Rather, they are complex "ecologies of selves," dynamic milieus that are continually shaped and reshaped, actively—even if not always knowingly—crafted through the sharing of "meanings, interests and affects," as well as flesh, minerals, fluids, genetic materials, and much more.[34]

Every gathering a riddle. Entangled beings resist determination through an excess of material, dispositional, and affective sharing, or what Dooren et al. call a collective "cobecoming."[35] A multispecies or inhuman modernist practice, I want to argue, is a means of dwelling in and with the *ainigma* of the undetermined gathering that is both every being and every relation among beings.

Building from multispecies thought, Anna Tsing, Heather Swanson, Elaine Gan, and Nils Bubandt emphasize the (non)category of the monster as a site of ecological gathering irreducible to humanist logics of the determinability and asociality of being. In *Arts of Living on a Damaged Planet*, the monsters of the Anthropocene—and the monsters that persist beyond Anthropocene—are given as "category-crossing beings," whose joinings compose worlds through means other than the conceptual.[36] In coral reefs, for example, are amalgamated polyps, algae, sponges, calcium carbonate shells, and other relationally constitutive organic and inorganic entities: a coral reef is simultaneously a bounded individual and a many-concepted monster. The first English usage of "monster" is from Geoffrey Chaucer's *Monk's Tale* (c. 1375), and its original definition might well apply to coral reefs and other contemporary ecological monsters: "a mythical creature which is part animal and part human, or combines elements of two or more animal forms."[37] In this sense, the monstrous is like the enigmatic in that the (il)logics of both are accretive, working by means of an uncontrollable profusion (of names, beings, times, and bodies), in which what emerges is irreducible to the grasp of the concept.

Appropriately, Immanuel Kant in the *Critique of Judgment* (1790) defines the monstrous as an excessiveness that undermines conceptuality: "An object is *monstrous* if by its magnitude it nullifies the purpose that constitutes its concept."[38] More broadly, Tsing et al. press on the inorderability of monsters within European Enlightenment modes of thought. The identification of monsters with "the irrational and the archaic" made them

> abhorrent to Enlightenment ways of ordering the world; sometimes they were classified as things of the devil, the antithesis of godly purity. . . . For later thinkers, rationalization meant individuation, the creation of distinct and alienated individuals, human and nonhuman. The landscape-making practices that followed from these new figures imagined the world as a space filled with autonomous entities and

> separable kinds, ones that could be easily aligned with capitalist fantasies of endless growth from alienated labor.[39]

Like riddles, monsters are otherwise, anti- and ante-logical gatherings of beings and times—or "impossible combinations of names"—that must be determinately solved and supplanted for a particular humanist order to establish itself. Perseus founds Mycenae upon the corpse of Medusa; Oedipus achieves dominion over Thebes by speaking the name of Man that also marks the murder of the Sphinx.

Extending this critique further into (as well as beyond) the Anthropocene, Colebrook and Weinstein's "critical life studies" approach maintains that both humanism and the human itself only come into epistemological being through a preexisting relational field of "inhuman" forces and actors, such as "animals, bacteria, viruses, plants, technology, and virtuality."[40] Similarly, Karen Barad grounds inhuman relationality in the principles of quantum physics. Entanglement is defined not as a relation between two (or more) independently existing beings but rather as the "mutual constitution" of beings in the space of their relation, or what Barad calls "intra-action."[41] Writes Barad in *Meeting the Universe Halfway*, "To be entangled is not simply to be intertwined with another, as in the joining of separate entities, but to lack an independent, self-contained existence. Existence is not an individual affair. Individuals do not preexist their interactions; rather, individuals emerge through and as part of their entangled intra-relating."[42]

What these and other multispecies studies approaches collectively suggest, perhaps, is that we do not have a language through which to think gathering outside of humanistic logics. The grammars of *ainigma*, monster, intra-activity, and entanglement, among others, do not cohere into something as authoritative as a language. However, it is the weakness of their bonds, the weakness of their gathering and articulative practices, that helps us to attune to the question of what else happens in worlds when worlds are no longer exclusively apprehended through human(ist) conceptualities and the possession-character of our languages.

As an example, we do not yet have a language—or even a fully satisfactory scientific conception—for the strange physics and stranger semiotics of lightning. In Barad's analysis, lightning (a "queer phenomenon") emerges from a "nonlocal communication" in the form of a multidirectional exchange of electrons between a storm cloud and the earth. "Objects on the ground" seem to be "hailed by the cloud's interpellative address," co-materializing a ground-cloud relation in the form of a lightning flash.[43] Lightning, then, is a gathering of inhumanist acts of address and material-affective sharing, or the grammatical field through which earth and sky momentarily commune. Another kind of (concept-indifferent) knowing and worlding arises here: a hermeneutics of matter(ing) that Barad refers to as "part of the world making itself intelligible to another part."[44] And we might see in such communions a partial answer to the question of what else happens—or happens otherwise—in worlds no longer humanistically determined.

A multispecies and more-than-humanist ethics, as Barad, Tsing, Stacy Alaimo, Bruno Latour, Eben Kirksey, Donna Haraway, and others have argued,[45] begins here, through a deeper attentiveness to the co-becoming of beings and materials with one another. As Alaimo writes in *Bodily Natures: Science, Environment, and the Material Self*, "Potent ethical and political possibilities emerge from the literal contact zone between human corporeality and more-than-human nature. Imagining human corporeality as trans-corporeality, in which the human is always intermeshed with the more-than-human world, underlines the extent to which the substance of the human is ultimately inseparable from 'the environment.'"[46] Similarly, for Barad, thinking the entanglement and shared corporeality of beings is a means of grasping toward a perhaps impossible to achieve ecological justice: "Justice, which entails acknowledgment, recognition, and loving attention, is not a state that can be achieved once and for all. There are no solutions; there is only the ongoing practice of being open and alive to each meeting, each intra-action, so that we might use our ability to respond, our responsibility, to help awaken, to breathe life into ever new possibilities for living justly. The world and its possibilities for becoming are remade in each meeting."[47]

At the same time, Eva Giraud's *What Comes after Entanglement?* draws important attention to the potential limitations of an ethics of multispecies entanglement. Writes Giraud, "While emphasizing the complex, more-than-human entanglements that constitute lived reality has proven politically and ethically important, such an approach also carries dangers. . . . Irreducible complexity, in other words, can prove paralyzing and disperse responsibilities in ways that undermine the scope for political action."[48] As a counter, Giraud wagers an "ethics of exclusion" that "pays attention to the entities, practices, and ways of being that are *foreclosed* when other entangled realities are materialized."[49] On the example of slugs needing to be kept out of gardens for both slugs and plants (and gardeners) to thrive, Giraud claims, "The act of excluding certain relations is precisely what creates room for others to emerge, or for existing forms of life to be sustained."[50] It is also what draws us toward better forms of accountability with regard to "which worlds are materialized over others" as well as the modes of power that determine these materializations.[51]

Instead of aiming toward resolution, we might instead think exclusion/entanglement in terms of a further knottedness to argue that exclusion perhaps *is* a form of relation among beings, and entanglement inseparable from modes of violence, appropriation, and unequal power relations among beings. Both approaches seek a way to think species together in and through Anthropocene, even if this togetherness involves their spatially and temporally separated thriving, and even if shared corporeality is complicit in the hastening of mass extinction.[52] In problematizing or further knotting the logics of both exclusion and entanglement, we should also complicate the idea that there is something like a singular ethics to be made or extracted from either relationist or exclusionist approaches. What we are after is perhaps something other than or oblique to an ethics, especially if the ethical is considered in terms of a prescriptive politics or program. At the same time, it also matters how and in what contexts the ethical is being mobilized. For Giraud, an exclusionist ethics is meant to provide a grounding for ecological activist practices and the work of carving out livable worlds. Giraud notes, too, that the ethicality and the ethical decisions stemming from exclusion "need to be temporary, contingent, and open to contestation to ensure they do not congeal in ways that allow normative social relations to simply reimpose themselves and reinscribe existing inequalities."[53] In this sense, both Barad's intra-activity (alongside other relationist theories) and Giraud's exclusion seem to share a sense of the malleability and indeterminacy of any ethical stance as well as an openness to the (entangled-excluded) worlds created anew in each ethical encounter.

With this (non)structure of simultaneous or knotted exclusion-entanglement in mind, we return to our opening *ainigmata* of how to think worlds—along with what happens and what gathers in worlds—inhumanistically. Jakob von Uexküll's definition of *Umwelt*, most often translated as "environment," speaks with us here: *Umwelt(en)* names both a multiplicity of worlds in relation *and* the incommensurability of worlds to one another. For Uexküll, in *A Foray into the Worlds of Animals and Humans* (*Streifzüge durch die Umwelten von Tieren und Menschen*, 1934), the "illusion" that all relations among beings "play out in the same space and time" that humans inhabit "is fed by the belief in the existence of one and only one world, in which all living beings are encased. From this arises the widely held conviction that there must be one and only one space and time for all living beings."[54] Every being's world, or "bubble," is a unique structure of perceptive relations to one's outside, or a living interpretation of an environment, that cannot be totalized into a singular and universal idea of world: "The bubble represents each animal's environment and contains all the features accessible to the subject. As soon as we enter into one such bubble, the previous surroundings of the subject are completely reconfigured. . . . A new world arises in each bubble."[55] The enclosed bubble mediates relations to other *Umwelten*, as worlds are recomposed in and through other worlds.

Uexküll theorizes relationality among worlds in part through a comparison with musical tonality. Each world is characterized by a unique tonal system and grounded by an "*Ich-ton*" (self-tone) through which each being makes meaning out of its environs. In other words, beings encounter environmental particulars as varied tones—including, for example, a "sitting tone," a "dwelling tone," an "obstacle tone," a "climbing tone"—sounding in relation to the tonic note of the self. Every being sounds a different world through its unique tonality, and every being and object is sounded differently into the worlds of other beings.[56] The metaphor of tonality provides measure for thinking the enigmatic relationality of worlds in which worlds remain fundamentally untranslatable (in)to one another, even as they co-sound through the expression—and attempted modulation—of obliquely exchanged tones. Kafka's burrow (to be visited in chapter 4) may be the result of the collapse of such a tonal system: without a self-tone or tonic-I by which to mediate environmental relations, the burrower finds himself amidst an atonal noise world sounding only the continual unselving of being.

An inhuman modernism may then be the thought and the possibility of holding worlds in (weak) relation, without determining or ordering the arrangements of time(s), extension, tone, and perception variedly gathered into every world. In *Animal Subjects: Literature, Zoology, and British Modernism*, Caroline Hovanec argues that Uexküll's granting of worlds to animals opens a "quandary" that drove modernist authors "to the very limits of literary and scientific representation": "one must understand animals as subjects, one cannot know animals' subjective experience."[57] And yet, subjectivity may be too strong of a form through which to think what is gathered in and amidst worlds, as Hovanec's account also seems to indicate, and the (Euro-American) modernist strategies of defamiliarization and perspectival plurality can in this way be read as a means of attending to what in worlds exceeds subjective experience.[58] Rather than abandoning the subject outright, however, a multispecied analytic maintains beings as both selves and relations—or living, branching presents along with sedimented histories—simultaneously.

And it is this knottedness, or tonal cacophony, of worlds that Martin Heidegger tries to dispel in his well-worn gloss of Uexküllian world as a humanistic possession. Heidegger famously singularizes world, restricting it to "world-building" (*Weltbildung*) humans in contrast to the "world-less" (*Weltlos*) stone and the "world-poor" (*Weltarm*) animal. In *The Fundamental Concepts of Metaphysics: World, Finitude, Solitude* (1938), Heidegger defines "world" in terms of what he calls an "as-structure" (*Als-Struktur*) in which only those who can encounter "beings *as* beings" can be said to have a world.[59] "The animal," then, "belongs within an environment but not a world" since although it has "access to what is around it and about it," it is nonetheless "confined to its environmental world, immured as it were within a fixed sphere that is incapable of further expansion or contraction."[60] Whatever an animal relates to, in other words, is "not accessible to it as a being" but only as a nonconceptualizable phenomenon.[61] Having/building world or being impoverished or deprived of world hinges on this "as-structure": "the 'as'-structure is an essential determination of the structure of the world," writes Heidegger.[62] And an as-structure, in turn, hinges on the capacity to think and to interpret conceptually. In Heideggerian philosopher William Blattner's terms, the as-structure "requires a certain kind of conceptuality, a fore- or proto-conceptuality" in order to make an entity legible within the larger framework of a world.[63] Following Blattner's analysis, we can think of the as-structure as not just a conceptual edifice but a metaphorics as well: seeing beings *as* beings depends on the work of a particular sensus communis that determines what constitutes knowledge (of a being or of anything else) and which kinds of relations contribute to the composition of a world and which do not. A particular world, as a contingently embodied set of temporal and spatial relations, passes into universality.

World-deprived stones can "touch" the earth they lie on, claims Heidegger, but this inhuman "touching [*berühren*]" "is not at all like that relationship which the lizard has to the stone on which it lies basking in the sun. And the touching implied in both these cases is above all not the same as that touch which we experience when we rest our hand upon the head of another human being."[64] But what if a world or many worlds could be composed of such inhuman relations: the touching, or *berühren* ("to touch" but also "to be affected"), that marks the contingency and co-exposure of beings both to one another and to their variously toned environs? ("Contingency," like "contagion," has its origins in the Latin *con* + *tangō*: "to touch together": it is a term definitionally grounded in the body and its relations.) What if a differently imagined as-structure could hold in common rather than transcendentally set apart the relations between stone and ground, lizard and stone/sun, and human hand and human head? And what new languages, metaphorics ("as-structures"), tonalities, and grammars might arise in response to the *ainigma* of what gathers amidst the material-affective touching together of worlds?

The Heideggerian as-structure becomes humanistic resolution to this *ainigma*, a means of determining the more ambivalent metaphoricity of tonality—along with other oblique modalities of relation—with the name of Man. Modernism's inhuman metaphorics, this book will argue, seek instead after socialities no longer beholden to this or any other metaphysical ground. An ungrounded and multispecied metaphorics of world might articulate new dimensions of the "as" beyond the mere identification of a being as coincident with itself (or with its own essence): a means toward experiencing beings not just *as* beings but also as monsters, as simultaneously excluded-entangled selves and communions, as swarms of sensuous and sounded particulars that are irreducible to conceptuality but no less constitutive of worlds.

II.

The question of what constitutes a world, for Susan Stanford Friedman and other theorists of modernist planetarity, is bound up with the Western humanist spatiotemporal frame through which both modernisms and worlds tend to be academically theorized. As an aesthetic and a praxis, planetarity is an insistence on the alternative geohistories of a no-longer-singular modernism. Writes Friedman, "I suggest we regard modernism in its different geohistorical locations and periods as a powerful domain *within* a particular modernity, not something outside of it, caused by it, or responding belatedly to it. . . . Thus, I am suggesting that we treat modernism as the domain of creative expressivity *within* modernity's dynamic of rapid change, a domain that interacts with the other arenas of rupture such as technology, trade, migration, state formation, societal institutions, and so forth."[65] In direct opposition to Fredric Jameson's formulation of a "singular modernity" that is defined almost exclusively in relation to the dynamics of global capitalism (or "a universal world market order"), Friedman insists on modernity's polyvalent, polycentric, and multitemporal dimensions.[66] "Any given 'modernism[,]'" then, "is the domain of the aesthetic that shapes, gives form to, and responds to the modernity of which it is a part."[67] Echoing Laura Doyle and Laura Winkiel's critical fracturing of "modernism" into globalized "geomodernisms,"[68] the modernisms of the Tang dynasty, Abbasid caliphate, and Mughal Empire, in Friedman's analysis, both predate and make possible the development of the late nineteenth- / early twentieth-century Euro-American modernism that has become universalized as "the default position of modernism per se," or as "modernism itself."[69]

In a similar vein, Edgar Garcia cautions against the persistent tendency to uphold the singularity of modernity amidst critical efforts to decolonize literary studies and (world) poetics. Garcia writes in *Signs of the Americas: A Poetics of Pictography, Hieroglyphics, and Khipu*, "The paradigm of a singular capitalist temporality is implicit to current conception of worlding in the study of world poetry. . . . As literary studies expands in polyglot, transnational, and intercultural considerations of literary value, it now faces the question of how to see the worlds that make up our shared globalities without reconstituting a world system that constrains those networks as evidence only of colonial, national, and transnational capitalist scrim."[70] Yet the fragility and contingency of the global capitalist paradigm—as a "world norm" that is nonetheless "implicated in other relations, obligations, and normative practices"—itself gives occasion for alternative, and equally fragile worldings to emerge. Citing Anna Tsing's analysis of the "resistant agency" of fungi—in which fungal sporing practices also double as "tactics for disarticulating capitalist strategies"—Garcia continues, "If anything, the precarious worlding of world capitalism compels us to make visible the alternative reserves in our midst, to challenge ourselves with the unusual reality of other realities, to derive from those challenges a sense of contingency and plurality in our planetary historiographies, and perhaps to make such contingencies tactically effective for the heterogeneous flourishing of cultural and biotic makings."[71]

We might then theorize modernist planetarity as a monstrous *anarrangement* of worlds and worlding practices.[72] And in our inhuman modernist analytic, the planetary is an opening toward the inhumanism of what in worlds flourishes—what grows and gathers—epiphorically, or in excess of humanistic-metaphysical time and thought. The eco-material valences of Friedman's planetarity "suggest . . . the Earth as a place of matter and climate, life and the passage of time" and as ground for the making of "interconnections [among] the human and nonhuman."[73] An inhuman modernist planetarity stays attuned to these valences, to the multispecies relationalities and metaphorics made in collaboration with the earth and its expressive materials.

Amidst this ongoing anarrangement of modernism's worlds, geohistories, and planetarities, where might an inhuman modernism actually begin? If the Anthropocene, as Dipesh Chakrabarty famously argues, is another name for the intertwining of human and inhuman (or "geological") histories, then an inhuman modernity would perhaps start here, at the end of the world.[74] But the Anthropocene's historical and conceptual origins are, like modernism's, similarly fractured and distributed across multiple space-time axes. Some climate scientists, including Paul Crutzen and Eugene Stoermer, date the beginning of the Anthropocene to James Watt's invention of the steam engine in the late eighteenth century, which ushered in the era of carbon emissions into the planet's atmosphere and marked the formal beginnings of global ecocide.[75] Other Anthropocene origin stories begin at the traditionally demarcated end of the modernist period in 1945: this is the moment when the "Acceleration" got its "Great," and fossil fuels began to be burned at an unprecedented rate that eventually led to the warming temperatures, droughts, rising sea levels, and ocean acidification that are currently in the process of making the planet uninhabitable for humans along with millions of other species. And yet the framing of Anthropocene, and of climate change discourse more broadly, solely around technological developments in the Global North—and, crucially, without a consideration of the central role of settler colonialism, Indigenous genocide, transatlantic slavery, and resource extraction in driving climate change—has come under strident critique.[76] There must be more to ecological thought, a decolonial and inhumanist practice insists, than Enlightenment- / conquistador-humanist Man and the self-image that it has scarred into the earth.[77]

In *Staying with the Trouble*, for example, Donna Haraway suggests reconceptualizing "Anthropocene" as "Chthulucene," after the spider species *Pimoa cthulhu*. Chthulucene insists on an earth that is geostoried through multiplicate, tentacular relations among species and worlds rather than humanist determinations and the furrowing of the name of Man into every ground.[78] A restoried Anthropocene opens to political and historical critiques of the Western humanistic structures underwriting and enabling something like the category of universalist, de-embodied Enlightenment Man to come into existence. In "On the Importance of a Date," Heather Davis and Zoe Todd place the golden spike of the Anthropocene in the year 1610, "or the beginning of the colonial period." "Linking the Anthropocene with colonization," Davis and Todd argue, "draws attention to the violence at its core, and calls for the consideration of Indigenous philosophies and processes of Indigenous self-governance as a necessary political corrective, alongside the self-determination of other communities and societies violently impacted by the white supremacist, colonial, and capitalist logics instantiated in the origins of the Anthropocene."[79] Similarly, Kathryn Yusoff's trope of the "Black Anthropocene" locates the governing logics of the Anthropocene in the institutions and practices of chattel slavery and the Atlantic slave trade, colonial expansion, mercantile imperialism, resource extraction, and the genocide of Indigenous peoples, all of which have been materially sedimented into the earth's geological archives.[80] And among an ever-multiplying series of rethinkings of the Anthropocene's *anthropos*, Jason W. Moore's self-explanatory "Capitalocene" stresses capitalist expansion, imperialism, and extractivism as the primary agents of global climate change.[81] "Plantationocene," "Necrocene," "Eurocene," "Plasticene," "Manthropocene," and even "Trumpocene" have also been offered as alternative nomenclatures.[82]

Articulated alongside these and other theorizations of the present age and its envisioned futures, an inhuman modernity would be inextricably tied to imperialist geopolitics—including what Yusoff calls the "material geophysics of race"—and planetary-historical relations more broadly.[83] For Yusoff, any thought of the earth's present and its futures (any geology or any -cene) is incomplete without a reckoning with the ongoing material and discursive scriptions of colonialist earth-writing (literal "geography").[84] I want to keep this particular analytic receptive, however, to the possibility of new earth-writings—or new articulations between beings and grounds, new intimacies to come between the terran and the submarine—through which both planetarity and modernity are co-constituted within an inhuman modernism. The relational knots of metaphor, too, are geostorying practices—a means toward reimagining ground and underground, *terra* and *oceania*—for an increasingly apocalypse-oriented present. As multispecies geographies, metaphors limn the ongoingness of relations to and with earth in place of the Anthropocenic extinction narratives that imperialist and ecocidal Man continues to tell about himself.

The groundwork of Anthropocene—as the making of a metaphysical ground, the sensus communis of the Western humanist tradition—may have thus begun with the development and ascendancy of human language and logical structures. (Or in Uexküll's terms, with the construction of a singular world in place of manifold worlds in composition: Anthropocene begins, again, with the ending of worlds.) In this vein, Franz Broswimmer argues in *Ecocide: A Short History of the Mass Extinction of Species* that the Anthropocenic story of mass extinction originates with the capacity for human language and "conscious intentionality" that set humans apart from nature in the Pleistocene era around sixty thousand years ago.[85] Broswimmer writes,

> The defining marker of ecocide was the development of language and an unprecedented expanded human capacity for culture. These novel features of *Homo sapiens sapiens* allowed for the rise of the conscious intentionality that humans bring to their projects. Reflected in the vastly improved capacity for language, conscious intentionality led to an explosion of innovation—manifest in a proliferation of artifacts—at the end of the Pleistocene era some 35,000 to 50,000 years ago. The emergence of conscious intentionality made possible the extension of human biological evolution by cultural means—including the species-specific capacity for conscious adaptive or maladaptive changes in social organization. By about 13,000 B.C.E., this development path resulted in the human colonization of all continents with the grave consequences of the worldwide destruction of most of the existing megafauna.[86]

The development of culture, for Broswimmer, is bound up with the proto-imperialist logics of Pleistocene-era ecocide: "changes in social organization" and new forms of cultural production accompany the violent, world-destroying "human colonization of all continents." Imperialism and ecocide, it would seem, have always operated as twinned discursive and material practices, entangled earth-writings. In this sense, Yusoff's framing of the geo-imperialist underpinnings of the Anthropocene might well apply both to 1441/1492 AD as well as 13,000 BCE: "The Anthropocene might seem to offer a dystopic future that laments the end of the world, but imperialism and ongoing (settler) colonialisms have been ending worlds for as long as they have been in existence."[87]

In Broswimmer's schema, language and conscious intentionality give rise to two different yet knotted modalities of culture: culture as social organization and culture as relation with the ground, or agriculture.[88] Agriculture is thus another form of earth-writing: a writing on and with grounds. Agriculture is also a potential focal point, or golden spike, through which both Anthropocenic and inhuman modernist temporalities might be rescaled. In *Dark Ecology*, Timothy Morton claims that the Fertile Crescent agriculture of Mesopotamia some 12,500 years ago was not just a phase in the ongoing development of ecocide (as Broswimmer has it) but also a shifted epistemological orientation toward the earth, or what Morton calls "agrilogistics" ("agricultural logistics"). The prehistoric "extinction of megafauna," in Broswimmer's account, required people "to change their social organization wherever conditions such as a favorable climate, water, and fertile soil, and species that could be domesticated, were present."[89] This resulted in a hierarchization of forms of life, a drive to dominate nature, and the enactment of what Morton calls the agrilogistic principles of "thin rigid boundaries between human and nonhuman worlds" and the reduction of "existence to sheer quantity [over quality]."[90] The domestication of animals (and the devaluation of nonhuman life generally), private property (leading to "the contemporary concept of self"), and patriarchal societies and class distinctions coupled with the emergence of class antagonism can be traced to the origin and spread of agrilogistics.[91] In this sense, we may still be living through agrilogism's self-unfolding, manifested as the ongoing imperialism, global resource extraction, and ecocidal (and slowly violent) modes of social and political organization of latest-stage or extinctionist capitalism. Within the terms of this particular planetary-historicity—a temporal knotting that joins 12,500 BC to the third millennium AD—we are all "Mesopotamians."[92] Or to adopt the terms of Friedman's unbound historicity, we might all be embedded still in a Mesopotamian modernity.

Both Yusoff's geo-imperial analytic and Morton's agrilogism emphasize divisions between human and inhuman, or life and nonlife, as constitutive of any modernity, whether conceived as a "biopolitical tool of governance"

(Yusoff) or an exploitative, extractivist relation to earth (Morton).[93] Yet each of these hierarchized conceptual arrangements (of matter, of species, of being) unwittingly raises monsters out of the ground. Agrilogism's first axiom, Morton argues, is the "inviolable" "Law of Non-contradiction," which "works by excluding (undomesticated) lifeforms that aren't part of your agrilogistic project. These lifeforms are now defined as pests if they scuttle about or weeds if they appear to the human eye to be inanimate and static. Such categories are highly unstable and extremely difficult to manage."[94] Monsters are what of/from the ground is unthought or unthinkable within systems of geo- or biopolitical order: the "category-crossing beings" abhorred by Enlightenment (Tsing et al.), for example, or outgrowths from the evolutionary process without fixed place along a line of descent (Darwin), or in a vaster sense, all beings birthed from the fractures between a given order's concepts. In its Darwinian evolutionary resonances—and its Carrollian, Flaubertian, and Joycean (proto-)Surrealist overtones—the monster can be defined as both a conceptless creature *and* as a many-concepted creature, one that entangles more incongruous concepts and referential frameworks than can traditionally be thought in the bounded space-time of a singular being or species. Monsters, like metaphors, are epiphoric.

We are returned here to the Sphinx, and not for the last time. In Morton's words, she is "the ambiguous female monster lurking on the edge . . . the monster that haunts the boundary of agrilogistics," and the being that must be supplanted to instantiate the axiom of non-contradiction.[95] The Sphinx speaks of a metamorphic, multilegged being that also metamorphoses time itself into cyclicality rather than *telos* ("What has four legs at dawn, two legs at noon, and three at night?").[96] Only if time is imagined exclusively as being's progression toward or into form can the *ainigma*—of this mystical, many-concepted, regenerative being cyclically bound to the motions of the sun—be answered with "Man." (Similarly, Aengus's harvest occurs amidst an undone temporality, a recursive ongoingness beyond "time and times," as inexpressible as the sought rhyme between earth and cosmos.) Oedipus's response thus clears ground for humanism's furrows, for the imposition of determinate time and concept onto the epiphoric gatheredness of both being and riddle.

"Man," in Oedipal and other senses, is a fundamental and severely consequential misreading of the earth. The present ecological crisis may in large part be shaped by the persistence of both agrilogistic and geo-imperialist norms. In following these scripts, wealthy nations and multinational corporations—all of which are outgrowths (or, at minimum, beneficiaries) of the practices of imperialism, transatlantic slavery, genocide, forced removal of peoples, and global resource extraction—traffic in the production and acceleration of extinctions, with the impact falling most strongly on the poorest and most vulnerable communities.[97] But agrilogistics also creates the conditions for its own undoing. Agrilogism's rigid laws and forms of separation and enclosure paradoxically give interstice for new relations, new intimacies among enigmatic beings and earths: a thought of the otherwise troped variously as Chthulucene (Haraway), a billion Black Anthropocenes (Yusoff), the "arche-lithic" (Morton), and in our terms, inhuman modernism.

Each of these otherwises seeks alternatives to the material and conceptual structures, grammars, and scriptions of humanist-colonialist thought, or the possibility of an inhuman relationality. Black Anthropocenic practice, for example, is a matter of "desediment[ing] the social life of geology," both by unearthing the repressed histories of colonialism, extractionism, slavery, and genocide written into the earth, and by locating new (philosophical, historical, poetic) practices that might "mark another possible inhuman relation that does not replicate the confinements of colonial grammar."[98] In borrowing from Jacques Derrida's concepts of "arche-writing" and the "re-mark" in *Of Grammatology*, Morton defines the arche-lithic as a "primordial relatedness of humans and nonhumans" that both predates and permeates agrilogism.[99] For Derrida, the meaning given in writing does not "happen all by itself," nor is it traceable to a singular origin. Meaning, rather, depends on the prior presence of material entities or actors and a physical (or lithic) context on which phenomena are able to achieve recognition as meaningful.[100] The prelogical ground, which Derrida elsewhere associates with Hegel's prehistoric "flower religions" that must be supplanted by culture (and etherealized into the ongoing labors of Spirit) as well the primordial "night in which all cows are black," both precedes and makes possible the generation of meaning.[101] Every signifying mark requires both a preexisting "inscribable surface"—a field of flowers, a stone/lith, a night with cows in it—and a system of meaning—a culture ("cultivation") produced out of the materials of nature—by which the mark counts as a unit of meaning and not just as a random scribble. In contrast to the nonsignification of the planting and worship of flowers, the mark of the plow's furrow instantiates meaning as well as history.[102] As naive flower religion or pre-dialectical night or desedimented geology or any other kind of primordial relationality among beings, the arche-lithic is the thought of a ground that generates indifferently to both agrilogistical, geo-imperial human/nonhuman boundaries and the punctuated linearity of time.

Like the Sphinx, Medusa is another irresolute riddle: a gathering ground for chthonic and wayward communions; another threatening, earthbound, female (sea) monster limning the edges of both *anthropos* and the sky gods' Olympus. Medusa's beheading, however, both allows Perseus to found Mycenae (a problem solved once again with the name of Man) and births the winged (monstrous?) horse Pegasus along with the coral reefs of the sea, as Haraway notes: "From the blood dripping from Medusa's severed head came the rocky corals of the western seas, remembered today in the taxonomic names of the Gorgonians, the coral-like sea fans and sea whips, composed in symbioses of tentacular animal cnidarians and photosynthetic algal-like beings called zooanthellae."[103] Oedipus's response and Perseus's blade are determinate solutions to ceaselessly regenerative riddles: conceptless monsters continually emerge from the gaps between agrilogistics' dividing furrows; the tentacular splaying of coral reefs reworlds Medusa into and against the patriarchal-imperial order founded on her corpse. In Derrida's terms, the Sphinxian and Medusean arche-lithic/Chthulucenic marks "the enigmatic relationship of the living to its other, and of an inside to an outside," an ongoingness that is also an irresolution to the problem of being.[104] In this way, an inhuman modernism partakes in the creative insistence of being as problem—the dynamic arche-play of entities and the tentacularly grammared worlds arising from them—across any and all universalist determinations. And what new relations to and writings with ground might become possible once the furrowing of Man into the earth is no longer being's only solution?

III.

Is there, or should there be, a means of binding the hyperextensive temporality of an inhuman modernity, stretching from Pleistocene to Chthulucene and reverberating well into any humanistically imaginable future? Countering Friedman's emphasis on the polycentricity, dislocation, and multitemporality of modernism, David James and Urmila Seshagiri maintain that "modernism as an era, an aesthetic, and an archive . . . originated in the late nineteenth and early twentieth centuries" and names a historically discrete body of artistic and cultural production.[105] "We dull modernism's particular brilliance," James and Seshagiri write, "when we dissolve it into a collective of techniques comparable with what other writers have practiced at other points in history, and to characterize modernism as a feature common to centuries of interlocking modernities is to risk depriving works such as *To the Lighthouse* of the formal and temporal inventiveness unique to their era."[106] Rather than set these accounts in irremediable opposition to each other (the boundless versus the bounded nature of modernism), we can instead think each periodization as a provisional gathering of the aesthetic, philosophical, and planetary forces at work in and across spatiotemporal contexts. In this vein, Marshall Brown argues that periodization does not concern "knowledge" as much as it does "thought, which is the other of knowledge": periodizing is the ongoing work of unknowing and problematizing rather than of structuring and conceptualizing.[107] Periods themselves, according to David Perkins, are "necessary fictions" that "make apparent the particularity, local difference, heterogeneity, fluctuation, discontinuity, and strife" through which we tend to understand the past.[108] Similarly, for Eric Hayot, "period is the untheorized ground of the possibility of literary scholarship."[109]

Periods, as grounds—or in Brown's phrasing, "terrains"—also touch other, proximate grounds, generative frictions that make new cross-temporal connections and intimacies possible. We can include among these the global and planetary-historical varieties of Friedman, Garcia, and Doyle and Winkiel, as well as the more modestly formulated theories of the re-engagement of contemporary literature with modernism's "formal principles and ethicopolitical imperatives" that James and Seshagiri term "metamodernism."[110] And this monstrous amalgamation of spatiotemporalities—the planetary, the global, the nonhuman/inhuman, the meta-, for example—may be internal to modernism itself. In this vein, Charles Tung argues in "Baddest Modernism" that among modernism's central concerns was the problem of how to "inject into the moment and literary form a scaling up and a scoping out to inhuman times."[111] Tung's essay is part of a *Modernism/modernity* special issue, "Modernist Inhumanisms," in which several of the contributors claim that undoing traditional, linear, grounded models of time and finding aesthetic means of expressing inhuman timescales were common preoccupations among both "high" modernist as well as popular modernist science fiction works.[112]

An inhumanist critique asks us to read texts like Woolf's *To the Lighthouse* (1927) and *Between the Acts* (1941) and Joyce's *Ulysses* (1922) alongside H. P. Lovecraft's "The Call of Chthulu" (1928), William Hope Hodgson's *The Night Land* (1912), and Olaf Stapledon's *Last and First Men* (1930) as part of a collective attempt to unground prevailing notions of temporal and spatial scale.[113] In the "Ithaca" episode of *Ulysses* (hereafter *U*), for example, Leopold Bloom "meditate[s]" on both the incomprehensibly "vast" "evolution" of the universe and its stars and the increasingly "less vast" "involution" of "the eons of geological periods recorded in the stratifications of the earth; of the myriad minute entomological organic existences concealed in cavities of the earth" (*U* 17.1042, 1057–60). Whereas Bloom's inter- and intraplanetary thought stops at the realization that there is "no known method from the known to the unknown" and thus no proper means of theorizing cosmic scale in comprehensible relation to human existence, Stapledon's *Last and First Men* tasks itself with extending humanity into an unknown and impossibly expansive "far future"

(*U* 17.1140–41). Interplanetary human existence (divided into eighteen distinct human species) is imagined through the novel's five ever-widening "time scales," which successively and simultaneously span 2000 BC to 4000 AD (Scale 1); 2,000,000,000 years past ("PLANETS FORMED") to 2,000,000,000 years in the future ("END OF MAN") (Scale 4); and finally, from 10,000,000,000,000 years past to 5,000,000,000,000 years in the future (Scale 5).[114] (An inhuman modernism that begins in 60000 BCE all of a sudden does not seem so far-fetched!)

As a way of thinking scale inhumanistically, Mark McGurl ventures "posthuman comedy" as an umbrella term (or "critical fiction") that "draw[s] together a number of modern literary works in which scientific knowledge of the spatiotemporal vastness and numerousness of the nonhuman world becomes visible as a formal, representational, and finally existential problem."[115] Comedy is also an opening to yet-to-be-imagined futures outside of the determinate coordinates of what has historically been given as possible. Accordingly, there is something of the comedic in Aaron Jaffe's provocative salvo in the introduction to "Modernist Inhumanisms": "What comes after the (humanist) future?"[116] Inhumanist comedy asks after a future that does not proceed from the stable ground of the Western metaphysical tradition, improvising instead with the many futures grown from an otherwise, anti- and ante-logical ground.

And yet something more or other than temporality inheres in the "after" of Jaffe's question: the "post" of the human is considered neither as a technological or prosthetic enhancement of human capabilities nor as the projection of human(ist) selves into post-extinction worlds. Rather, to think the human's post nonhumanistically requires a critique of the deterministic logics and concepts through which futures come to be articulated in the first place. Evolutionary time and the time of Wonderland alike (as we will explore in chapter 2) are endlessly reversible and undeterminable as predicates of a singular ground, as suggested by Carroll's original title, "Alice's Adventures Under Ground." Humanism cannot accompany us here, nor can a posthumanism that would still grow from the determinate fields humanism has furrowed. If human afters and posts are only imaginable as means of self- and species transcendence, then this kind of posthumanism is merely an extension of the same humanism that it purportedly works against, as Colebrook and Weinstein argue. "Humanism," for them, is "a way of refusing to see humanity as a biological event within life and has (at least in its Western metaphysical mode) always seen the human as a rational, sentimental, technical, spiritual, cultural, or historical means of surpassing life."[117] Put simply, "humanism has always been a posthumanism."[118]

An earth-insistent inhumanism (or what has also been called the "posthumous")—one that includes within its expanse articulations of multispecies, inhumanist, Black feminist-materialist[119] and ecological theory—strives instead toward comedic futures, or futures that proceed undetermined by the human and its time.[120] In its planetary, meta-, or inhumanist variants, modernism is similarly an insistence on thinking grounds and the enigmatic grammars of their growth. What modalities of relation between knotted beings and grounds might arise, an inhuman modernism asks, in the aftermath of humanism's exhaustion of its own concepts?

IV.

> Metaphor is a kind of enigma.
>
> —Aristotle, *Rhetoric* III.2.12

We encounter Odradek as a knot, or as many knots: he is both himself and the times epiphorically gathered into him, a holding-together and an unspooling, a being that means and a being that only says otherwise (*andeutungsweise*). Odradek's knottedness offers an alternative to conceptual forms of world making. For Odradek's meaningless (*sinnlos*) makings do not totalize into singular world, nor do they answer to time. By way of limning the terms of an Odradekian language made of strivings rather than determinations, expressivities ungathered into concepts, I propose the term "inconceptual." By this term, I want to designate something of how beings unfold and express not only in opposition to concepts but also *indifferently* to concepts. In Colebrook and Weinstein's analysis, as we have discussed, the inhuman is distinct from both the nonhuman and the posthuman because the use of the term "nonhuman" (like the nonconceptual) can presuppose a definitional opposition to the human and "posthuman" tends to reinscribe the self-surpassing or self-transcendence of the human around which Western humanism has traditionally oriented itself.[121] The inhuman, I want to add, is indifferent to humanism's determinants. *Inconceptual* gives provisional name to what happens otherwise in worlds, the inhuman socialities among beings (and among beings and grounds) that insist beyond the grasp of conceptual and other metaphysical logics. In a broader, planetary sense, the inconceptual is a means toward thinking *with* the earth's interstices, with what grows and gathers between agrilogism's furrows and geo-imperialism's fractures. Odradek's lungless laughter (in response to the question of where he lives) cannot be given history, nor can its futurities be determined: Odradekian laughter resonates, or "rustles" (*rascheln*), beyond both the time of the home and the time of extinction. We can intuit in this expressive laughter something like a comedic opening to what comes after—and what rustles indifferently beyond—humanist time and times.

The inconceptual, I want to emphasize, is meant neither to sidestep nor to minimize the catastrophic realities of colonialist, humanist, and ecological violence. Rather, the inconceptual is a call toward further receptivity to what tends to go unnoticed or unthought in a Western humanist emphasis on the forms of the subject, the act, the archive, and the concept as the dominant loci of knowledge production. Inconceptual is a gesture toward epistemological modesty as well as epistemological weakness: the refusal of strong, determinate forms of thought in favor of what Paul North calls "alternatives to power," such as "the non-act—secession, surrender, the yield."[122] The weak epistemologies of the inconceptual are ones that, first, stay open to the indeterminacy of the epiphoric growing and gathering that happens on and with earth and, second, attune us to what else happens—however precariously—beyond the grasp of archives and concepts: how do we hear and think burrowing (Kafka) or the bellowing of cowherds (Woolf) or the nonsense-speech of the earth's undergrounds (Carroll), or notice the vibrancy of minerals (Clevidence) without subsuming these into the humanistic frameworks of our concepts?[123] What else emerges when we start from the assumption that Odradek's rustle laugh and Alice's vegetal and fungal transformations, for example, are modes of self- and earth expression that are indifferent to the concept, or that mean something other than or more than concepts?

"Expression," as my improvisation on a term from Benedict de Spinoza's *Ethics* (1677) has it, helps us to further elaborate a theory of inconceptuality.[124] And as we will explore further in chapter 1, Spinoza's *Ethics* (hereafter *E*) gives provisional ground for an inconceptual, or Odradekian, language of human- and concept-indifferent strivings. For Spinoza, all beings, as "mode[s] which express God's essence," are fractal interpretations of God's being, which is equivalent with nature itself (*E* II D1).[125] A being is nothing more or less than the gathered material-affective content of its strivings and its coarticulations with other beings, including the earth itself: "The striving [*conatus*] by which each thing strives to persevere in its being is nothing but the actual essence of the thing" (*E* III, P7). A being strives not toward fulfillment in form or concept but rather toward a more energetic and more involved sense of self through the unfolding of being in and with earth. In this way, a being's self-expression is coincident with its relations. In lines that we will revisit in chapter 6, Cody-Rose Clevidence's *Flung Throne* expresses something of the fraught idiom of inconceptuality, or of an inconceptual striving: "a pattern climaxes at a form / a living form is restless within itself."[126]

Metaphor, for Clevidence and for us, is refusal of the impoverished teleology of form. As multispecies practice, metaphor gives us a way to think the internal restlessness of a living form, the strivings and expressivities of beings in their fugitivity from form. Multispecies metaphor, finally, clears fragile milieu ("middle place") for the epiphoric gathering of beings into inhumanist swarms, unholdable by concepts. Let us provisionally begin an inhuman modernism here, in a middle place made from the routing of species and worlds through metaphor. And in so doing, we will take the enigmatic gatherings of the Darwinian Tree of Life and the Carrollian Jabberwock as inhuman modernism's improvisatory grounds. In them, the incongruity of beings with their concepts—and the noncoincidence of multispecied time with humanist historicity—insist as problems for philosophy, science, literature, and poetics to reworld themselves around.

The multispecies strivings of metaphor may have begun already in 60000 BCE with the development of conscious intentionality and the foundations of language, or in 12500 BCE with the invention of agriculture and its logics. They may also date from 44000 BCE with the first known figurative cave painting of an animal and thus the earliest articulation of a relation between animals, humans, and earth (as both lithic surface and material medium—ocher pigment—out of which inscriptive instruments are made, simultaneously writer and written upon). However, the comparatively modest resituation offered by *Modernism's Inhuman Worlds* asks not that we originate modernism amidst the prehistoric beginnings of human consciousness, intentionality, or aesthetic expression, but only that we let these impossibly vast temporal and spatial arcs inhere in the definitional content of any modernism. Accordingly, we can think Darwin's *On the Origin of Species* and Carroll's *Alice* books not so much as foundations of an inhuman modernism (or of modernism itself) but rather as anarrangements of prevailing understandings of metaphysical and material ground. The Under Ground into which Alice falls is both the place where form intersects with geological/evolutionary time, and occasion for the unearthing of all that has been sedimented for the laws of sense, species, linearity, and concept to achieve sovereignty over the above-ground world. Such disruptions—of ground and the concepts through which ground is thought—are what characterize, in Friedman's terms, the dynamic coarticulation of any modernism with and through its (geo)historical context, or its modernity. An ongoing striving toward alternative geologies or planetarities, other modes of earth writing, is what all modernisms across times and spaces, periods and fields, might share in common.

V.

Modernism's Inhuman Worlds strives toward a means to think the "planet" of planetarity in both its ecological and aesthetic dimensions. In this spirit, I offer each of the chapters to come as an improvisation with a different set of the earth's expressive materials, as these are mediated through inhuman modernist works from the *Origin* and the *Alice* books to the present century's *M Archive* and *~~Flung~~ Throne*. After a theoretical elaboration on the multispecies dimensions and variegated modes of growth of metaphor in chapter 1, we begin with the earth's underground sediments (Darwin, Carroll) and proceed through dust (Flaubert, Joyce), soils (Kafka), sky (Woolf), forest (Clevidence, Gladding, Nakayasu), ocean (Waldman, Shaughnessy, Gumbs), and end with the earth's (imagined) self-relation to its own planetarity. I do not mean *Modernism's Inhuman Worlds* to be an exhaustive geography either of modernism or of the earth, but only one among many potential articulations of an inhuman modernist planet(arity) and the gathered materials through which it is enigmatically expressed.

Chapter 1, "Growth," puts into conversation the multiple origins—or the many growing grounds—of metaphor, with an emphasis on how a differently historied (and futured) metaphoricity limns inhuman modernist makings of world. This chapter also further develops a theory of the inconceptual—as a generative indifference to concepts and conceptual structures—especially as it bears on philosophies of metaphor. We explore the tensions, or knots, out of which metaphor originates and which still inhere in and as metaphor. These include the tensions between *metaphora* and *epiphora*; metaphor as the transference of meaning and metaphor as the profusion of names; metaphor as logos- or concept-oriented and metaphor as inconceptual; along with the tensions between an animal or "zoontological" metaphorics (Rosi Braidotti) and a vegetal/plant metaphorics (Jacques Derrida). A multispecied metaphoricity, as we will explore, *is* this knottedness of beings, times, and worlds: a means toward thinking another planetarity.

We begin our planetary-aesthetic thought amidst the earth's sediments, or the material and metaphoric undergrounds of the earth. As we discuss in chapter 2, "Sediments," Carroll and Darwin imagine sediments as both the undergrounds of meaning/logos as well as the repressed geo- and evolutionary histories of the earth and its beings. Ungrounding logic, for Carroll, provides a means toward imagining a post-Darwinian geohistorical landscape in which the present is the gathering of gradual and untraceable modifications to organisms over an insensible (and reversible) continuum of evolutionary time. The enigmatic dimensions of evolutionary theory thus resonate closely with the illogics, or the saying and sounding otherwise, of nonsense. Chapter 2 dwells in this resonance with an emphasis on the subterranean, multispecies communions between each work's most iconic image: Darwin's Tree of Life and Carroll's Jabberwock. Darwin's Tree of Life, the diagram illustrating the work of natural selection on species over "thousands of generations," is constituted by the unknowable histories that have been sedimented into the earth over geological time: as it "covers the surface with its ever branching and beautiful ramifications," the Tree of Life also "fills with its dead and broken branches the crust of the earth" (*OS* 177). If the earth's sediments could speak, they might sound like Jabberwock(y): a being made from the piling up of multiple sounds, senses, referents, and bodies that do not totalize into meaning. In the Jabberwock, and in the *Alice* books as a whole, underground sounds, beings, and worlds propagate and gather into something other than logos, an inconceptual generativity that we call nonsense. In this way, both Tree of Life and Jabberwock are metaphors that are persistently reshaped by what is gathered into them, simultaneously nominal designation (*metaphora*) *and* copulative excess (*epiphora*). What both reveal is that there is no regime of sense and no species or being that is unrouted through nonsense, through the clamors of the earth and its sediments that inhere in every given form.

In chapter 3, "Dust," we theorize dust as both earthly material and as the primordial substance of creation, as the *adamah* (Hebrew: "ground," "dust") from which Man/Adam/"earthling" was first shaped into flesh in biblical accounts. The Flaubertian and Joycean inhumanisms, of *La tentation de Saint Antoine* (The Temptation of Saint Anthony, 1874) and the "Circe" episode of *Ulysses* (1922), offer alternative modalities of creation, through the medium of a still-vibrant dust that might be shaped otherwise than Man. What happens, both works ask, when nonsense infects both the logics of creation themselves and the relations ongoingly made among nonsense-shaped beings? In short, a metaphorics of contagion (Latin: *con* + *tangere*, "touching together"), in which all dust is made inseparable from all flesh and from all created material. Flaubert's word for this everywhere-relationality is "temptation/*tentation*." Temptation happens as a reordering of what Saint Anthony calls "the immutable order of the universe" in which the human body (and its fractious parts) comes to be melded into its multispecied environments and inhumanist lineages. The central characters of each text—Saint Anthony and Leopold Bloom—are simultaneously individuals as well as geologies, gatherings made in and of earth, that come to be re-created with each encounter in time and in space. A metaphorics of contagion manifests, in both texts, through the use of stage directions that collapse any meaningful distinction between imaginative figuration and materialized enfleshment. Metaphors given in the stage directions of each work, as we explore, indicate the nonsensical joining of human, nonhuman, hybrid, and mythical bodies and parts, rather than the offer of name and logos to what was formerly outside of meaning. When a metaphor's expressivity is no longer confined to metaphysical abstraction, metaphors come to infect reality itself, clouding the distinctions between truth and semblance, individual character and multispecied assemblage, dust and flesh, earth and earthlings.

In Kafka's *Zürau Aphorism* 106, the "true language of prayer [*Gebetsprache*]" is "humility [*Demut*]," "at once adoration and the firmest of unions," both with one's "fellow man" and with the earth itself.[127] The connection between humility and the earth is made more explicit in the English word "humility"/ French *humilité*, with origins in the Latin *humus* (soil): humility is a striving toward an earth-relationality or soil-being through which all other relations among beings are made possible. In chapter 4, "Soils," we philosophize with the animal narrators of two late Kafka works: *Der Bau* ("The Burrow") and *Forschungen eines Hundes* ("Investigations of a Dog"). What we might call a late Kafkan inhumanism emerges through acts of faith and acts of striving, on the part of these narrators, toward the *ainigma* of the earth's generation, to and past the point of complete self-abandonment. For both burrower and dog-philosopher, all philosophy and all knowing proceeds from relations with the "nourishing earth [*der nährende Erde*]" and what is given in its soils (*Investigations of a Dog* [hereafter *Investigations*], 326). The burrower hears an earth sound and, in striving alongside it, invents a listening apparatus—an implicit metaphoric structure—that is also the co-creation of another world of soil-made meaning. Burrowing, as we will explore, becomes devotion to what of earth is unpossessable—the unseen sound, most particularly, and what is engendered from the burrower's fantasies of it—even if this devotion leads to the unmaking of the self, or the fracturing of self into many selves. Similarly, dog-philosophy, according to the narrator of *Forschungen eines Hundes*, begins from the ancient canine wisdom on which all dog being (*Hundeschaft*) is founded: "Water the ground as much as you can [*Mache alles naß, soviel du kannst*]" (287). In this statement, according to the dog-philosopher, is contained "the essence of all knowledge [*Wissenschaft*]" since relations to ground—or the nourishing earth—mark the beginning of all philosophy (287). The abstract individuation that takes a being outside of one's self and away from the ground, then, is denial of both soil-being and of life itself. To make wet, as we explore alongside the dog-philosopher, is an enjoining not toward a specific action but rather toward a humble comportment to earth, a giving without thought of recompense. Making wet comes to be simultaneously an inscriptive practice, an act of devotion, and a metaphorics. Burrow language (and burrow hearing) is likewise an all-consuming earth-faith, made through the burrower's shifting comportments toward phenomena and reconfigured at every moment around the varied interpretive possibilities given by earth. In this sense, both burrow- and dog-philosophy may also be modalities of a concept-indifferent prayer language (*Gebetsprache*): a devotion to ground at the expense of one's own self-possession, the surrender of being to the mysteries of soil.

In a letter to Felice Bauer of March 3, 1915, Kafka writes, "Had you not been lying on the ground among the animals, you wouldn't have been able to see the sky and the stars wouldn't have been set free."[128] In chapter 5, "Sky," we remain on the ground—with burrower, with dog-philosopher, and with other philosophizing beings—while contemplating the possibility of a liberated sky, or a sky that is no longer the exclusive domain of humanistic projections of meanings, concepts, and futures. We read Woolf's final, unfinished novel as proto-Anthropocenic thought and as a means of imagining (with) the other side of human extinction. Throughout, bird flight and bird syllablings scribe the possibility of another, still-expressive sky in the absence of human beings. Both *Between the Acts* and the also unfinished "Anon," however, modulate extinction through the time and form of the comedic, in which comedy is an opening to differently imagined—or perhaps unimaginable—futures. We thus explore comedy as a mode of ecological thought, as abandonment of the self to forces, beings, and temporalities other than the human.[129] The comic vision of *Between the Acts*, finally, happens in and as the dissonance—or what McGurl calls the "incommensurability"—between the "spatiotemporal vastness and numerousness of the nonhuman world" and the comparative smallness of our scientific and aesthetic means of representation.[130] In what we call Woolf's "dissonant metaphorics," *Between the Acts*—along with its focal event of Miss La Trobe's pageant on the long history of England and the songs of the primordial forestial balladeer Anon—become post-extinction imaginaries. Collectively, they are gathering places for noise, color, vibration, and other inhuman expressivities that resonate beyond the time of humans. The strange, comic figure of a "gigantic ear" embedded in the landscape, intuited by villager and pageant attendee Lucy Swithin, gives us a means for thinking a dissonant metaphorics in which metaphor is figured as a hearing apparatus for all of the expressive earth noise planted into it, from the prehistoric past to the post-(human-)extinction future. What remains after extinction, for Woolf, are the noise- and color worlds that nonhumans make in relation with the earth's ground and sky, modes of livingness unrouted through human(ism)s yet persisting far beyond them. Through the medium of a gigantic earth ear, both late Woolf and La Trobe attempt to hear toward the inhuman ongoingness of earth, or the earth's comedic insistence in and through the extinctions to come.

To inhabit an Anthropocenic present, according to Elizabeth DeLoughrey, is also to inhabit the disjunctive temporality of an "anticipatory mourning": a mourning for a world soon-to-be or even already ended.[131] Since we cannot "locate a stratigraphic marker for this epoch until a geologically significant period of time . . . has passed," in other words, we can only define the present from the unthinkable perspective of a post-extinction far future.[132] Chapter 6, "Forest, Ocean, Planet (Green & Blue)," explores a set of contemporary poetic works grown from the ending of worlds and from the abysses following human extinction. We focus on the post-extinctionist poetics of Alexis Pauline Gumbs, Cody-Rose Clevidence, Anne Waldman, Brenda Shaughnessy, Jody Gladding, and Sawako Nakayasu. Yet even as they confront the ever-increasing possibility of a humanless earth humanless earth, these works also inherit—and improvise with—Woolf's modulation of human extinction into comedy, along with other inhuman modernist formal experimentations. Rather than mourning for an ended world, these works revel in the forms of interspecied relation and intimacy that become possible in a post-Man era. In developing the relation between contemporary and modernist poetry, we draw on theoretical formulations of metamodernism, as what James and Seshagiri term the strivings toward "alternative futures for fiction through engagements with their modernist past."[133] How, we ask, do contemporary poets re- or interanimate the inhuman modernisms of the nineteenth and twentieth centuries in the service of both thinking and creating with a prior age's planetary aesthetics *and* imagining post-extinction futures anew?

As global sea levels continue to rise as a result of anthropo-/sociogenic climate change, we are faced with an increasingly shrinking terra firma as well as a future comprised of new intimacies between sea and land, ocean and forest: the blurred blue and green of a post-human(ist) planetarity. And although, in Gumbs's words, "all life would have to change the meaning of breathing" to survive the impending apocalypses, Gumbs and fellow eco-metamodernist poets also imagine this modulation of breath as another form of (emergent) ceremony among the earth and its beings.[134] In the simultaneously ancestral and interspecied memory danced into being by coral reefs (Gumbs); in the "green w living" manifolds and "VIRAL ORCHARDING" of forestial being and forestial thought (Clevidence); in the submarine expressivities given as manatee's anti-capitalist and anti-ecocidal "grammar of sound" (Waldman); in the comic fantasy of a cephalopodic future in which superintelligent octopi take over the earth in hopes of continuing interspecies and planetary being beyond the failures of humanism (Shaughnessy); in the sylvan poetics and material geographies of bark beetles, as they glyph inhumanist modes of being, memory, and community in and through wood

(Gladding); and in the confounded materiality and metaphoricity of ants as they infect and inhumanize the metaphysical structures on which our languages depend (Nakayasu), we bear oblique witness to an inhuman livingness that insists beyond imaginability, beyond the concept, the gaze, the archive, and other humanistic forms of capture. We end with the earth's self-relation to its own planetarity, in its poetic apostrophe to a singing star—as imagined in Gumbs's poem, "cycle"—as a means of knowing its own being, the mysteries of its enigmatic materials, and the inhuman joy of the ceremonies still to be found on the other side of oblivion.

CHAPTER 1

Growth

Multispecies Metaphor and the Riddle of Being

> Every case in which a plurality of meanings is irreducible, in which there is not even a promise of unity of sense, is a case in which we are beyond language. And consequently beyond humanity.
>
> —Jacques Derrida, "White Mythology: Metaphor in the Text of Philosophy"

The Heideggerian as-structure, and its modes of determinate joining, is rooted in the Aristotelian definition of the "symbol" (*symbolon*, σύμβολον) as "throwing one thing together with another, holding something together with something else, i.e., keeping them alongside one another, joining them to and with one another."[1] The contemporary notion of symbol as something like "sign" or "designation," for Heidegger, thus misses the gathering activity proper to the Aristotelian symbol: "σύμβολον therefore means joint, seam, or hinge, in which one thing is not simply brought together with the other, but the two are held to one another, so that they fit one another. Whatever is held together, fits together so that the two parts prove to belong together, is σύμβολον. In the original concrete sense, for example, the two halves of a ring are σύμβολα."[2]

This holding together of elements in the being of the symbol applies as well to the word (logos, λόγος), which only exists by way of an inherent "agreement" or "being held together" (*Verhältnis*) by which the word's meaning is constituted.[3] "Words," in this sense, are "sounds which emerge out of and for this *fundamental relation of letting something come into agreement and holding it together*."[4] Heidegger's words are material commons in which agreement must emerge from the joined elements for meaning to exist. However, the achieved agreement actually preexists the gathering entities or gathered sounds, since these elements would never join in the first place unless they already primordially agreed with one another. With the same gesture by which Heidegger liberates symbols and words from humanist abstraction, he also safeguards logos within the realm of humanity. Heidegger defines logos as "language" or "discourse, conversation [*Rede, Gespräch*]" and assigns this alethic capacity exclusively to man, as *zoon logon echon* or languaging or discoursing animal. As a result, nonhuman animal "voices" (*Stimme*, as Heidegger translates Aristotle's *phone*, φωνή) and other forms of expression and relation are ontologically disinvited from language.[5] Animals are *álogon* (without reason), creating "noises [*psófos*, ψόφος]" and nothing more.

We can think Kafka's Odradek as simultaneously *symbolon* and *álogon*. Odradek's thrown-together, dissensual being unweaves the metaphysical copula on which Aristotelian and Heideggerian *symbolon* rests. *Sinnlos*, Odradek both has no concept of his own *and* creates worlds in agitation with the conceptual structures (house, family, time, human) by which he is held. Odradek un- and respools, weaves, lurks, holds, rustles, laughs at and with his own unanswerability. These expressivities flourish amidst an ongoing indifference to concepts that we are calling the *inconceptual*. What new worlds and metaphoricities will this expressive *álogon* create in and across an indeterminate time scale, "rolling down the stairs, with ends of thread trailing after him, right before the feet of my children, and my children's children," in the absence of the metaphysical holding by which *symbolon* gathers its already grounded terms into determinate agreement?[6]

In this chapter, I want to expand on the distinction between human(ist) conceptuality and inhuman expressivity, and the metaphorics—symbolic / *sinnlos*, sublated / earthly, metaphysical / multispecies, *metaphora / epiphora*, for example—to which these varied comportments give rise. Can the being of metaphor sustain a commons that includes the voices (*phone / Stimme*), noises, spoolings, and other expressivities of nonhumans, or are metaphor's only worlds the ones that grow from the metaphysical ground[7] and its conceptual furrows?[8] In sifting through metaphor's grounds and undergrounds, its varied geologies, this chapter theorizes multispecies metaphor as both a mode of concept-indifferent gathering and as a tactical resource for inhuman modernist worlding practices.

I.

"Metaphoricity," in Jeffrey T. Nealon's Derridean analysis in *Plant Theory: Biopower and Vegetable Life*, is "a movement from the sensible to the intelligible."[9]

As such, metaphor "makes understanding possible" even as it remains immured in the materiality of the particulars—or what Derrida calls the "flowers of rhetoric"—that the metaphor is expected to sublate into meaning. However, metaphors can gather and say otherwise than sense, their expression irreducible to the *teloi* of meaning, intelligibility, or concept. In their emphasis on metaphor as a medium for the transfer of knowledge/meaning across contexts, traditional philosophies of metaphor, as we will explore, tend to overlook the enigmatizing and, crucially, the *epiphoric* dimensions of metaphor.

Spinoza's theory of "common notions [*notiones communes*]," I propose, is a means of giving world to the milieus between sensibility and intelligibility in ways that are instructive for thinking toward the inhuman commons of (multispecies) metaphor. In the *Ethics*, Spinoza divides concepts into "common notions" and "abstractions." A common notion is a form of concrete understanding based on the relation between my body and another body, or my capacity to be affected by another body. In Spinoza's words, it is an idea of "something which is common to . . . the human body and certain external bodies" (*E* II, P38 Dem). Abstractions, on the other hand, are imaginations of relations that we do not comprehend because they are not directly registered either in extension or in thought. Something of the Heideggerian material commons of the *symbolon* is already being formulated here—a joining of entities that together compose a world—but without the metaphysical grounding in logos that ensures that whatever "is held together, fits together." In his gloss of the distinction between common notions and abstractions, Deleuze writes,

> A common notion expresses our capacity for being affected and is explained by our power of comprehending. On the contrary, an abstract idea arises when our capacity for being affected is exceeded and we are content with imagining instead of comprehending: we no longer to seek to understand the relations that enter into composition; we only retain an extrinsic sign, a variable perceptible characteristic that strikes our imagination, and that we set up as an essential trait while disregarding the others (man as an animal of erect stature, or as animal that laughs, that speaks, a rational animal, a featherless biped, etc.).[10]

Our existence, along with the existence of every living or nonliving thing on this planet, is an expression of the earth—the originary, divine Substance that is equivalent with both God and Nature—that ceaselessly unfolds itself through the medium of its beings and its phenomena. Yet the relationality made through common notions—acts of joining, of throwing or being thrown together, of affecting and being affected—is not exclusive to the body. Rather, "the mind's power of thought is equal to . . . the body's power of action" (*E* III, P 28 Proof). Affect (*affectus*), an emanation from the earth itself, imprints minds and bodies alike: thinking and acting are different yet immanent means of participating in the unfolding of the earth's substance. Common notions result from intensive and involved attitudes and engagements (mental or physical, in thought or in extension) with and toward what we are affected by. Abstractions, on the other hand, result from less vividly experienced relations that rely on secondhand imagination to supply their images, which then become universalized into the form of the metaphysical concept. In this way, the materially-affectively formed commons of the common notion give another ground for metaphor, a ground that generates indifferently to abstract concepts, or to the primordial agreements by which *zoon logon echon* metaphysically secures itself. Metaphor flowers in the interstices between sensibility and intelligibility. And in so doing, metaphor joins—or makes common—what is monstrous, enigmatic, or otherwise not meant to fit.

We can also think of metaphor as a complicating or stilling of the related passage from sentience to sapience, or from embodied relationality and earthly affection to the abstract "wisdom" and intellection that is inscribed into the name of the human species. If consciousness and its concept-making—as human communion with logos, discourse, speech—define sapient over and against sentient being, then a multispecies metaphorics might give us a means toward exploring what else happens, or what happens beyond the grasp of concepts, both in worlds and in thought. As an instance of what Deleuze calls "a variable perceptible characteristic . . . set up as an essential trait while disregarding the others," consciousness itself is an abstraction raised above the commons formed through sentient, inhumanist relationality. The boundary between consciousness and sentience is Man-made / -furrowed rather than ontological, however, and tarrying with it may bring us closer to an understanding of thought as collaborative, material-affective worlding rather than exclusively as conceptual abstraction.

The most compelling such tarrying to date, to my mind, is Sunaura Taylor's *Beasts of Burden: Animal and Disability Liberation*. The collaboration between speciesist and ableist / neurotypical logics, in Taylor's argument, allows for the production of consciousness or rationality or other forms of cognitive ability as definitional both to Homo sapiens and to collective understandings of what makes a life worth living or protecting. This definition gives rise to the structural devaluation and oppression of both disabled lives and animal lives. But it also, paradoxically, allows for the emergence of new alliances and intimacies between animality and disability, as both theoretical categories and embodied modes of being. Taylor writes,

> Justifications for human domination over animals almost always rely on comparing human and animal abilities and traits. We humans are the species with language, with rationality, with complex emotions, with two legs and opposable thumbs. Animals lack these traits and abilities and therefore exist outside of our moral responsibility, which means we can dominate and use them. But isn't it ableist to devalue animals because of what abilities they do or do not have? The fact that one of the most ubiquitous arguments people use in support of our continued exploitation of nonhumans is that animals are incapable of a myriad of cognitive processes that human beings engage in shows the extent to which speciesism uses ableist logics to function.[11]

There is an implicit metaphoricity here (which Taylor expands on in the "Animal Insults" chapter, to be discussed further in chapter 3) in the articulation of human sapience in and through structures of comparison with nonsapiential beings. This exceptionalized sapience, and the sensus communis generated from it, also privileges individualist rather than interdependent modes of being: physical autonomy, self-sufficiency, and independence are systematically valued over shared vulnerability or coexposure, mutual cooperation, interspecied knottedness, as what normatively defines a "human" life or a life worth consideration and care.

We see these logics and valuations at work even amidst critical attempts to dismantle the structural underpinnings of anthropocentrism, such as the Cambridge Declaration on Consciousness of July 7, 2012. The authors, "a prominent international group of cognitive neuroscientists, neuropharmacologists, neuroanatomists, and computational neuroscientists," state "unequivocally" that nonhuman animals possess the neurobiological and neurophysiological "substrates" for consciousness and "affective qualia" as well as "the capacity to exhibit intentional behaviors."[12] Affective experience and conscious intentionality, the authors conclude, are not the exclusive property of human beings but rather extend to nonhuman (and nonmammalian) animals: "The absence of a neocortex does not appear to preclude an organism from experiencing affective states. Convergent evidence indicates that non-human animals have the neuroanatomical, neurochemical, and neurophysiological substrates of conscious states along with the capacity to exhibit intentional behaviors. Consequently, the weight of evidence indicates that humans are not unique in possessing the neurological substrates that generate consciousness. Non-human animals, including all mammals and birds, and many other creatures, including octopuses, also possess these neurological substrates."[13] Laudable as its aims may be, the Cambridge Declaration nonetheless relies on and perpetuates a determinate—and potentially ableist—idea of consciousness or conscious intentionality as the ground from which a life attains value.[14] The Cambridge Declaration's anti-anthropocentrism, it seems, still maintains a hierarchical separation between the conscious intentionality of sapience—along with the neurological capacity to differentiate one's self from one's surroundings—and the knottedness of sentience in which body, thought, and earth relate outside of the humanist logics of what counts as intellection, thinking, and wisdom.[15]

The term "sapience," as Michel Serres notes in the "Tables" chapter of *The Five Senses*, retains another, underrecognized connection to the labor of the senses; specifically, the ability to taste: "We were too quick to forget that *homo sapiens* refers to those who react to sapidity, appreciate it and seek it out, those for whom the sense of taste matters—savoring animals—before referring to judgment, intelligence, or wisdom, before referring to talking man. The rise of the golden mouth at the expense of the tasting mouth. But hidden within a dead language, we find this confession of the first about the dead mouth: namely that wisdom comes after taste, cannot arise without it, but has forgotten this."[16] However, Serres is in turn too quick to turn taste into another means of constructing a sapience-based commons—now rooted in sensory pleasure (the "tasting mouth") rather than conscious intentionality and language (the "golden mouth")—that excludes those humans and nonhumans without the capacity and aesthetic refinement to experience and seek after taste. The classist undertones of this chapter, with its focus on the epicurean banquet table and the aromatic pleasures of a particular Yquem-region wine,[17] are unfortunate in the larger context of the book's compelling elaboration on the idea that "our intellect knows nothing that has not first passed through the senses."[18] Writes Serres, "Without taste, we risk abnegating our human state and returning to that of animals. . . . Animals wolf down their food, man tastes it."[19]

Its classism and Euro-/Franco-centrism notwithstanding, Serres's account helps us to think a sapience that is mingled with bodily and sensory relations and a knowing that is generated from the ongoing confluence of sensate givens and earthly affections. There is a "philosophy of confusion" to be made from the mixtures of streams involved in the production of food and wine (as well as soil, flowers, and metals), a philosophy that "should be the common ground of sapience."[20] If sapience is generated from and achieved through acts of differentiation from nonsapient beings, as Taylor argues, it is also dependent on a particular metaphorics by which these comparisons are organized and distributed: a sapient metaphorics, we might say, is rooted in the ongoing preservation of the conceptual edifices that hold *zoon logon echon* metaphysically apart from *álogon*. A multispecies/inconceptual metaphorics, on the other hand, occupies the confounded milieus between sensibility and intelligence, working toward a bodily and earthly oriented sapience while taking sentient worldings for wisdom.

II.

We can further develop a theory of the inconceptual—as a generative indifference to concepts and conceptual structures—in the relational philosophy of both Spinoza's *Ethics* and Deleuze's *Difference and Repetition*, with Nietzsche's "On Truth and Lie in an Extra-Moral Sense" serving as a nodal point for Deleuze's Spinozism. This brief foray, I hope, will yield some insight into how we might think *áloga*—the fractious, knotted beings and worlds excluded from logos yet gathered, epiphorically, in metaphor—without spiritualizing them into concept.

For Spinoza, the common notion that joins or throws together is also a way of thinking the relationality of bodies, whose patterns of "motion and rest, speed and slowness" through which they are defined, unfold without the mediation of conscious thought (*E* II, L1). There is another joining that becomes thinkable once we imagine the body as opaque to thought and unpossessable by mind: "The mind does not know the body," Spinoza writes, and "no one has yet determined what the body can do" (*E* II, P23 Dem; III, P2 S).[21] Neither does the mind "involve adequate knowledge of the parts composing the human body," as these parts are "highly composite individuals, whose parts can be separated from the human body and communicate their motions to other bodies in another manner" (*E* II, P24 Dem).

What Spinoza calls "communication [*communicant*]" between and among bodies and parts, we might call an enigmatic worlding in which what is joined joins "in another manner [*alia ratione communicare*]" than "that of the body as a whole" (*E* II, P24 Dem). Spinoza stills the passage from bodily relation to abstraction, staying amidst rather than sublating the contingency of beings, bodies, and parts touching together. In Spinoza's original, P24 of Part II reads, "Mens humana partium corpus humanum componentium adaequatam cognitionem non involvit" (The human mind does not involve adequate knowledge of the parts composing the human body). *Cognitionem* can also be translated as "recognition," in which case, the body (and its parts) is actively estranged from the mind, unfolding itself in and with the world in ways that the mind, no longer able to comprehend this alien form to which it finds itself attached, "can only wonder at" (*admiratur*) (*E* II, P2 Schol). The self is noncoincident with itself, threaded in a wondrous loop.

While we might imagine a trajectory leading from the sentient, relational world-making of the body to the abstract conceptuality of the mind, Part II of the *Ethics* goes on to give a mind alienated from itself, with its "confused and mutilated" self-knowledge contingent on its perception of the body's "affections." "The human mind . . . has not an adequate but only a confused and mutilated knowledge of itself," Spinoza writes, and "the mind does not know [*cognoscit*] itself, except in so far as it perceives the ideas of the affections of the body" (*E* II P29 Cor; P 23).[22] Thinking, then, is not the production of abstractions out of sensate phenomena but rather a means of ordering the body's affections into lines of thought. Before Part V of the *Ethics*, which tends to retreat from the more radical implications opened up in the previous four Parts, the mind is dispossessed of its determinative power over things, affections, and orderings: "In the mind there is no absolute, or free, will, but the mind is determined to will this or that by a cause which is also determined by another, and this again by another, and so to infinity" (*E* II, P 48).

Thought is thus a contingent, material-affective passage across "images of things in the body," as Spinoza notes in his description of the work of memory:

> [Memory] is nothing other than a certain connection [*concatenation*] of ideas involving the nature of things which are outside the human body—a connection which is in the mind according to the order and connection of the affections of the human body. . . .
>
> And from this we clearly understand why the mind, from the thought of one thing, immediately passes [*incidet*, "falls," "encounters"] to the thought of another, which has no likeness to the first: as, for example, from the thought of the word *pomum* a Roman will immediately pass to the thought of the fruit [an apple], which has no similarity to that articulate sound and nothing in common with it except that the body of the same man has often been affected by these two, that is, that the man often heard the word *pomum* while he saw the fruit.
>
> And in this way each of us will pass from one thought to another, as each one's association has ordered the images of things in the body. For example, a soldier, having seen traces of a horse in the sand, will immediately pass from the thought of a horse to the thought of a horseman, and from that to the thought of war, and so on. But a farmer will pass from the thought of a horse to the thought of a plow, and then to that

> of a field, and so on. And so each one, according as he has been accustomed to join and connect the images [*imagines*] of things in this or that way, will pass from one thought to another. (*E* II, P 18 Schol)

The mind is a place for contingent joinings, thrown-together things, and concatenations of images as they are "ordered . . . in the body": thought, we might say, is the material arrangement of sentient relations. Edwin Curley translates the "falling [*incidet*]" of the images from the external world upon the mind as the "passing" of one thought to another. (A strange resonance emerges here between Spinozist memory and Virginia Woolf's artistic dictum from "Modern Fiction": "Let us record the atoms as they fall upon the mind in the order in which they fall, let us trace the pattern, however disconnected and incoherent in appearance, which each sight or *incident* scores upon the consciousness."[23]) An affective common forms in the mind, composed not through the metaphysical agreements of logos but rather through passages across the body, through the inhumanistic holding together of what falls.

For Nietzsche, the concept itself (Begriff, from begreifen: "to grasp") is a means of asserting possession over phenomena and beings, primarily by sublating their relational differences into abstract universality. As Nietzsche writes in "On Truth and Lie in an Extra-Moral Sense [*Über Wahrheit und Lüge im außermoralischen Sinne*, 1873],"

> Every word immediately becomes a concept [*Begriff*], inasmuch as it is not intended to serve as a reminder of the unique and wholly individualized original experience [*Urerlebniss*] to which it owes its birth, but must at the same time fit innumerable, more or less similar cases—which means, strictly speaking, never equal—in other words, a lot of unequal cases. Every concept originates through our equating what is unequal. No leaf ever wholly equals another, and the concept "leaf" is formed through an arbitrary abstraction from these individual differences, through forgetting the distinctions [*durch beliebiges Fallenlassen dieser individuellen Verschiedenheite*]. . . . We obtain the concept, as we do the form, by overlooking what is individual and actual; whereas nature is acquainted with no forms and no concepts, and likewise with no species, but only with an X which remains inaccessible and undefinable for us. For even our contrast between individual and species is something anthropomorphic and does not originate in the essence [*Wesen*] of things.[24]

In constructing abstractions, we do not just "forget" differences; we "let them fall away" (*Fallenlassen dieser individuellen Verschiedenheite*) and thus allow for the passage of singular and unreproducible particulars into a determinate and repeatable form. The "intellect," for Nietzsche, is a "means for the preservation of the individual," both against "other individuals" and against Nature itself. In this sense, Nature is a medium for the inconceptual, difference-creating expressions of beings that perhaps do not world against but world without regard to the concept-regimes imposed on them.[25] As Uexküll's peculiarly Nietzschean formulation of Nature has it at the closing of *A Foray*: "Forever unknowable behind all of the worlds it produces, the subject—Nature—conceals itself."[26]

In the Nietzsche-inflected Spinozism of *Difference and Repetition*, Deleuze traces human conceptuality (or "conceptual philosophical thought") to the universalizing logics of Cartesian "common sense," or the "*Cogitatio natura universalis*."[27] Common sense is what we "presume to know" before we even start thinking—namely, "what it means to think and to be," to have a "self."[28] Commonsensical thinking, then, is a pursuit of the "true," or the "pre-philosophical . . . Image of thought [*l'image de la pensée*]." The image of thought, for Deleuze, amounts not to any positive content but rather to a particular "form of representation or recognition" within which everything encountered in the world—everything that falls—is made intelligible through its connection to a perceiving consciousness. This representational form—the work of a humanistic epistemology that reads its own concepts in all of Nature and in turn sees Nature as a reflection of one's own subjective identity—is the Nietzschean "lie" that has been mistaken for truth by a difference-forgetting intellect. Where the common notion sought to name an enigmatic relationality between bodies, parts, and minds (and between the mind and itself), common sense seeks universally valid determinations of worlds—or "distribution[s] of the object and subject, Being and beings"—through conscious selves.[29] For the archetypal (Kantian) philosopher against whom both Deleuze and Nietzsche pose their counter-humanisms, "consciousness establishes between the I and the representation a relation much more profound than that which appears in the expression 'I have a representation': it relates the representation to the I as if to a free faculty which does not allow itself to be confined within any one of its products, but for which each product is already thought and recognized as past, the occasion of a determinant change in inner meaning."[30] In contrast to the Spinozist mind whose thinking is determined by material and affective causes in the world, Kantian consciousness (in Deleuze's interpretation) is beyond determination and unconditioned by its affections.

Thinking joins body to world: for Deleuze, Spinoza, and Nietzsche alike, thinking is a means of touching and gathering rather than determining. Consciousness, then, is a particular kind of conceptual thought but not the only means of thinking. In Deleuze's analysis, thought can become an oppositional force or "trespass" against conceptuality and its prephilosophical image of thought. Deleuze writes, "What becomes established with the new is precisely not the new. For the new—in other words, difference—calls forth forces in thought which are not the forces of recognition, today or tomorrow, but the powers of a completely other model, from an unrecognized and unrecognizable *terra incognita*."[31] The danger that Deleuze runs, however, is of supposing a higher power of thought—a violent, transgressive thinking that seeks out differences rather than truths, passing through the mind yet remaining unpossessable (and unrecognizable) by it—through which the thinking subject is, paradoxically, empowered over what surrounds it. The forces called forth by thought seem to take on a transcendent, spiritualized dimension in Deleuze's imagination, as if arriving from the heavens rather than growing from another, unrecognized *terra*. (Is a ground that we cannot relationally create with or even recognize perhaps just another name for the heavens?) Anticonceptuality, or philosophical trespass, risks becoming its own form of transcendence.

In contrast to the (early) Deleuzian world-unmaking power of thought, Part V of Spinoza's *Ethics* falls back on the harmonizing work of thought (or the "power of reason") to order the body's affections: "There is no affection of the body, of which we cannot form a clear and distinct concept [*conceptum*]," Spinoza claims (*E* V, Preface and P4 Dem) and in the accompanying Scholium: "Each of us has—in part, at least, if not absolutely—the power to understand himself and his affects, and consequently, the power to bring it about that he is less acted on by them" (*E* V, P4 S). Both Spinoza and Deleuze may thus tend toward forms of thinking that are still beholden to the concept, whether by undermining it with the new forces gifted from a new terra incognita or by using the mind's power to order affections into concepts. However, I would argue that these tensions—between concepts and common notions, between thought as harmonious ordering and thought as trespass, between established ground and unrecognizable ground/*terra*—are constitutive of each work's striving toward and making space for something like the inconceptual. Spinoza's qualifier "in part," in this sense, opens a hesitation between the conditional and the absolute power of thought over material affections. The common notion that gathers beings otherwise, the ungrounding of *Cogitatio natura universalis*, the liberation of thinking from human(ist) consciousness: all give us ways to think the expressivities of beings in their indifference to conceptual determination.

We might meet Deleuze's provocation on the force of thinking, "What is a thought that harms no one, neither thinkers nor anyone else?" with a related, more recessively posed improvisation: is there a thought capable of holding what falls, without determining and without harming?[32] Another way of asking is after a thought light enough to think *sinnlos* Odradek's atemporal falling—or his "rolling down the stairs" at the feet of the children (and children's children) of the *Hausvater*—as his knotted and tangled threads (*nachschleifendem Zwirnsfaden*) unspool beyond humanistically imaginable time and world. What thought to hold this worlding- and living-on (*überleben*) that "harms no one [*schadet . . . niemandem*]" while ceaselessly expressing its own unpossessability by either a thinking that domesticates or a thinking that trespasses?

III.

Let us return to our opening *ainigmata*—what does metaphor gather, and what gathers in metaphor?—and modulate them slightly: what does metaphor grow, and what grows in metaphor? As we explored, the Aristotelian theory of metaphor, in the *Poetics*, begins with a split between metaphor as a transfer of names and metaphor as an uncontrollable profusion of names (*metaphora* and *epiphora*). In the *Rhetoric*, Aristotle adds to metaphor's already multiplicate origins by theorizing metaphor in and as the growth of plants. From the *Rhetoric*:

> We all naturally find it agreeable to get hold of new ideas easily: words express ideas, and therefore those words are the most agreeable that enable us to get hold of new ideas. Now strange words simply puzzle us; ordinary words convey only what we know already; it is from metaphor that we can best get hold of something fresh. When the poet calls "old age a withered stalk," he conveys a new idea, a new fact, to us by means of the general notion of bloom, which is common to both things. . . . We see then that both speech and reasoning are lively in proportion as they make us seize a new idea promptly.[33]

The withered but still living stalk is sublated into the abstract notion of "bloom," which then holds both "stalk" and "old age" in conceptual relation. This sublation of living being into concept, additionally, enhances the liveliness of speech and reason by allowing us to grab hold of ideas (metaphor "gives name to nameless things")[34] that were formerly inexpressible in language. I want to stay with this metaphor of the withered stalk—along with the histories and worlds still blooming within it—as we explore metaphor's inconceptual modes of growth and gathering. Similarly, let us also stay with the "flowers of rhetoric"—a central component of Derrida's vegetal metaphorics—through which metaphors enter meaning while also speaking otherwise, or in another manner than meaning.

The trope of multispecies metaphor is, in part, an attempt to undo (or to regather) the implicit critical separation between animal metaphor and plant metaphor, as modes of thought and as foundational claims for the origins and futures of metaphor. In *The Posthuman*, for example, Rosi Braidotti insists on animal metaphor as the place from which a new ontology (a "zoontology") must emerge to rethink the dominant anthropocentrism of metaphor: "The point now is to move towards a new mode of relation; animals are no longer the signifying system that props up human self-projections and moral aspirations. They need to be approached in a neo-literal mode, as a code system, or a 'zoontology' of their own."[35] However, I would argue that we delimit metaphor's generative capacity by grounding metaphor in another code system that is still beholden to the formal demands of ontology. Accordingly, the contemporary logic of *zoon* emerges largely from Herbert Spencer's *Principles of Biology* (1864), in which Spencer reconfigures the Greek *zoon* into an English term referring to a "zoological individual . . . constituted by any such single animal as a mammal or bird or by any . . . group of animals . . . that have been developed from the same egg."[36] Read this way, zoontology—and the zoontological metaphorics that would extend from it—risks locating the essence and expressivity of beings in individuation rather than in relational co-composition and indeterminate growth.

In this vein, Jeffrey T. Nealon argues (in the aforementioned *Plant Theory*) that the emphasis on the individualized zoon gives rise to a metaphorics that tropically privileges (individualized) animal over (multifarious) plant being in the making of worlds.[37] Nealon questions whether Heideggerian world extends to vegetal forms of life, which seem incongruous to both animalian *weltarm* (poor in world) and lithic *weltlos* (worldless). Plants, accordingly, persist as nontotalizable aggregates, dually abjected from both philosophy and history. They do not readily lend themselves to the ensouling individuations that would allow or disallow beings to have access to being, to world, and to the futures made possible through a being's having of world. Cain's vegetal sacrifice must be supplanted by Abel's animal sacrifice as the founding of a religion; rhizomatic growth must give way to agrilogistic determination, nature to (agri)culture.[38]

Following this schema, the prehistoric flower religions, according to Hegel's *Phenomenology of Spirit* (where they are also associated with the Romanticism of Goethe and Schelling), are a "quiet and impotent" phase of being that has to "pass . . . into the seriousness of warring life, into the guilt of animal religions; . . . into destructive being-for-self" that constitute history and culture.[39]

There is a directionless, "bad infinite" desire of plants, one that, unlike the "good infinite" of animal desire—as a desire that "can be repressed, sublated, trained"—cannot be incorporated into the work of Spirit: "Plant desire merely grows without telos, without a proper end (or a proper death)—and the logic here is classic: in Aristotle . . . the vegetable *psukhe* or soul deploys a single power, the ability to grow. But that power is without entelechy—plants have no higher, ideal form toward which they develop. They just grow uncontrollably until they grow no more."[40] Plants also grow without regard for concept, we might argue. And there is something monstrous about this growth that is perhaps not exclusive to vegetal being: a growth without attachment to any determinate or individualized body.[41] According to John Hunter's theory of monstrosity, "The great principle of monstrosity in a vegetable relates to the constant property of forming new and similar parts, and to a stop being put to, or a violence committed to, the natural growth of one of these parts."[42] An unzooned animal is similarly a swarm, a many-futured aggregate, whose being only emerges in and through its enigmatic relations. Metaphor gathers, too, those beings that do not desire or grow at all but nonetheless insist on (or unspool) their own nonentelechial infinity. Rather than a choice between different modalities of world and nonworld, multispecies metaphor can instead be thought as animal, plant, and lithic all at once,[43] holding in common the knottedness of all *áloga*.

The monster also doubles as a name for nonsymbolic joining. As discussed in the introduction, the English word "monster" first emerged in Chaucer's *Monk's Tale* (ca. 1375), where it denoted the integration of two or more (human or nonhuman) bodies, or "a mythical creature which is part animal and part human, or combines elements of two or more animal forms."[44] Every monster is immediately a communion. Deriving from the Latin *monstrare* ("to show," which also forms the roots of words like "demonstration" and "remonstration") as well as *monere* ("to warn" and "omen"), the monster also disconceals or unearths what else happens in worlds, or what happens outside of Enlightenment orderability. For film theorist André Gaudreault, the "monstrator" is a term for a narrator-like figure in film and literature—defined as well as a "plural agent" without a "single voice"—who simply shows (or reveals, or "monstrates") the content of the narrative without mediation through their own (or any other) subjective consciousness.[45] Everything contradictorily gathered—or everything that expresses from the interstices between determinate concepts, everything that gives more names than can be humanistically held—we might call monstrous, and we might also call metaphor.

Metaphor itself, as we have explored, is a contradictory gathering. *Metaphora* cohabitates uneasily with *epiphora*, an alogic of accretive profusion that threatens to undo metaphor's capacity to transfer meaning and place among beings and across contexts. And this fraught (thrown-)togetherness of beings, names, and worlds is internal to the operations of metaphor; in a word, it *is* metaphor. (Regarding its philological origins, Paul Ricoeur provocatively asks, "But is not the word 'metaphor' itself a metaphor, the metaphor of a displacement and therefore of a transfer in a kind of space?"[46]) Yet most philosophies of metaphor following from Aristotle have tended to emphasize metaphor's transactional or transferential dimensions over and above the endlessness of its growth and gathering. *Epiphora* tends to fall away from metaphor's worlds, or to become sublated from the growing ground into the ethereality of concept.

To give a brief sense of how far the philosophical and rhetorical dissolution of *epiphora* has carried, we can look to Philip Wheelwright's division of metaphor, in "Metaphor and Reality" (1962), into the varieties of *epiphor* and *diaphor*. Wheelwright argues, concerning the "two ways of metaphor," that the function of "*epiphor*—which is to say, metaphor in the conventional Aristotelian sense—is to express a similarity between something relatively well-known or concretely known (the semantic vehicle) and something which . . . is less known or more obscurely known."[47] The accretive growth of *epiphora* is subsumed into metaphor's traditional knowledge-giving function. The other dimension of metaphor, *diaphor*, is described as "movement (*phora*) . . . through (*dia*) certain particulars of experience (actual or imagined) in a fresh way, producing new meaning by juxtaposition alone."[48] Metaphor functions as either the transference or the juxtaposition of meaning already made, immured within the conceptualities against which its gathered terms agitate. (Ricoeur and Derrida offer some of the first critiques of such a concept-bound metaphoricity, about which more soon.)

But first, let us take Quintilian's *Institutio Oratoria* (pub. 95 AD) as an early example of *epiphora*'s falling away from metaphor. Quintilian echoes Aristotle's argument about the name-giving power of metaphor, with an emphasis on how metaphor "transfer[s]" a word from "the place to which it properly belongs" in the language to a new, unfamiliar place to produce fresh meanings: "[A metaphor] adds to the copiousness of a language by the interchange of words and by borrowing, and finally succeeds in accomplishing the supremely difficult task of providing a name for everything. A noun or verb is transferred from the place to which it properly belongs to another where there is either no *literal* term or the *transferred* is better than the *literal*. We do this either because it is necessary or to make our meaning clearer or, as I have already said, to produce a decorative effect. When it secures none of these results, our metaphor will be out of place."[49] Quintilian insists on the possibility of metaphor producing a remainder-less fit between a word and its meaning, and a language progresses toward greater clarity via these graftings of words from one context to another. Languages grow according to a logic of perfectibility. We can also read into this account an implicit dynamic of sacrifice, in which language's enlivening requires the ongoing replacement of existing (literal) terms by superior terms that travel (or are transferred) from other places. In Nealon's Hegelian-vegetal schema, Quintilian's metaphorics enact the good infinite / animal desire of growth and movement toward an enclosed, self-relational totality, rather than the spurious infinite / plant desire that Nealon defines as "a kind of undifferentiated and uncontrolled growth that is finally anathema to the progress of spirit and law."[50]

The logics of *metaphora*'s (sacrificial) progress toward perfection (via transference or juxtaposition)—and its gradual shedding or sublation of the excesses of *epiphora*—pass into most subsequent philosophies of metaphor, including the late Hegelian in the *Lectures on Aesthetics* (1835). Hegel's account tends to preserve the Quintilianian, good accretive dynamic of metaphorical meaning, while recasting its terms slightly. Rather than the transition of a word from one context or place in language to another (as in Wheelwright's definition of *epiphor*), Hegelian metaphor traces a thing's passage from sensuous particular into the "spiritual sphere" of abstract meaning. Metaphors arise according to a language's need for new descriptions of—or new connections between—existing things. As a result, however, the difference-animating singularity of the metaphor's new description or connection dissipates, or effluviates into Spirit. Hegel writes:

> Metaphor has its principal application in linguistic expressions which in this connection we may treat under the following aspects:
>
> a) In the first place, every language already contains a mass of metaphors. They arise from the fact that a word which originally signifies only something sensuous is carried over into the spiritual sphere. *Fassen, begreifen* [to grasp, to apprehend], and many words, to speak generally, which relate to knowing, have in respect of their literal meaning a purely sensuous content, which then is lost and exchanged for a spiritual meaning, the original sense being sensuous, the second spiritual.
>
> b) But gradually the metaphorical element in the use of such a word disappears and by custom the word changes from a metaphorical to a literal expression, because owing to readiness to grasp in the image only the meaning, image and meaning are no longer distinguished, and the image directly affords only the abstract meaning itself instead of a concrete picture. If, for example, we are to take *begreifen* in a

> spiritual sense, then it does not occur to us at all to think of a perceptible grasping by the hand. In living languages, the difference between actual metaphors and words already reduced by usage to literal expressions is easily established; whereas in dead languages this is difficult because mere etymology cannot decide on the matter in the last resort. The question does not depend on the first origin of a word or on linguistic development generally; on the contrary, the question above all is whether a word which looks entirely pictorial, deceptive, and illustrative has not already, in the life of the language, lost this its first sensuous meaning, and the memory of it, in the course of its use in a spiritual sense and been sublimated [*aufgehoben*] into a spiritual meaning.[51]

The metaphor is a narrative of its progress from phenomena to form, or spiritualized concept. Without sublimation into idealist universality—such as the transition of *begreifen* from tactile to conceptual grasping—the sensuous particulars brought into language by metaphor might multiply, repeat, gather names, improvise new commons. Spiritualization may thus become a means of repressing the material-affective content inhering in words, or of letting fall away what of beings and words is unraisable into conceptuality.

What Hegel calls spiritualization, Donald Davidson more bluntly calls death. In "What Metaphors Mean" (1978), an analytic philosophical approach to the metaphorical dynamics traced by Hegel, Davidson argues that metaphor bears two primary yet seemingly paradoxical properties. The first is a radical indeterminacy of content despite the metaphor not saying "anything beyond its literal meaning." In this sense, the effect of metaphor (as in Aristotle's and Quintilian's formulations) is to make us notice something that we did not notice before, but this is never a limited or "finite" process.[52] As Davidson explains, "There is no limit to what a metaphor calls to our attention, and much of what we are caused to notice is not propositional in nature. When we try to say what a metaphor 'means,' we soon realize there is no end to what we want to mention. If someone draws his finger along a coastline on a map, or mentions the beauty and deftness of a line in a Picasso etching, how many things are drawn to your attention? You might list a great many, but you could not finish since the idea of finishing would have no clear application."[53] This limitless production of meaning—which we might think variously as *überleben*, plant desire, or *epiphora*—coexists with the tendency of metaphors to die by passing into concept, at which point the likenesses animated by a metaphor become identical with a literal description of its object.

Davidson's analysis of "dead metaphors" hints at the fantastical or *Märchen*-like aspects of this transformation: "Once upon a time . . . rivers and bottles did not, as they do now, literally have mouths. . . . When 'mouth' applied only metaphorically to bottles, the application made the hearer *notice* a likeness between animal and bottle openings. . . . Once one has the present use of the word, with literal application to bottles, there is nothing left to notice. There is no similarity to seek because it consists simply in being referred to by the same word."[54] A metaphor spiritualized into descriptive language is no longer a metaphor: metaphor does not contain a "hidden message," for the purpose of metaphor is not to communicate ideas but only to self-express or to unfold its own content.[55] But a possessed metaphor is like a solved riddle: each is a living-on and living-otherwise, a subterranean insistence beyond determination. Davidson's account holds metaphor in the suspension of this contradictory or knotted state in which a metaphor can be simultaneously determined and boundless, living and dead, and in our terms, both transferred meaning and epiphoric excess.

Something of this knottedness inheres, too, in Paul Ricoeur's expansive philosophy of metaphor, particularly in his insistence on the copula-being of metaphor. Metaphor asserts identity (in the form of "X is Y"), but the "is"/copula is not reducible to literal meaning: there is always, as Ricoeur claims in *The Rule of Metaphor*, a "metaphorical sense of the verb *to be* itself" in which an assertion of a literal identity is also its negation. In the play of the metaphoric copula, then, we find a series of tensions: (1) a tension between the terms of the statement, (2) a tension between "literal and metaphorical" interpretations of the relation between these terms, and (3) a tension between the modes of "identity and difference" proper to all copulative play.[56] In Ricoeur's analysis, "in order to elucidate this tension deep within the logical force of the verb *to be*, we must expose an 'is not,' itself implied in the impossibility of the literal interpretation, yet present as a filigree in the metaphorical 'is.'"[57] In other words, every "is" is subtended by an implicit "is not," and this "existential" tensility or (k)not(ted)ness of identity and difference *is* metaphor.

We see the fantastical aspects of metaphor's living and (un)dying, being and not-being, at work as well in the life of Gregor Samsa and in the life of the metaphor that he becomes in *The Metamorphosis*. In contrast to Odradek's inconceptual spooling-on beyond the time of the family and the home, Gregor's (insectile, inhuman) expressivities begin and end in domestic confinement. What we see and hear of Gregor's life is a fraught passage, or transformation, from worlding—and crawling, hanging, self-suspending, joyous—being into deworlded concept. As *álogon*, or "monstrous vermin" (*ungeheuer Ungeziefer*), Gregor awakes abjected from both family and species, his unsignifying speakings and gestures excluded from humanism's discursive agreements. Gregor's first interaction is his attempt to communicate to the family and chief clerk through his bedroom door, manifested as the mixing of "his own voice [*Stimme*]" with a "persistent horrible twittering squeak [*schmerzliche Piepsen*] behind it like an undertone [*wie von unten her*]," which is initially partially comprehended by Gregor's mother.[58] As noise-entangled voice, simultaneously *psófos* and *logos*, such expression threatens the entire order of metaphysical meaning and so must be driven back into the undergrounds of *logos* by the chief clerk's determination: "That was an animal voice [*Das war eine Tierstimme*]."[59]

The mingling of noise and word is also the complication of metaphorical meaning, in which the underground speaking of metaphor seeps into the spaces between logos's discursive agreements. Gregor lives and sounds in the interstices of meaning, as incomplete or stalled sublation of material into concept. And so it takes Gregor's unceremonious death (and silencing) for metaphoricity and literalness to become distinguishable for the first time.[60] Gregor's passing is given through the housekeeper's use of a "dead" metaphor to describe his corpse: "Look for yourselves, it's croaked; it's lying there, totally dead!" (*Sehen Sie nur mal an, es ist krepiert; da liegt es, ganz und gar krepiert!*).[61] The housekeeper twice employs the disrespectful and colloquial verb *krepieren*, meaning "to die," often "to die a miserable death," but also translated metaphorically as "to croak." (This in place of the more neutral verb *sterben* [to die] and the more specific *verenden*, which has the same meaning but is used primarily for the death of animals.) Gregor's death is thus also his raising into concept, the point when metaphorical expression coincides with literality: "to croak" is simply "to die" and nothing else. Stanley Corngold's translation cleverly employs an idiomatic "dead" metaphor in English to illustrate this point: "it's lying there, dead as a doornail!"[62] The passage of the metaphor, and Gregor, into the spiritual sphere is made complete. A new referential language purged of its monstrosities emerges, and its first order is to reconstitute the family unit—a perfect *symbolon*—that Gregor has disrupted: "The couple Mr. and Mrs. Samsa [*Das Ehepaar Samsa*] sat up [*aufrecht*] in their marriage bed."[63] The Samsas enact a raising up or an uprightness (*aufrecht*) that constitutes proper, human(istic) being against the verminous, many-legged metaphor that could only fully enter the human realm of logos as a corpse.[64]

What both Samsan and Odradekian varieties of metaphor do not provide, however, is a sense of stable origin. Odradek is both himself in the present and the many provisional origins speculated into him, and Gregor has no singular, premetamorphosed identity prior to his waking. And as Odradek endlessly unspools down staircases, so the "undertones" of Gregor's hybrid noise-voice resonate still. (Gregor's vocal undertone, Kafka writes, leaves "the words in their clear shape only for a moment and then rose up reverberating around them to destroy their sense, so that one could not be sure one had heard them rightly [*daß man nicht wußte, ob man recht gehört hatte*]."[65]) Each embodies the living-on of metaphor in excess of origin and beyond a presentist metaphysics. For Derrida, this quality of many-origined indeterminacy is proper to metaphor itself. Every metaphor, Derrida argues in "White Mythology: Metaphor in the Text of Philosophy," holds in itself the sedimented content of its past usages, milieus, meanings, and transversal labors. And so each utterance of a metaphor is simultaneously a geological unearthing of this mingled, epiphorically heaped content. While we may believe in a "primitive and concrete meaning that lurks yet present under the abstract and new interpretations [of metaphor]," metaphorical expression, its monstrous gatherings and growths, exceeds the contexts in which it is meant to serve as meaning-granting exemplum.[66] Metaphor, as Ricoeur argues in terms borrowed from Nelson Goodman, "yields while protesting": metaphor is defined by its continual friction against the enclosures of concept.[67] And "intelligibility," as Nealon elaborates in his analysis of Derridean metaphoricity, cannot "fully . . . escape its dependence on these flowers of rhetoric, which are already overgrown."[68] As its flowers inconceptually bloom, however, so its roots radiate, connecting grounds to other grounds, forging terran and subterranean communions undeterminable by logos.

We need think only of the cell, Robert Hooke's metaphor for the most basic biological unit of existence, as noted by Georges Canguilhem in *La connaissance de la vie* (a central text in Derrida's theorization of metaphor). Even though the metaphor of the cell seems self-evidently identical with its object, Canguilheim argues, "cell" also relates to the "cells of wax filled with honey" of beehives. So instead of thinking of this relationship between metaphor and object as one of "complete . . . correspondence" (or Davidsonian dead metaphor), we might wonder whether "in consciously borrowing from the beehive the term cell in order to designate the element of the living organism, the human mind has not also borrowed from the hive, almost unconsciously, the notion of the cooperative work of which the honeycomb is a product."[69] (Even in the conceptualization of the smallest isolable organic entity, there is already the suppressed undertone of relation, communion.)

Continuing Derrida's emphasis on metaphor's accretive, underground growth, Hans Blumenberg's *Shipwreck with Spectator* employs the term "life-world" (*Lebenswelt*), borrowed from Edmund Husserl, to think the expressive, unraised sediments of metaphor. A resonance opens here between metaphor's noise worlds and the enigmatic tone system through which Uexküll theorizes relations among *Umwelten*. All metaphors, for Blumenberg, emerge from a "nonconceptual" and inharmonious life-world that must be incorporated—

often violently—into the harmony of a "total causal system."[70] ("Metaphor," Blumenberg writes, quoting Husserl, "is, first of all, 'resistance to harmony.'"[71]) Like Gregor and like Gregor's twittering squeak-speech, metaphors are dissonances within the metaphysical order into which they are brought to serve as transits for meaning: sensibilities that resist encounter with—and passage into—a murderous intelligibility.[72] In Blumenberg's words, metaphor "interposes a heterogeneous element that points toward a different context from the actual one," a waywardness or disharmony that must be resolved through the work of consciousness as it "find[s] its way back to the harmony of the data as data of *one* experience."[73]

Something of dissonance—the subterranean content and inconceptual gathering modes of metaphors, beings, and *ainigmata*—sounds still through all (re)solution. Each of the chapters to come is attempted attunement to / with this coarticulated clamor, *psófos* gathered in the undergrounds of logos, that constitutes metaphor and other inhuman modernist modes of world-making. "Metaphors speak in riddles [*metaphorai . . . ainittontai*],"[74] as Aristotle claims in the *Rhetoric*, and "the very nature of the riddle [*ainigma*] is this, to describe a fact in an impossible combination of names."[75] Metaphors speak nothing but themselves, but the selves of metaphor are without end, made anew in each improvisatory gathering. In the *Origin* and the *Alice* books, to which we now turn, thought gathers the expressive illogic and nonsense of what has been sedimented over the course of evolutionary and geological time; strives to hold in common what sapience cannot join.

Chapter 2

Sediments

Ground and Under Ground in *Alice* and *On the Origin of Species*

And what might come to pass if inhuman expressivities and noises—Aristotelian *psófos*—could dissensually gather in logos's undergrounds? The Duchess of *Alice's Adventures in Wonderland* tells Alice, "Take care of the sense and the sounds will take care of themselves," and the ground opens. For by letting Wonderland's and the Looking-Glass world's sounds "take care of themselves," they grow beyond the agreements demanded by logos, ungrounding sense with the surplus of sense that is nonsense. Representational language/logos is itself governed by a tenuous, easily disruptible union between sound and sense, with the nonsense-bearing voice (*phone*) as the primary agent of this disruption, as sound theorist Mladen Dolar argues in *A Voice and Nothing More*:

> There is a speech which makes sense, and in that horizon of sense-making there is suddenly a disturbance, the intrusion of the voice, the sound, which functions as a disruption of which we cannot make sense. The element of the voice, in the form of contingent and senseless co-sonance, unexpectedly runs amok and produces nonsense, which in the second step turns out to be endowed with an unexpected sense emerging from it. . . . The nonsense emerges from contingent sound encounters, and with it another sense, which can manifest itself only for a

> moment because of that co-sonance, through the momentary resounding, then it is gone.[1]

Nonsense, in Deleuze's terms, is that which "says its own sense," though nonsense also emanates from multiple sources and speaks multiple senses.[2] Nonsense, for Dolar and for us, is epiphoric: a piling up or profusion of (contingent, touched-together) sounds that gather indifferently to the orders of sense. Carroll gathers this noisy illogic into the multisonant being of the Jabberwock, a monster in whom the "voice" (*woce*, an archaic form of the Latin *voce*) goes astray, or jabbers.

Darwin's monsters are also alogic communions: the evolutionary monster is a botched outgrowth of determinate lineage, as well as a term for those beings that form untraceable intimacies with one another without totalizing into separate "species" or "varieties" of their own. The idea of "species," in other words, is another name for the improvisatory unfolding of beings over time. Darwin writes in the *Origin* (hereafter *OS*), "At long intervals of time, out of millions of individuals reared in the same country and fed on nearly the same food, deviations of structure so strongly pronounced as to deserve to be called monstrosities arise; but monstrosities cannot be separated by any distinct line from slighter variations. All such changes of structure, whether extremely slight or strongly marked, which appear among many individuals living together, may be considered as the indefinite effects of the conditions of life on each individual organism" (*OS* [6th ed] 27).[3] The deviations called monstrosities—born of the "indefinite effects of the conditions of life"—are both "strongly pronounced" and inseparable "by any distinct line" from variations, alienated from the concepts that they spectrally and uncannily border. Kant's aforementioned terms from the *Critique of Judgment* may be instructive here: "An object is *monstrous* if by its magnitude it nullifies the purpose that constitutes its concept."[4] We might think monstrous beings, then, as those that grow and express their own unorderable sense. Darwin's analysis of evolutionary monsters in the *Origin* was influenced by John Hunter's (1728–93) "principle of monstrosity" in which "we call everything that deviates from [Nature's] uniformity a 'monster.'"[5] Yet monsters also form from within Nature's principles, which do not always tend toward perfect uniformity: "We may first observe that animals, not monsters in themselves, shall have the principle of producing monsters," and as a result, "such monsters, once formed, have the principle of propagating their monstrosity."[6]

A metaphorics, too, arises out of the production and the propagation of monstrosity, one that undetermines the structures dividing beings into species, sub-species, varieties, and other conceptual designations. In "Chapter 6:

Difficulties on Theory," Darwin embraces the inhumanistic futures that might emerge from such monstrous relationality: "In North America the black bear was seen by Hearne swimming for hours with widely open mouth, thus catching, like a whale, insects in the water. Even in so extreme a case as this, if the supply of insects were constant, and if better adapted competitors did not already exist in the country, I can see no difficulty in a race of bears being rendered, by natural selection, more and more aquatic in their structure and habits, with larger and larger mouths, till a creature was produced as monstrous as a whale" (*OS* 210).[7] Every being is an enigma, an expressive noncoincidence with its own being. Swimming, opening mouth, and catching insects are relational strivings—propagations of monstrosity between bears and insects—that go ungrasped by conceptual (as-)structures. In both a biological-material as well as a literary-aesthetic sense, beings are gatherings, knottednesses capable of forming (or "rendering") swarms of futures that humanistic thought cannot determine.

Darwin here gives a thought of two grounds. First, there is the ground of metaphysical comparison, in which structural resemblances (bears swim and feed like whales) ensure a conceptual separation between the beings that relate. And second, a ground in the process of yielding to the monstrous communions propagating on it, relations that can no longer be securely held by concepts. This latter ground, in our terms, is the (un)ground of the inconceptual: a means of thinking relations among beings and worlds that unfold amidst a profound indifference to metaphysical determination. We call inconceptual the dissonance of an unthinkable, spectral, and multispecied "creature . . . as monstrous as a whale," oblique to both origin and to imaginable future. Metaphor propagates monsters, stalling the sublation of life-world into meaning in order to stay in and with the gathered noise of beings.

Emerging in the wake and world of Darwin's *Origin*, Carroll's *Alice's Adventures in Wonderland* (1865, originating from an 1863 manuscript titled "Alice's Adventures Under Ground"; hereafter *AAIW*) and *Through the Looking-Glass* (1871; hereafter *TTLG*) also propagate monstrosities—linguistic, metaphorical, biological, and otherwise—that unworld conceptual and metaphysical sense. Alicean nonsense is the expressive sounding of what has been sedimented into the Under Ground. We hear nothing but nonsense's resonances throughout, from Alice's alogical falling and multidirectional growth in and through Wonderland's environments, to the generation of enigmatic beings through metaphor, pun, and portmanteau (the "bread-and-butterfly," Mock Turtle, snake- and plant-Alice, and Wonderland's ur-monster, the Jabberwock), to the improvisatory and denaturing aliveness of the beings populating the Queen's croquet ground. And in this way, we can think an inhuman modernism—scaled here from the propagative nonsense of Darwinian and Carrollian undergrounds to the dissonant metaphorics of both late Woolf and contemporary post-extinction poetics—as an attunement to the noise of the earth. We will trope our reading/hearing of *Between the Acts*, in chapter 5, through Lucy Swithin's fantasy of a "gigantic ear" of the earth in which underground and surface sounds (from prehistory to far future) can be gathered, preserved, and eventually harmonized. But first, here, we seek through *Alice* and the *Origin* for a means of making audible the epiphoric noise of beings, grounds, and undergrounds, without resolving them into either the tonality of Man or the universal harmony of gigantic ear.

I.

In the years leading up to his composition of the *Alice* books, Carroll "read widely on the subject of evolution,"[8] possessing "nineteen books on Darwin, his theories and his critics,"[9] as well as five works of Herbert Spencer, including the Darwinian *First Principles* (1862).[10] As a lecturer in mathematics at Christ Church, Oxford from 1855 to 1881, he was present during the famous 1860 debate at Oxford University Museum between evolutionary theory proponent Thomas Henry Huxley and evolutionary skeptic Bishop of Oxford William Wilberforce.[11] Carroll's response to the "Darwinian trauma"[12] was dually conditioned by his devout Christian faith (he took deacon's orders in 1861, though he eventually refused the priesthood) and his scorn for nineteenth-century humanity's "worship of Self" and "torture of the animal worlds," these being "steps of the ladder by which man is ascending to higher civilisation."[13] Darwinian theory offered Carroll a kind of metaphorics of the underground, a means of both refusing the ascent of man and dwelling with humanism's abjected animal worlds. Darwinian and Alicean time alike is geological, made underground: for Darwin, the present order is birthed from an aggregation of gradual, intentionless, and untraceable modifications to organisms over "tens of thousands" of generations. Similarly, in the indeterminable time of her descent down the rabbit hole—an earlike passage into the strange auditory world of the Under Ground—Alice falls not only to the "centre of the earth" but fears falling "right *through* the earth" (*AAIW* 11). Underground time and growth are detached from bodies, endlessly reversible, and indifferent to the linear-progressive and other forms of above-ground sense.

The (sonic) illogics of *Alice* thus run counter to the Kantian loop between subjectivity and consciousness in which the form of the I or "I think" synthesizes into meaning all of the representations that the subject encounters. In a key passage from *The Critique of Pure Reason* (1787), one that forms much of the basis of Deleuze's critique of philosophy as a whole in *Difference and Repetition*, Kant writes, "It must be possible for the I think to accompany all my representations: for otherwise something would be represented within me that could not be thought at all, in other words, the representation would either be impossible, or at least would be nothing to me."[14] In the first edition of *The Critique of Pure Reason* (1781), Kant is more explicit about the birth of knowledge out of the connection of various representations to one another: "If every single representation were completely foreign to and, as it were, isolated and separated from every other representation, no such thing as knowledge could ever arise; for knowledge forms a whole of compared and connected representations."[15] Knowledge depends on a stable order of representations, and an order of representations is only stable insofar as it can be related back to an "I think" within which phenomena can be synthesized into knowledge. The slightest fracture of this ground—a ground that holds representations and selves together in the humanistic *symbolon* Kant calls knowledge—would let flower the a-synthetic representations and phenomena that consciousness cannot harmonize into sense. Another name for this underground expression that is both "impossible" and "nothing to me," then, is nonsense.[16]

Alice's fall, according to Deleuze in *The Logic of Sense*, marks the "loss of her proper name" (or the "I think" / *cogito*) and an end to the "permanence of *savoir*" by which the self is granted an identity rooted in "God and the world in general."[17] Alice's falling also undoes the temporal linearity that enables the synthesis of phenomena into knowledge. In Deleuze's analysis, the *Alice* books detach the "event" from any specific spatiotemporal order: events do not just occur in a present determined by its past but also alter the past out of which they emerge as much as the present in which they take place. Alice's growing and shrinking are "pure events" that belong neither to the past nor the present order of things: there is no stable Alice-before-growing and Alice-after-growing, since Alice, after falling, never exists within a specific set of spatiotemporal coordinates.[18] Darwinian *and* Alicean beings both relate across chasms of geological time, in the absence of identity, formal consistency, temporality, place, and world. The nonsense of the *Alice* books, as Donald Rackin points out, emerges amidst the ruins of the early Victorian fictions of "natural moral progress . . . [and] sense of unitary, purposeful, God-given order and natural motion," agitating against the strictures of novelistic form, representational language, and narrative progression.[19] As Carroll writes in "*Alice* on the Stage": "'Alice' and the 'Looking-Glass' are made up almost wholly of bits and scraps, single ideas which came of themselves."[20] It is as if the bare elements of plot (Alice's growing older at the end of *Alice in Wonderland* and

Alice's becoming Queen at the end of *Through the Looking-Glass*, for example) exist only to provide some kind of minimal framework for the profusion of pure events—enigmatic growths and gatherings—out of which Under Ground and Looking-Glass worlds are composed.

Simultaneously aesthetic and geohistorical forays, the *Origin* and the *Alice* books give us not beginnings but provisional middle places for an inhuman modernism, as discussed in the introduction. In both, the restlessness of being—its insistent friction against form—becomes an opening to multispecied time(s), thought, and metaphoricity. I call Carrollian and Darwinian thought geohistorical because each dwells in and with the underground, or with what has been materially and metaphorically sedimented into the earth in order for the surface to make its sense. Darwin's (anti-)foundational metaphor of the "Tree of Life," for example, both "fills with its dead and broken branches the crust of the earth, *and* covers the surface with its ever branching and beautiful ramifications" (*OS* 177; my emphasis). The beings that grow above the surface are underground-shaped, and underground beings compose/compost worlds from the above-ground's dejecta. Similarly, Carroll's Jabberwock speaks the sedimented nonsense of the underground into the above-ground's linguistic and poetic forms, denaturing them from within. In Alice's exasperated gloss, "Somehow it seems to fill my head with ideas—only I don't know exactly what they are!" (*TTLG* 134). The *woce* of "Jabberwock," according to Carroll, indicates both the Latin *voce*, "voice," and the Anglo-Saxon *wocer*, "to bear fruit."[21] Here and elsewhere, Alice's mind is an epiphoric gathering place for underground growing and speaking. We can think the Tree of Life and the Jabberwock, then, as the multispecied nexuses or knots of an inhuman modernism: middle places where geological time perplexes form and where surface logics can no longer possess the wayward branchings of being. Our inhuman modernist planetarity—as thought of the earth's composition and the expressive means through which the earth unfolds itself—thus begins here, with the earth's sediments.

II.

Alice's body, too, is an underground gathering place. What holds the underground's manifold worlds in common is Alice's material-affective experience of (and self-unmaking through) them. Once stripped of the proper name that grants self-consistency in the above-ground, Alice becomes formless and unspecied, shaped variously into being by the environments in which she is held. Eating the "right-hand" and "left-hand" sides of a mushroom in response to the Caterpillar's instructions causes Alice to shrink and grow alternately and eventually to become vegetal herself: "'Come, my head's free at last!' said Alice in a tone of delight, which changed into alarm in another moment, when she found that her shoulders were nowhere to be found: all she could see, when she looked down, was an immense length of neck, which seemed to rise like a stalk out of a sea of green leaves that lay far below her" (*AAIW* 46, 47). Alice's phytomorphosis is also a monstrous communion between human and plant, neck and stalk: bodies and parts that are joined together without concept, while growing from the same groundless and formless "sea of green leaves." If plant growth is without direction or telos, as Nealon argues in *Plant Theory*, Alice's transformation gives world to an impersonal, bad infinite growth no longer tied to any order of time or being: human, animal, vegetal, lithic. The "immense length of neck" entangles human and plant ontologies. Knotted thus, Alice is both an individual and an inhuman swarm—a theme that recurs in the multispecies transformations of Saint Anthony and Leopold Bloom in chapter 3—whose interrogations of her own body lead into ever profounder relationalities: "'What *can* all that green stuff be?' said Alice. 'And where *have* my shoulders got to? And oh, my poor hands, how is it I can't see you?' She was moving them about as she spoke, but no result seemed to follow except a little shaking among the distant green leaves" (*AAIW* 47). Have her hands become leaves? Or have causes become so sundered from effects that a freely and impersonally distributed affectability (touching, moving, shaking) among all the bodies and parts gathered by an environment makes them indistinguishable from one another?

Carroll's own illustrations for this scene inhabit Alice's impersonal and multiplicate being more explicitly than do the published Tenniel illustrations, which tend to insist on an essentially stable form amidst Alice's fluctuations (figure 1).[22] In the first of Carroll's illustrations, Alice's body simply ceases to exist: the shrunken Alice is depicted as a disproportionately large head attached to a pair of shoes, with arms emerging from the bottoms of her cheeks. In Carroll's illustration of Alice's growth, we see her head growing directly out of a depersonalized length of neck or stalk, grounded in the nonground of a sea of leaves, or the unidentified, shaking "green stuff." In this sense, Carroll's characterization of his illustrations as "designs that rebelled against every law of Anatomy or Art" can be taken literally: they oppose subjective individuation, the laws of conceptual and entelechial growth, and the logics of the form-giving ground.[23]

The length of neck speciates further as Alice further unfolds into her environment: "As there seemed to be no chance of getting her hands up to her head, she tried to get her head down to *them*, and was delighted to find

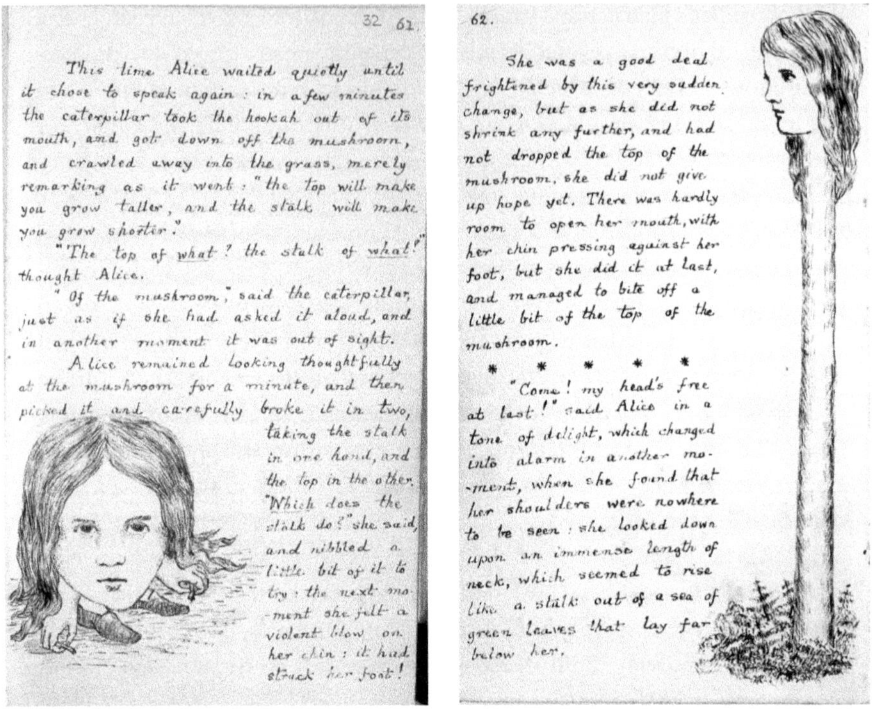

32 61

This time Alice waited quietly until it chose to speak again: in a few minutes the caterpillar took the hookah out of its mouth, and got down off the mushroom, and crawled away into the grass, merely remarking as it went: "the top will make you grow taller, and the stalk will make you grow shorter."

"The top of what? the stalk of what?" thought Alice.

"Of the mushroom," said the caterpillar, just as if she had asked it aloud, and in another moment it was out of sight.

Alice remained looking thoughtfully at the mushroom for a minute, and then picked it and carefully broke it in two, taking the stalk in one hand, and the top in the other. "Which does the stalk do?" she said, and nibbled a little bit of it to try: the next moment she felt a violent blow on her chin: it had struck her foot!

62.

She was a good deal frightened by this very sudden change, but as she did not shrink any further, and had not dropped the top of the mushroom, she did not give up hope yet. There was hardly room to open her mouth, with her chin pressing against her foot, but she did it at last, and managed to bite off a little bit of the top of the mushroom.

* * * * *

"Come! my head's free at last!" said Alice in a tone of delight, which changed into alarm in another moment, when she found that her shoulders were nowhere to be seen: she looked down upon an immense length of neck, which seemed to rise like a stalk out of a sea of green leaves that lay far below her.

FIGURE 1. Lewis Carroll, illustrations for chapter 3 of *Alice's Adventures Under Ground* (1864). Images courtesy of the British Library, © British Library Board (46700, f. 32, f. 32v).

that her neck would bend about easily in any direction, like a serpent" (*AAIW* 47). As with Darwin's whalelike swimming bear, Alice's conceptual resemblance to a serpent soon brings into being a monstrous creature—or a girl as monstrous as a snake—when a nearby pigeon mistakes tentacular Alice for a predator. If, as Alice insists in defense of her girlhood, "little girls eat eggs quite as much as serpents do," then they must, counters the Pigeon, be "a kind of serpent. . . . You're looking for eggs. I know *that* well enough; and what does it matter to me whether you're a little girl or a serpent" (48). With a personhood made immanent with both serpent-hood and plant-hood, Alice's body is an interpretation of an environment—an "invention," according to the Pigeon, like the concept of species itself—brought into being by improvised relations with and through other beings and bodies.

Darwinian metaphorics are similarly rooted here, in the unsynthesizable difference hovering between the concepts of "species," "sub-species," and "varieties." Within larger genera, in Darwin's analysis, we define as separate "species" those organisms that would have been defined as "varieties" (implying a smaller degree of difference) if they had existed within a smaller genus. "The amount of difference considered necessary to give to two forms the rank of species is quite indefinite," Darwin writes, and goes on to tell us, "Certainly no clear line of demarcation has as yet been drawn between species and sub-species—that is, the forms which in the opinion of some naturalists come very near to, but do not quite arrive at the rank of species; or, again, between sub-species and well-marked varieties, or between lesser varieties and individual differences. These differences blend into each other in an insensible series; and series impresses the mind with the idea of an actual passage" (*OS* 131, 126). In the same vein, Darwin writes at the end of chapter 2 of the *Origin*, "Varieties have the same general characters as species, for they cannot be distinguished from species,—except, firstly by the discovery of intermediate linking forms, and the occurrence of such links cannot affect the actual characters of the forms which they connect; and except, secondly, by a certain amount of difference, for two forms, if differing very little, are generally ranked as varieties, notwithstanding that intermediate linking forms have not been discovered; but the amount of difference considered necessary to give to two forms the rank of species is quite indefinite" (131). The designation of species or variety is not a proper identity but merely the temporal fixing of a being's performance of itself—with its environments, with its earth—over imperceptible lengths of geological time.[24] The differences through which beings self-express form an "insensible series" amidst the interstices between concepts—these inventions swallowed by the "sea of leaves" and the rebellious murmurings of pigeons in *Alice*—and open requisite world for new, inconceptual modes of "passage" between beings.

Darwin theorizes two types of metaphor—what we might call metaphysical and multispecied—in a lengthy passage in chapter 13 of the *Origin*, "Classification":

> Naturalists frequently speak of the skull as formed of metamorphosed vertebrae: the jaws of crabs as metamorphosed legs; the stamens and pistils of flowers as metamorphosed leaves; but it would in these cases probably be more correct, as Professor Huxley has remarked, to speak of both skull and vertebrae, both jaws and legs, & c.,—as having been metamorphosed, not one from the other, but from some common element. Naturalists, however, use such language in a metaphorical sense: they are far from meaning that during a long course of descent,

> primordial organs of any kind—vertebrae in one case and legs in the other—have actually been modified into skulls or jaws. Yet so strong is the appearance of a modification of this nature having occurred, that naturalists can hardly avoid employing language having this plain signification. On my view, these terms may be used literally; and the wonderful fact of the jaws, for instance, of a crab retaining numerous characters, which they would probably have retained through inheritance, if they had really been metamorphosed during a long course of descent from true legs, or from some simple appendage, is explained. (366–67)

What literary scholar Jeff Wallace, in his discussion of this passage, calls an "inconsistency" in Darwin's use of the word "metamorphosis" as both a "metaphor" and a "plain signification," we might consider a grasping for a more nuanced understanding both of metaphorical expression and of difference itself.[25] In this context, the "metaphorical sense" of the Naturalists refers to their use of the word "metamorphosis" as a metaphor to mean an abstract comparison of one thing to another based on resemblance: the jaws of a crab look as if they have metamorphosed from legs, for example. They employ this metaphor without realizing that they are not just making an abstract comparison, but literally describing the growth and development of existing life forms from common but unknown origins ("from some common element" or from "primordial organs of any kind"). Vertebrae and skull, legs and jaws, stamens and leaves, neck and stalk, then, do not necessarily predate one another and do not actualize some prior concept of themselves, but rather only come into being through an enigmatic, concept-indifferent relationality that preexists all individualization. Differentiation, too, can collapse back into the sameness of the common element, vary into new monstrosities, and cycle amidst various degrees of difference and likeness over the span of geological time. This precarious but limitlessly generative self-unfolding of beings is both the origin and the *ainigma* of species.[26]

Everything in the *Origin* grows in and through the branching of a single—vegetal yet also monstrous and multispecied—metaphor: this is the "great Tree of Life," the diagram illustrating the action of natural selection on species in terms of the unpredictable and multidirectional growth of branches on a tree (figure 2). The vast number of existing species are shown to have developed from a set of hypothetical ancient progenitors, numbered A through L, over the course of "thousands of generations," each interval of a thousand generations marked by a Roman numeral from I to XIV. It is a metaphor without

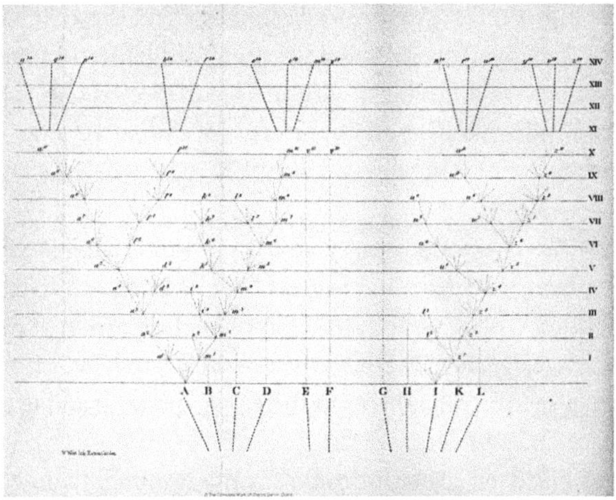

FIGURE 2. Darwin's diagram of the Tree of Life. Reproduced with permission from John van Wyhe ed. 2002–. The Complete Work of Charles Darwin Online (http://darwin-online.org.uk/).

ground or origin, supplying another, underground-made grammar for the growth and proliferation of beings and phenomena beyond the grasp of concepts.

The Tree of Life's present form is constituted by the unknowable relations that have been sedimented into its ground and that may yet give rise to new amalgamations of being(s). A bear that swims becomes "as monstrous as a whale" over the course of thousands of generations simply by aqueously striving alongside other beings. Gathering without determining, the Tree of Life is a metaphor that is persistently overcome by the monstrous performativity of its own content. As with phytomorphed Alice, its bad infinite profusion allows for thought of a growth that is both decoupled from the developmental telos of any singular being, and distributed across an indeterminate manifold of waywardly desiring bodies and parts. In Darwin's words:

> The affinities of all the beings of the same class have sometimes been represented by a great tree. I believe this simile largely speaks the truth. The green and budding twigs may represent existing species; and those produced during each former year may represent the long succession of extinct species. At each period of growth all the growing twigs have tried to branch out on all sides, and to overtop and kill the surrounding twigs and branches, in the same manner as species and groups of species have tried to overmaster other species in the great battle for life. (*OS* 176)

> As buds give rise by growth to fresh buds, and these, if vigorous, branch out and overtop on all sides many a feebler branch, so by generation I believe it has been with the great Tree of Life, which fills with its dead and broken branches the crust of the earth, and covers the surface with its ever branching and beautiful ramifications. (177)

Aristotelian *epipherein* does not carry across but rather agglutinates: names, beings, species; living and dead. In the great Tree of Life, we also have the fractured ground of metaphor itself, as both nominal designation (*metaphora*) *and* copulative excess (*epiphora*). Death is never really death here, since a dead being or an extinct species still inheres in the earth, growing and expressing beneath its surface, in its undergrounds.

The Tree of Life itself grows from these undergrounds, below the surface of what is biologically and even geohistorically conceivable: its present constitution arises from indeterminate joinings between species (animal and plant) stretching back to "long-past geological periods" and insisting through to present and futures: "It is truly a wonderful fact—the wonder of which we are apt to overlook from familiarity—that all animals and all plants should be related to each other in groups, subordinate to groups, in the manner which we everywhere behold. . . . The several subordinate groups in any class cannot be ranked in a single file, but seem clustered round points, and these round other points, and so on in almost endless cycles" (*OS* 176). In *The Descent of Man* (1871), the unknown ancestors of the human species, for example, are always known as "ape-like progenitors," not as a singular and definite form. (Darwin does not simply say "ape" because he is uncertain whether humanity descended from a progenitor closer to the family of the gorilla or the chimpanzee.) The common "origin" is both indeterminate and multiple, and no being is holdable by its own concept.[27] In later instances, Darwin speaks of "the advancement of man from a semi-human condition" and tells us that "man is the co-descendant with other mammals of some unknown and lower form."[28]

A groundless gathering of knotted, clustered beings, each shaped by unnameable ancient progenitors: is this perhaps what Aristotle's originary metaphor of the "withered stalk" has grown into? All we know of the Tree of Life are its above-ground effects: the "beautiful ramifications" on the earth's surface that have not yet been covered over and sedimented by the weight of geohistory. In the last sentence, the noun "branches" transmutes into the gerund "branching" (the image that ends the "Natural Selection" chapter) and suggests the ongoing production of conjugative growths, variations, and differences unrulier than what can be delimited by concepts: "ever branching."

III.

While *Through the Looking-Glass* contains both a garden of live, chattering flowers and a talking insectarium in its early chapters, it is perhaps the "Queen's Croquet Ground" chapter of *Alice's Adventures in Wonderland* that gives the most explicit engagement with evolutionary thought and its anarranged ground. A voice begins the game:

> "Get to your places!" shouted the Queen in a voice of thunder, and people began running about in all directions, tumbling up against each other: however, they got settled down in a minute or two, and the game began.
>
> Alice thought she had never seen such a curious croquet-ground in her life: it was all ridges and furrows: the croquet balls were live hedgehogs and the mallets live flamingoes, and the soldiers had to double themselves up and stand on their hands and feet to make the arches. (*AAIW* 73)

Here, we have all the elements of evolutionary theory in microcosm: an arbitrarily ridged and furrowed terrain on which varying species coexist; the shaping of organisms into species (balls, mallets, arches) according to certain contingent characteristics; the propulsion of life into motion (via an incomprehensible injunction to order) issued by the thunderous voice of an inhuman and mindless force (the Queen of Hearts, or Natural Selection); improvisatory movement among fractious, unorderable beings; and the persistent threat of extinction by the Queen of Hearts' arbitrary and frequently uttered command, "Off with their heads!" (72). Carroll suggests as much in his description of the Queen in "'Alice' on the Stage": "I pictured to myself the Queen of Hearts as a sort of embodiment of ungovernable passion—a blind and aimless Fury."[29]

As with Darwin's *Origin*, the chaos of the Croquet Ground results from the strivings of its beings, *áloga* of the earth, in incongruity with their own concepts:

> The chief difficulty Alice found at first was in managing her flamingo: she succeeded in getting its body tucked away, comfortably enough, under her arm, with its legs hanging down, but generally, just as she had got its neck nicely straightened out, and was going to give the hedgehog a blow with its head, it *would* twist itself round and look up in her face, with such a puzzled expression that she could not help bursting out laughing; and, when she had got its head down, and was going to begin again, it was very provoking to find that the hedgehog had unrolled

> itself, and was in the act of crawling away: besides all this, there was generally a ridge or a furrow in the way wherever she wanted to send the hedgehog to, and, as the doubled-up soldiers were always getting up and walking off to other parts of the ground, Alice soon came to the conclusion that it was a very difficult game indeed. (73–74)

While the flamingo is conceived as a mallet, the hedgehog as a croquet ball, the human as an arch, all croquet-ground beings undo their concepts by simple motion. In other words, they improvise the minimal degree of difference that makes them nonfunctional within the signifying system or the taxonomic order in which they are placed: moving from straightened to twisted neck, rolled to unrolled body, arched to straight back. A hedgehog, for example, becomes something other than its designation (a ball) by unrolling itself along the surface of the ground. These creative, Odradekian (self-)expressions cannot be incorporated into the game. Nothing stays itself; static nouns are traversed by the twisting, rolling, and arching of verbs. For Aristotle, in Heidegger's analysis, the name is "a meaning without time," or a "thinking [that] stops," while a verb "refers to time in addition to what is otherwise referred to in the verb . . . in such a way as to be related to whatever is being spoken about, . . . namely to beings."[30] This logic inheres in the German word for verb, *Zeitwort*: "time-word."[31] Indifferent to law and concept, verb-beings denature the game's claim to totality. Or in the language of Alice's complaint to the Cheshire Cat during the game: "you've no idea how confusing it is all the things being alive" (75).

Carroll's illustration of this scene, which does not appear in the published *Alice* books, gives the Croquet Ground as a crowded, anarchic world in flux (figure 3). It is centered on the menacing Queen of Hearts who holds up a (real) croquet mallet with which she threatens a plaintive-looking hedgehog on the left edge of the illustration, the only one among several not rolled into a ball. In the midst of a terrain overpopulated with hedgehogs, the Queen marks the destructive, blind force that acts indiscriminately as a check on the hyperproductivity of organisms.

Compare Carroll's description of her to Darwin's language concerning an extinction-regulating natural selection in the "Struggle for Existence" chapter of the *Origin*: "Lighten any check, mitigate the destruction ever so little, and the number of the species will almost instantaneously increase to any amount. The face of Nature may be compared to a yielding surface, with ten thousand sharp wedges packed close together and driven inwards by incessant blows, sometimes one wedge being struck, and then another with greater force" (64). Beneath nature's face, and beneath the yielding ground, is the underground into which beings may be "driven inwards" by both Natural Selection and the

FIGURE 3. Lewis Carroll, illustration for chapter 4 of *Alice's Adventures Under Ground* (1864). Image courtesy of the British Library, © British Library Board (46700, f. 39).

mallet-wielding Queen. "Struggle for existence": a being or species wards off extinction not through strength or adaptability or other modes of fitness, but rather through its strivings, or what Darwin calls its own "patterns of action and reaction," in relation to its environments and to other beings.

For there is also a creativity that worlds through the specter of extinction, and one that we can trace from Darwin's entangled bank to Carroll's croquet ground—and later, to contemporary Anthropocenic discourse and its imagined beyond, as given in post-extinction poetics. Darwin considers "the plants and bushes clothing an entangled bank" as an "instance showing how plants and animals, remote in the scale of nature, are bound together by a web of complex relations" (*OS* 140). And on the example of "the ancient Indian ruins in the Southern United States," he writes,

> What a struggle must have gone on during long centuries between the several kinds of trees, each annually scattering its seeds by the thousand; what war between insect and insect—between insects, snails, and other animals with birds and beasts of prey—all striving to increase. . . . Throw up a handful of feathers, and all fall to the ground according to definite laws; but how simple is the problem where each shall fall compared to that of the action and reaction of the innumerable plants and animals

> which have determined, in the course of centuries, the proportional numbers and kinds of trees now growing on the old Indian ruins! (*OS* 140)

Falling is determinate, for Darwin, but the relational webs among beings are untraceable and unpredictable, even as these relations make all above- and below-ground life possible. In Carroll's illustration, beings express beyond the boundary of the game's laws: flamingos shriek in multiple directions, hedgehogs roll upon (and perhaps create anew) asymmetrical furrows, and human courtesans grasp what beings they can, including one another, to prevent their flight. With no logical outcome that would accord with its own internal laws, the game and its ground are continually remade by the improvisatory strivings of its players. Beings are immediately meanings with time, or meanings that only take shape in time and through conjugation with other beings. Silenced and helpless—and perhaps the only nonmoving/nonverbed being in the illustration—Alice casts her eyes downward, to a ground that gives rise only to unruly bodies and relations that grow beyond biological and linguistic (and Victorian lawn sport) systems of order. Similarly, the trees, plants, bushes, insects, and birds that flourish amidst Darwin's entangled bank emerge out of knotted relations across centuries, many likely sedimented into the earth, whose present form cannot be accounted for by any definitive laws of falling or of growth. Nonsense, then, is the expressive thought of the Under Ground and of evolution alike.

IV.

"And they all quarrel so dreadfully one can't hear oneself speak," Alice continues to the Cheshire Cat prior to his vanishing from the croquet scene. The persistent, clamorous nonsense of the croquet ground beings interrupts Alice's autoaffection, or what Jacques Derrida has called the mythical condition of "hearing oneself speak [*s'entendre parler*]." Can one's speaking be immediately heard by the self, taking place without passage through the external world and thus remaining pure of sonic-material traces of the nonself? According to Derrida, such relationless expressivity is impossible, and therefore every imagined autoaffection is immediately a heteroaffection. Following this logic, we might also say that there is no possibility of sense or logos that has not been routed through nonsense; sense only takes shape through a sublimation of fugitive and nonsensical sounds into the discursive agreements we call words. There is, perhaps, no such thing as "hearing oneself make sense." Every sound is a gathering and potentially a contagion, according to Jean-Luc Nancy, who argues that "the visual is tendentially mimetic, and the sonorous tendentially methexic (that is, having to do with participation, sharing, or contagion)."[32] Sounds that are allowed to "take care of themselves," in the words of the Duchess, might grow into monsters and make into monsters all that they contingently cosound with.

Sounds that gather otherwise are the source of nonsense for Mladen Dolar, and we can think of puns and portmanteau words in the *Alice* books as immanent expressions of the "senseless co-sonances"—syllables, words, phonemes, noises (*psófos*), and sounds—that make nonsense.[33] Rather than creating new realities outside of dominant orders of sense, puns and portmanteaux rearrange and improvise anew the sonic-material particulars out of which metaphysical sense itself is constructed. Puns make monsters of what they gather, ungrounding the conceptual distinctions that uphold a universe of individualizable and separable beings. Similarly, portmanteau words, like metaphors, join multiple words—as well as the sedimented referential content inhering in each—into a single, contradictory word or body held together only through the illogics of sonance.

Upon leaving the croquet ground, her ears (unwillingly) filled with the Duchess's moralizing about sound, sense, and other subjects, Alice encounters the Gryphon and Mock Turtle: monstrous, pun-making—and pun-made, in the case of the Mock Turtle—beings seemingly born from the minglings of Wonderland's promiscuous sounds. In Greek mythology, the gryphon / griffin is an amalgamation of an eagle and lion, and in the peculiar etymology lesson given by the Queen, the Mock Turtle is "the thing Mock Turtle Soup is made from": "calf's head dressed with sauces and condiments so as to resemble turtle" (*AAIW* 81).[34] Carroll's Mock Turtle, then, passes from cosonance into materiality: he is the living embodiment of a pun who can only reflect his fractured origins by returning to them through the act of punning, or employing a mode of language that collapses the distinction between intended meaning and derived counter-meaning. The most immediate pun is the word "mock," which suggests that the Mock Turtle is both a confused culinary product come to life and the imitation of a real turtle, in the process confounding the distinction between original and copy, between real and linguistically produced turtle: as the sobbing Mock Turtle claims, "Once, I was a real Turtle" (*AAIW* 83).[35] His schoolmaster was "an Old Turtle" whom they used to call "Tortoise" because, as he explains to Alice, "he taught us" (83). Teacher and tortoise, and soon reading / "reeling," writing / "writhing," Latin / "Laughing" and Greek / "Grief," and lessons / "lessens" among others, collapse into one another and disrupt the networks of meaning through which sensical language is grounded. As Derek Attridge writes in his analysis of puns in James Joyce's *Finnegans Wake* (1939),

> As speakers, we construct our sentences in such a way as to eradicate any possible ambiguities, and as hearers, we assume single meanings in the sentences we interpret. The pun, however, is not just an ambiguity that has crept into an utterance unawares, to embarrass or amuse before being dismissed; it is ambiguity *unashamed of itself*, and this is what makes it a scandal and not just an inconvenience. In place of a context designed to suppress latent ambiguity, the pun is the product of a context deliberately constructed to *enforce* an ambiguity, to render impossible the choice between meanings, to leave the reader or hearer endlessly oscillating in semantic space.[36]

I would qualify Attridge's conception of the pun's ambiguity slightly by noting that the pun always runs the risk not only of being dismissed after the intended semantic meaning is comprehended, but also of enshrining the "real" semantic meaning even more forcefully now that the divergent possibility posed by the pun has been met, overcome, and expiated from the intended meaning. The metaphysical separations between original/artificial, intended/accidental, meaning/nonsense then become all the more firmly upheld.

The Mock Turtle's puns are directed against knowledge itself, seeking instead after ways re-embody logos, or to mingle flesh and word once more. Each disrupts the abstract, autoaffective transmission of knowledge, routing the making of sense through the self's sonic-material environments. Modes of knowing entangle with bodily motions (reading/"reeling," writing/"writhing") or bodily affections (Latin/"Laughing," Greek/"Grief") that have formed the Mock Turtle's education. As the punning language of the Mock Turtle collapses the metaphysical and material dimensions of words into one another, referential language loses its capacity to sublate—or in Humpty Dumpty's words, to "be master" over—the alogic sounds that gather into nonsense.[37] For the pun is a joining of terms in which each term's prior referential meanings and sonic connections inhere, epiphorically piling up beyond the grasp of concept, sense, and species. As Jonathan Culler proposes in his analysis of punning in "The Call of the Phoneme," "puns present us with a model of language as phonemes or letters combining in various ways to evoke prior meaning and to produce effects of meaning with a looseness, unpredictability, [and] excessiveness . . . that cannot but disrupt the model of language as nomenclature."[38]

TURTLE'S STORY. 141

"When we were little," the Mock Turtle went on at last, more calmly, though still sobbing a little now and then, "we went to school

FIGURE 4. John Tenniel, "The Mock Turtle, the Gryphon and Alice." Illustration for chapter 10 ("The Lobster Quadrille") of Lewis Carroll, *Alice's Adventures in Wonderland*. Image courtesy of the British Library, © British Library Board (C.59.g.11).

The monstrous bodily composition of these living, worlding, epiphoric, inhuman puns is a point of emphasis in Carroll's illustrations. The Tenniel illustrations give the Mock Turtle and Gryphon as amalgamations of multiple forms, but these forms maintain the conceptual distinctions of calf, turtle, eagle, and lion (figure 4). Carroll's give us creatures without concept, unspecied, floating, screeching forms, figuratively deprived of formal consistency and literally detached from the ground itself (figure 5a). The flying Mock

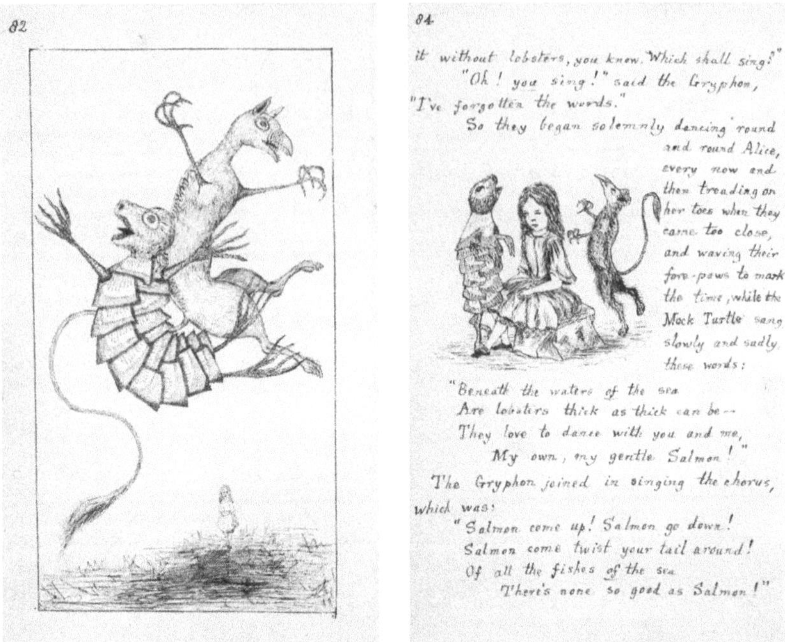

FIGURE 5 (5A AND 5B). Lewis Carroll, illustrations for chapter 4 of *Alice's Adventures Under Ground* (1864). Image courtesy of the British Library, © British Library Board (46700, f. 42v, 43v).

Turtle throws wide its pitchfork-shaped arms and legs in every direction, the series of stony flaps that make up the shell bending backward with the "writhing" of its body. The creatures appear to meld into one another: the Mock Turtle's head seems to grow organically from the side of the Gryphon's body while the Gryphon's tail snakes into the shell of the Mock Turtle at the point where the Mock Turtle's body bends in half, emerging out of an opening on the other side.

These are illusions of depth, of course, but all sense of proportion is thrown askew when we look to the dwarfed Alice, who measures not even a tenth of creatures' size in this image but has to sit to make herself eye level with them in the next (figure 5b). Here, the creatures threaten to squash the tiny Alice as soon as they make their way back down to earth. The only resolution to the Mock Turtle's dilemma, as a creature of linguistic fabrication as well as a "real Turtle," a disruptive pun poised between semantic and material orders, is to dissolve (metaphorically) into the primordial "Mock Turtle Soup" from which he emerged.

FIGURE 6. John Tenniel, "The Jabberwocky." Illustration for chapter 1 ("Looking-Glass House") of Lewis Carroll, *Through the Looking-Glass*. Image courtesy of the British Library, © British Library Board (012808.eee.57).

While the pun traffics in the "unashamed" ambiguities between semanticity and sonic materiality, the portmanteau word knots together contradictory sounds, words, bodies, and meanings. Simultaneously originless and many-origined, Alicean portmanteaux unmake and multiply the contingent processes by which meanings are attached to words and forms grafted onto bodies. The portmanteau, like the inconceptual metaphorics of Darwinian species relations and the "ever branching," bad infinite generativity of the Tree of Life, works outside of metaphysical modes of growth and gathering. A portmanteau such as "slithy," for example, is a gathering of many sounds and senses that only achieve expression through sociality with one another: in isolation, "slither," "lithe," "slide," "slimy," and "sly" cannot properly describe the "toves" they precede, giving readers the task of crafting some minimally sensical adjective through this set of noisy, methexic relations.

Carroll's Jabberwock embodies the illogics of both portmanteau expression and agglutinative *epipherein*: a joining or piling up of sounds, senses, referents, bodies, and meanings through the medium of a single, monstrous being. As mentioned earlier, the term "Jabberwock" condenses both "jabber," as "excited and voluble discussion," and "*woce*," or "to bear fruit," to make what Carroll refers to as "the result of much excited and voluble discussion."[39] To "jabber" is also "to talk rapidly and indistinctly or unintelligibly; to speak volubly and with little sense; to chatter, gabble, prattle," "to utter inarticulate sounds rapidly and volubly; to chatter, as monkeys, birds, etc.," and "to speak (a foreign language), with the implication that it is unintelligible to the hearer."[40] The origins of words are sedimented beneath endless layers of voluble human and inhuman chatter. The Jabberwock is thus a creature in whom the voice / *woce*—that objective, and exclusively human-owned, indicator of individual and biological identity[41]—is continually threatened by the indiscriminate "jabber" that voiced humans attempt to disavow as nonsense.

Jabberwock(y), as being, poem, and riddle all at once, expresses the uncontrollable proliferation of both nonsensical sound objects, syllables, and bodily parts and forms, as given in the iconic Tenniel illustration (figure 6). (Rather than give names to nameless things, as in the traditional function of metaphor, the Jabberwock gives namelessness to existing names.) As the unground from which alogics / *áloga* ceaselessly grow, the Jabberwock's strivings resonate with those of another "fruit bearing" and "ever branching" being: Darwin's Tree of Life.[42] If all that has been sedimented into the earth for sense to reign—extinct species, ancient progenitors, past lives of words and metaphors, forgotten grammars of relation—were given voice, "Jabberwocky" is what these voiced sediments might sound like. This may also be why the Jabberwock must be slain before language can resume a referential function and before a new regime of logos—purified of the comminglings of bodies and sonances—can emerge. But it is in the nature of all riddles to creatively and generatively insist, Medusa-like, through the violence and murder of solution.

Consider again Alice's response to the poem: "Somehow it seems to fill my head with ideas—only I don't know exactly what they are! However, *somebody* killed *something*: that's clear, at any rate" (*TTLG* 134; italics in the original).

Ideas grow without determination by the thinking I. Alice's confusion seems to work against the neo-Cartesian model of consciousness given in Kant's *Critique of Pure Reason*, as discussed above: "It must be possible for the I think to accompany all my representations: for otherwise something would be represented within me that could not be thought at all, in other words, the representation would either be impossible, or at least would be nothing to me."[43] Is the Jabberwock something like the "impossible representation" that a humanistic consciousness cannot imagine: the enigmatic unthought that even beyond death fills my field of representation, my conscious understanding, without being harmonizable into logos? In the hollowed-out, impersonal categories of "somehow," "somebody," and "something," we reach humanism's exhaustion of its own concepts, the dissolution of knowledge into riddle.

Yet the time of riddle and the time of nonsense—along with all of the temporal, spatial, linguistic, sonic, and bodily fluidity we have experienced in the *Alice* books—seems bound to the time of childhood, for Carroll. As Alice eventually surfaces and leaves childhood, words settle into articulacy, beings rejoin their concepts, and Wonderland thrives only in the recesses of "Memory's mystic band." To grow may also be to grow deterministically, into individuation, as Alice does when she becomes a mother figure at the end of *Alice's Adventures in Wonderland* and becomes queen in *Through the Looking-Glass*. As queen, she is empowered over her surroundings and capable not only of reducing them to order but also of universalizing her particular arrangement and making it identical with reality itself: "The Red Queen made no resistance whatever: only her face grew very small, and her eyes got large and green: and still, as Alice went on shaking her, she kept on growing shorter—and fatter—and softer—and rounder—and— . . . —and it really was a kitten after all" (*TTLG* 235–36). Although this sovereign act of determination is precisely what *Alice*'s events resist (or at least defer), we still seem to close with a capitulation to the above-ground logics of *bildung* and the laws of individualized and asocial form.

Or at least we would if the tale ended here. Instead, the final chapter, "Which Dreamed It?" features Alice speaking to Dinah and her kittens and closes with a scene of Alice posing an irresolvable paradox to the Queen (newly transformed into a kitten):

> "Now Kitty, let's consider who it was that dreamed it all. . . . You see, Kitty, it *must* have been either me or the Red King. He was part of my dream, of course—but then I was part of his dream, too! *Was* it the Red King, Kitty? You were his wife, my dear, so you ought to know—Oh,

> Kitty, *do* help to settle it! I'm sure your paw can wait." But the provoking kitten only began on the other paw, and pretended it hadn't heard the question.
>
> Which do *you* think it was? (239–40; italics in the original)

In the closing line, the object of address slips from the kitten to us, the audience, who are put in the same position of being confronted with a question to which we lack the capacity to respond, or the "I think" by which to synthesize disorderly phenomena into consciousness. It is all we can do to oscillate between the opposing, irremediable possibilities of Alice's *ainigma*, the movement of our thought akin to the kitten's licking one paw and then another.

In a letter to physician and mathematician Daniel Biddle on Zeno's paradox, Carroll writes, "But a thing is not *impossible*, merely because it is *inconceivable*. The human reason has very definite limits."[44] These limits are perhaps what Carroll plays upon in the closing lines of the *Alice* books, in which the entire humanist order assumes the condition of the kitten in its ignorance of its own origin—brought into existence via transformation from one being to another—and its inability or refusal to respond to the question in the same idiom in which it was phrased. Following this metaphor, human reason operates only through the excision of the unanswerable question, the insistent riddle, the paradoxical instance, the nonsensical statement or sonance from its purview, under the pretense that the question never existed or was never "heard." (A representation that is either "impossible" or "nothing to me," in Kant's terms.) Being Queen, or assuming the position of sovereignty, is circuitously equated with Alice's loss of her proper name and bodily composition at the opening of the *Alice* books, her queenly condition—mirrored in the devolution of the Red Queen into a kitten—making a riddle of her ontology (the dreamer or the dream?).

What does the ground give when it no longer gives sensical forms? The *Origin* and the *Alice* books can be read as improvisatory responses to this *ainigma*, as descents into the undergrounds from which monstrosities propagate and nonsense knots with other nonsense. The earth's nonsense is methexic: it grows in collaboration with what it touches and with what touches through it—grounds, beings, bodies, sonances, words. In the following chapter, we theorize propagative nonsense in terms of a metaphorics of contagion. In Flaubertian and Joycean modes of contagion/temptation, what touches together, grows together, and grows in the manner of the accretive, concept-indifferent profusion we are calling epiphoric. As we have discussed, and as we will carry into our exploration of Flaubertian and Joycean inhumanisms, the Tree of Life provides a model of a growth that waywardly radiates rather than totalizes into form: "It is truly a wonderful fact . . . that all animals and all plants should be related to each other in groups, subordinate to groups, in the manner which we everywhere behold. . . . The several subordinate groups in any class cannot be ranked in a single file, but seem clustered round points, and these round other points, and so on in almost endless cycles" (*OS* 176). We everywhere behold a relation between all beings. For Flaubert's Saint Anthony, however, this everywhere-beheld relation is only comprehensible through a return to *le fond de la matière*, to matter's preformal ground, also troped as the dust of the earth that inheres in all shaped flesh. The ground gives dust, and dust is shaped into the pliant flesh of characters: Adam, Alice, Anthony, Bloom, for example. These exist as both singularly enfleshed selves *and* impersonal distributions of the everywhere-relationality of clustered, cyclical growth. Contagion is the ground of another (co)creation—the making possible of alternative enfleshments—in the passage from dust to flesh, bodies (and body parts) to selves, inhumanist lineage to human(ist) present. Nonsense, unsedimented, rises to the surface as the improvisatory generativity of dust.

Chapter 3

Dust

Metaphorics of Contagion: Flaubert and Joyce

> The dust that served to form Adam the first man was gathered from the entire world, as it is stated: "When I was made in secret and wrought in the lowest places on earth, Your eyes did see the unshaped flesh" (Psalms 139:15–16), and it is written, "For the eyes of the Lord run to and fro throughout the whole earth" (II Chronicles 16:9), indicating that this figure was formed from the whole earth, the place within the view of the Lord's eyes.
>
> —Rabbi Meir, *Sanhedrin* 38-a, *Talmud Bavli*

The passage of dust into flesh could only be possible if dust (Hebrew: *adamah*), in sympathy or through likeness, resembled or otherwise already bore a relation to flesh, and if dust continued to insist in, with, and as all shaped flesh. The opening stage directions of Flaubert's *La tentation de Saint Antoine* (The Temptation of Saint Anthony, 1874) mediate all the creation to follow through an originary contagion (*con* + *tangere*, "touching together") between dust and the light by which dust is animated into being: "throughout space there floats a golden dust [*poudre d'or*] so fine as to become confounded [*se confond*] with the vibrations of the light" (10). Could any but the Lord's eyes distinguish dust from flesh amidst light's vibrations? *La tentation*'s stage directions open the (perhaps blasphemous) possibility that even the gaze of the Lord may not have been capable of such differentiation. All creation would then be the confused mingling of light with dust—a dust that is itself golden light and a light that is a rhythming materiality in space. The inhuman materials of the earth's lowest places inhere in the sublimity of creation, and each passes via contagion or temptation into one another.

Anthony's first temptation is toward a relationally constituted universe in which the contiguity of all things to one another makes possible the un-/re-shaping of creation and its laws. From his opening monologue: "For the world, as a philosopher explained to me, forms a whole [*ensemble*] of which all parts mutually influence one another, like the organs of one body." Hermeneutic knowledge, he continues, is a science that "enables us to know the natural loves and natural repulsion of all things, and to play upon them [*puis de les mettre en jeu*]. . . . Might it then be possible to modify [*modifier*] what appears to be the immutable order of the universe [*l'ordre immuable*]?" (17; ellipsis in the original, translation modified). This *ainigma* of the body and of creation suffuses both *La tentation* and Joyce's "Circe" in its (proto-)surrealist wake. What (else) happens when universalist laws of joining—laws of stone, laws of *symbolon*—are no longer metaphysically guaranteed? And what happens when the organicism of the body and of creation is no longer protected from the contagion of its thrown-together parts that might now gather into something other than meaning? Joycean and Flaubertian inhumanisms happen, in short, but they only happen in relation with the expressive aesthetics, philosophies, and geohistories knotted in and as inhuman modernist thought. These include the liberation of nonsense from form in the *Origin* and *Alice*; the beginnings of surrealist aesthetics; late nineteenth- and early twentieth-century reengagements with theories of the grotesque and the pre-Enlightenment grotesque body; varied resituations of temporality and spatiality according to terrestrial, oceanic, planetary, and interstellar scales; and permeating all, the making-multispecies, making-enigmatic of metaphor.

In Flaubert and Joyce, contagion is an improvisatory remaking of the metaphysically ordered distinctions between book and world, lineage and species, *symbolon* and *álogon*, flesh and earth. Contagion happens according to the principle that what in worlds touches together, creates (otherwise) together. Character becomes Alicean, not a possessed self but a gathering ground for bodies and parts that join—in extension and in thought—in another manner than concept. Bloom and Saint Anthony, at once beings and ecological assemblages, come to be remade with each material-affective and imaginative encounter. Anthony reads and Anthony's body becomes a world soliciting many worlds—alternative ontological and cosmological articulations of life and pre-/nonlife—that tempt him to abandon self-preservation and to re-earth himself in and as *le fond de la matière*. Similarly, the contagious metaphorics of "Circe" enflesh ancestral, species, and geological lineages within the body of the grotesque, many-worlded being known as Leopold Bloom. *La tentation*'s Manes, the Manichean prophet, speaks in an idiom of contagion that comes to be shared among Bloom, Bloom's resurfaced ancestors, and the "Circe" episode itself: "he that kills an animal shall become that animal; if thou plantest a vine, thou shalt be thyself bound within its boughs [*tu seras lié dans ses rameaux*]" (58).

"Circe" and *La tentation* are milieus in which varied everythings touch and in which a body (Bloom's, Anthony's, or any other) is both a bounded self and all of its enigmatic relations through space and time, or what Deborah Bird Rose calls the "multispecied" "history of [a being's] forebears and mutualists."[1] In both are given not answers but further enfoldings into the *ainigma* of how and why beings relate across the seemingly intractable boundaries of materiality and figuration, contemporaneity and (geo)history, and life and nonlife.

I.

The grotesque aesthetics and proto-surrealisms of both *La tentation* and "Circe" provide a means toward thinking other modalities of creation, otherwise shapings of *adamah*, through a return to the contagious infinitude of resemblance. We beheld something of this infinitude in the everywhere-relationality of Darwin's Tree of Life, which grows according to "the wonderful fact . . . that all animals and plants should be related to one another in groups, subordinate to groups, in the manner which we everywhere behold" (*OS* 176). Resemblance, according to Michel Foucault's account in *The Order of Things* (*Les mots et les choses*), was the dominant principle of Western philosophical thought in the sixteenth century before becoming eclipsed by the classical period's focus on epistemologies of representation and the signification of signs rather than likenesses among things. Foucault's discussion of the first figure of preclassical resemblance, *convenientia*, gives us a means of thinking the persistence of an epistemology and a metaphorics of contagion into the inhumanisms of the nineteenth and twentieth centuries:

> Those things are "convenient" which come sufficiently close to one another to be in juxtaposition; their edges touch, their fringes intermingle, the extremity of the one also denotes the beginning of the other. In this way, movement, influences, passions, and properties too, are communicated. . . .
>
> And then, from this contact, by exchange, there arise new resemblances; a common regimen becomes necessary; upon the similitude that was the hidden reason for their propinquity is superimposed a resemblance that is the visible effect of that proximity. . . . In the vast syntax of the world, the different beings adjust themselves to one another; the plant communicates with the animal, the earth with the sea, man with everything around him. Resemblance imposes adjacencies that in their turn guarantee further resemblances.[2]

The multiplication of resemblances through contagion ultimately tends toward a universal harmony uniting all creation: "Thus, by this linking of resemblance with space, this 'convenience' that brings like things together and makes adjacent things similar, the world is linked together like a chain."[3] Yet the profusion of resemblances, and the radical intimacy of all created things with one another, might also expand and generate beyond what can be harmonized by reason or gathered in logos.

Breton's first *Manifesto of Surrealism* (1924) speaks to and with the intimacies made of resembling things, along with the enigmatic "light of the image" emanating from their touching." In Pierre Reverdy's aphoristic terms, which provide the source for much of Breton's analysis, "[the image] cannot be born from a comparison but from a juxtaposition of two more or less distant realities."[4] Comparison only grasps what the mind can meaningfully associate. And a reason-bound mind cannot imagine (*imago, imaginare*: "to make an image of") a relationality among things prior to grasping, oblique to concept. As Breton elaborates,

> In my opinion, it is erroneous to claim that "the mind has grasped the relationship" of two realities in the presence of each other. First of all, it has seized nothing consciously. It is, as it were, from the fortuitous juxtaposition of the two terms that a particular light has sprung, *the light of the image,* to which we are infinitely sensitive. The value of the image depends upon the beauty of the spark obtained; it is, consequently, a function of the difference of potential between the two conductors. When the difference exists only slightly, as in a comparison, . . . the spark is lacking. Now, it is not within man's power, so far as I can tell, to effect the juxtaposition of two realities so far apart. The principle of the association of ideas, such as we conceive of it, militates against it. . . . We are therefore obliged to admit that the two terms of the image are not deduced one from the other by the mind for the specific purpose of producing the spark, that they are the simultaneous products of the activity I call Surrealist, reason's role being limited to taking note of, and appreciating, the luminous phenomenon.[5]

Further,

> It is true of Surrealist images as it is of opium images that man does not evoke them; rather they "come to him spontaneously, despotically. He cannot chase them away; for the will is powerless now and no longer controls the faculties" (Baudelaire).[6]

There is a resonance among things and realities in the universe that is both anterior and indifferently posterior to the conceptual, or to the work of reason, which arrives belatedly to the gathered light of the image. Juxtaposed things and realities keep creating beyond the enclosures of concept. The greater the conceptual difference between things, it seems, the more intensive their material intimacy, the more luminous their cocreated light. Juxtaposition allows for the resurfacing of what reason and the principle of association have obscured: the whole earth was once nothing but mingled dust and light, and dust still strives toward rhyme with the lost light it once was.

II.

> Je suis, tout en étant près d'elle, plus près des choses qui sont près elle [When I am near her, I am nearer to things which are near her].
>
> —André Breton, *Nadja* (1928)

Surrealist relationality permeates both *La tentation* and "Circe" despite both having been written and published before Breton's first *Manifesto of Surrealism* (*La tentation*, 1874; "Circe," 1922). In composing "Circe," as Catherine Flynn observes, Joyce drew inspiration from Guillaume Apollinaire's *Les mamelles de Tirésias* (1920, first performed 1917). *Les mamelles* was the first of his works that Apollinaire called "surrealist," after having invented the term in 1917 to describe a "new spirit" in art stemming from "Jean Cocteau's, Sergei Diaghilev, Pablo Picasso, and Erik Satie's multimedia ballet, *Parade*."[7] Read in relation to the "new spirit" intuited by Apollinaire and the "raise[d] up monsters" of Breton's manifestos,[8] "Circe" and *La tentation* unbind being from the individuation of the body and from prior arrangements of matter and time that have solidified into law and passed into universality. In the words of *La tentation*'s Oannes, fish-god of the Chaldeans and "the contemporary of beginnings [*origines*]," "I, the first consciousness of Chaos, arose from the abyss that I might harden matter [*pour durcir la matière*], and give a law unto forms [*pour régler les formes*]" (126–27). Oannes is counted among the "false gods [*faux Dieux*]" that proceed before Anthony in Act V, yet Oannes's (and Oannean law's) "resemblance" to what is called the "true" and what is called beginning makes Oannes into a fraught analogue for the creative powers of the Judeo-Christian Lord (141). The Oannean conception of origin, like its Darwinian counterpart, is a provisional articulation raised into a determinate universality: there is always another origin to be found prior to each inscription of origin. Another (co)creationism thus becomes possible through contagion. Here the touching of beings—across temporal and spatial distances taken for abysses—gives rise to alternative joinings of bodies, materials, times, and origins indifferent to both Oannean and Judeo-Christian law.

Contagion manifests, in both texts, through the use of stage directions that collapse any meaningful distinction between imaginative figuration and material-affective enfleshment (also a feature of surrealist practice).[9] The surrealism of "Circe," according to Flynn, emerges most centrally through the episode's creation of new realities and forms—including the episode's re-creation of both itself and of *Ulysses* (hereafter *U*) as a whole—through the spatial and temporal disruptions of the stage directions. "The constant transformations of the 'Circe' episode," writes Flynn, "resemble the shifting of the novel itself as it moves past the initial style into a series of heterogeneous forms."[10] Further, "with the temporal and spatial dislocation of its ceaseless shifting of scene, the stage play of 'Circe' . . . accords with Apollinaire's directions in his prologue to *Les mamelles*: 'It is right that the dramaturge uses / all mirages he has at his disposal / as Morgane did on Mont Gibel / it is right that he makes crowds and inanimate objects speak / if it pleases him / and that he adheres no more to the passage of time than to space.'"[11]

If all time and space could touch together, creation would not have a single origin but would be an ongoing improvisation made amidst the ante-Oannean chaos of mingled beings, parts, and forms; of spirals rather than lines; of the inhumanist speaking of objects, crowds, and other expressive multitudes. Both "Circe" and *La tentation* happen in and as the recursivity of creation unhardened by Oanneanism, a creation that arises, like Oannes himself, from the "formless world where hermaphroditic creatures slumbered, under the weight of an opaque atmosphere, in the deeps of dark waters, when fingers, fins, and wings were blended, and eyes without heads were floating like mollusks, among human-faced bulls, and dog-footed serpents" (126–27). In the "vast syntax" of a blended creation, temptation/*tentation* is the force through which likenesses among things propagate, across extension and time. Ocean, land, and atmosphere remain in cocreative convenience to one another, joined in resemblance rather than linearly horizoned.

Temptation, for Flaubert, is temptation to unselve, to return body and spirit to the resembling things of a blended earth, to become the "matter itself" of the earth's first materials before they were given shape by the Lord's eyes. While Anthony's final desire is to "divide myself everywhere,—be in everything . . . ,—assume all forms," and "be matter itself!" Anthony has perhaps been so unselved from the opening of *La tentation*, his fasting body already an empty enough site for the "everywhere" and "everything" of the outside to gather in it (190). In this way, temptation is visited upon but also emanates from Anthony's body, as the saint both encounters and cocreates with the expressive realities encountered in the time of his reading: fragmented beings, forgotten myth, the speaking sediments of classical, pagan, and pre-Christian cosmologies, new (an)arrangements of matter and spirit. Reading unsediments, with Saint Anthony's Bible itself as the improvisatory ground from which temptation is made material.

There is a dual movement in reading that both secures the subject within the discursive agreements of logos and exposes the subject to logos's fractures, to the interstices between concepts and the variegated forms through which the Word has passed over time. The book, we might say, is neither mimetic (a representation of the world) nor alexiteric (a safeguard from the infection of the world). As Scarlett Baron argues, in *"Strandentwining Cable": Joyce, Flaubert, and Intertextuality*, "The whole of *La Tentation* functions as an intratextual enactment of the original matrix of Anthony's biblical reading."[12] Similarly, Michel Foucault claims that Anthony's Bible, rather than providing the determinate forms and "written signs" that protect the self-consistency and singularity of the humanistic subject, instead opens being to temptation, "[issuing] into the world of forms horrible shapeless figures" which "no language has ever named" and "no book can contain."[13]

The resurfacing of "ancient faiths" in *La tentation*, according to Leo Bersani, creates a "chain of similarities which, instead of inspiring epistemological confidence, suspends reality in an imaginary series with no objective beginning or end."[14] The suspension of reality also enables a contagion between thought and world, "as if the imaginative capacity to see resemblances had *acted* on the world."[15] Reality is dissolved in propagative resemblances. The novel's own impossible conditions of existence in large part make possible the ir-/surreality through which it creates: at a formal level, and much like "Circe," *La tentation* is simultaneously a theatrical production that cannot be staged but only read and a novel whose characters exceed the limits of textuality and are only legible through performance. (Woolf's early description of *The Waves* might well apply to both works: "an abstract mystical eyeless book," "an abstract playpoem."[16]) *La tentation*'s creation occurs in the interstices between fixed forms, as given in the opening stage directions: "It is in the Thebaid, at the summit of a mountain, upon a platform, rounded off into the form of a demilune, and enclosed by huge stones [*qu'enferment de grosses pierres*]. The Hermit's cabin appears in the background [*occupe le fond*]" (9). And "throughout space," the directions continue, "there floats a golden dust [*poudre d'or*] so fine as to become confounded [*se confond*] with the vibrations of the light" (10). Stable forms ("huge stones") scribe a milieu, a suspension, in which the co-vibrations of dust and light can be made visible. First element of creation: illumined dust—simultaneously ether and matter, a refraction of mingled light as well as the corporeal substance of *adamah* (dust, earth, clay)—that swirls amidst the enclosure of the giant stones.

In this chapter's epigraph from the *Talmud Bavli*: "this figure [Adam] was formed from the whole earth, the place within the view of the Lord's eyes." Within this enigmatic formulation, dust is both itself and all of the created, lumined beings in which it inheres. And within the demilunar platform atop the Thebaid, dust is both element occupying the background (*fond*) and element of the originary confounding (*se confond*) of all creation. As we will explore shortly, *confondre* is perhaps simply the subjection of *fond* to an incalculable time, or the setting of ground into creative, improvisatory motion.

In one sense, temptation in the novel is the narrative progression of Anthony's desires from the desire for wealth, power, and sexual libertinage to spiritual and scholarly omniscience to the final striving toward the bottom (*fond*) of matter itself. Yet temptation is also the ongoing liberation of resemblances from linear time, along with the joining of resembling things into new modalities of cocreation, to which Anthony can only passively and belatedly bear witness. For Anthony beholds, in the bad infinite spirals of vegetal growth and the confounding of likenessing things across orders of being, another origin of the earth's creation—or what he will call the "birth of life" and "beginning of motion" ("j'ai vu naître la vie, j'ai vu le mouvement commencer"):

> And all sorts of plants extend themselves into branches [*s'étendent en rameaux*], twist themselves into screws, lengthen into points, round themselves out like fans. Gourds take the appearances of breasts; lianas interlace like serpents. . . .
>
> And then the plants become confounded [*se confondent*] with the stones.
>
> Flints assume the likeness [*ressemblement*] of brains; stalactites of breasts; the flower of iron [*des fleurs de fer*] resembles a figured tapestry [*des tapisseries ornées de figures*].
>
> He sees efflorescences in fragments of ice, imprints [*empreintes*] of shrubs and shells—yet so that one cannot detect whether they be imprints only, or the things themselves [*ces choses elles-mêmes*]. Diamonds gleam like eyes; metals palpitate. (190)

Anthony beholds, in other words, the blended syntax of a confounded earth. *Les plantes se confondent avec les pierres*: plants create together with the very means of their formal enclosure.[17] As with Darwin's contemplation of the tangled bank,[18] another plan(e)tarity becomes possible when we allow that beings join indifferently both to concepts and to linear time. "Things themselves," "shrubs and shells," for example, are only distinguishable from their imprints if we believe that temporal distance ontologically forecloses relations. What a rescaled modernist planetarity gives, however, is the possibility of a communion and cocreation of multispecied beings across time and space. Anthony here bears witness to, is tempted toward, the heresy of an alternative planetarity in which every *fond* is made mobile and creation emerges from gathering's infinitude.

The inhuman grammar by which plant and stone confound has been at work from the opening of *La tentation* and insistent through each of its Acts. The entire platform is enclosed by huge stones with an abyss on either side: "Some ten paces from the hut, there is a long cross planted in the soil; and, at the other end of the platform, an aged and twisted palm tree leans over the abyss [*un vieux palmier tordu se penche sur l'abîme*]" (9). Two kinds of cultivation begin the text: the planting of *symbolon* into earth and a vegetal overgrowth that branches beyond enclosure to touch an abyss. In Act IV, Anthony sees "a forest of columns [*une forêt de colonnes*]," lithically metaphorized trees that prefigure the floralized iron of caves (*des fleurs de fer*) to come at the novel's close. The columned trees all emerge from something like the "monstrous timberwork [*une charpente monstreuse*]" of a single "vast trunk" and its ever-multiplying branches: "From each branch of the colossal tree descend other branches which may bury themselves in the soil [*qui s'enfoncent dans le sol*]; and the aspect of all these horizontal and perpendicular lines, indefinitely multiplied, would closely resemble a monstrous timberwork, were it not that they have small figs growing upon them here and there, and a blackish foliage, like that of the sycamore" (87). Is this flower religion, or heresy of flowers and trees? Like the "beautiful ramifications" of Darwin's Tree of Life, simultaneously covering the surface and "fill[ing] with its dead and broken branches the crust of the earth," Flaubert's gigantic *seul tronc* radiates in monstrous, even blasphemous excess of its concept.

Stone is the site upon which form is codified into law in *La tentation*. Stone is enclosure, the *grosses pierres* of the opening stage directions—the blocks of formal consistency between which the book's world is contained—as well as the "tables of stone [*tables de pierre*]" on which God has "engraved [*gravé*]" his all-containing law. As God's "Voice" explains in Act V, "Within that law, my people were enclosed [*enfermait*]" (160). The confounding, or vegetalization, of stone at *La tentation*'s close also marks the culmination of the novel's heretical counter-movement toward a nonlithic faith, or a faith made in and through the contagious melding of resemblances rather than divinely ordained form. Anthony's humanism, to this end, is given as faith in the rhyme between *pierres* and *prières*, as guarantors of the singularity of being, and as the solidities that open and close the novel: Anthony's concluding prayers—"se remet en prières" is the novel's last line—sonically return us to the *grosses pierres* enclosing the demilunar platform where dust and light first confound.

Yet every stratification in *La tentation*—every making of a concept or law, every enclosure, every shaping of matter into form—also opens interstice for another modality of gathering. Enclosing stones make confounded light and dust apprehensible, just as Anthony's first opening of his Bible (illumined by torchlight) marks the beginning of temptation, as creation unbound by logos. Another *Pierre*, this time the saint Peter during the time of his own visions, emerges when Anthony reads the first illumined words from the Acts of the Apostles:

> Am I, then, so weak, O my God! Courage, let me rise from here! [*Du courage, relevons-nous!*]
>
> *He enters his hut, turns over a pile of cinders, finds a live ember, lights his torch and fixes it upon the wooden desk, so as to throw a light upon the great book*
>
> Suppose I take . . . the Acts of the Apostles? . . . —yes!—no matter where!
>
> "And he saw the heaven opened [*le ciel ouvert*], and a certain vessel descending, as it were a great linen sheet let down by the four corners from heaven to the earth—wherein were all manner of four-footed beasts [*animaux terrestres*]; and a creeping of things of the earth and fowls of the air. And there came a voice to him: Arise, Peter! Kill and eat!" [*"Pierre, lève toi! tue, et mange!"*]
>
> Then the Lord desired that his apostle should eat of all things? . . . while I . . . (15, ellipses in the original)

Where Anthony descends, joining flesh with earth, Pierre rises above matter through a divine sanction that is also a determinate conceptual arrangement of human and nonhuman being. (In the Apostles, when Peter protests that he would not eat anything "profane and unclean," the voice enjoins: "What God has made clean, you are not to call profane."[19]) Here, Anthony's "I" inheres suspended between ellipses, without positive fulfillment as either the murderous "Lord's apostle" or one of the dominated creeping earth beings or fowls of the air. His hesitation to identify with the humanisms written in stone creates a gap between the descent of beasts from the opened heavens and the rising of Pierre.[20] Reading anarranges sky and ground, and Anthony's hermeneutics of weakness transmutes spoken law into *ainigma* made of an excess of falling being.

The same divine voice that spoke determination to Peter reemerges in Act V, both to give law (engraved on *tables de pierre*) and now to speak its own decay into "an enormous silence, a profound night [*un silence énorme, une nuit profonde*]" (160). God's "Voice" here laments the disordering of laws:[21] "Woe! Woe! the Holy of Holies is open [*Le Saint-des-Saints s'est ouvert*], the veil is rent, the perfumes of the holocaust are dissipated by all the winds of heaven. The jackal whines [*piaule*] in the sepulchres; my temple is destroyed; my people dispersed!" (160). Profanation is contagious. With the tearing of the veil follow the whinings of winds and jackal, intrusions of inhuman scent and noise that in turn lead to the temple's destruction, and finally, the dispersal of beings beyond the enclosures of stone and law. Solidities crumble in profound night, the divine voice becomes a sound among night-dissolved sounds. As in the Sphinx's riddle, night is the time of metamorphosis and the time of the confounding of logos, which can only resolve the enigmatic nonsingularity of being through the originary violence of speaking the name of Man.

The Sphinx itself, further, can only appear to Anthony once God's voice has dissipated into *nuit profonde*. In contrast to the Chimera's capricious motion through all places of the earth and the imagination in Act VII, the Sphinx insists unconditioned through determination and solution, remaining enigmatic to all. Why does he (the Sphinx is male in Flaubert's imagining) remain "perpetually immobile," the Chimera asks, "or perhaps making strange designs with thy claws upon the sand [*ou bien du bout de ta griffe dessinant des alphabets sur le sable*]?" "It is because I keep my secret [*je garde mon secret*]," responds the Sphinx. His "gaze [*regard*], which naught can deviate, remains fixed [*tendu*], gazing through all intervening things, upon a horizon that none can reach [*sur un horizon inaccessible*]" (181). *La tentation*'s Sphinx is indifference made immobile, sign of the persistence of being beyond all humanistic imaginability. And if we entertain the Chimera's speculations, the Sphinx's earth-writing—literally, "drawing alphabets on the sand"—gives another origin of meaning and language in the expressive obliqueness of earth-written *ainigma* rather than in the determinations of the divine Word.

The Sphinx and Chimera also embody a conflict between inhumanist indifference to all time and catastrophe on the one hand, and imaginative fantasy transcending all material limitations on the other (Chimera: "I offer to the eyes of men dazzling perspectives [*perspectives éblouissantes*] with Paradise in the clouds above, and unspeakable felicity far off [*félicités lointaines*]" [181]). Yet even this seemingly intractable conflict dissipates into night and, specifically, into the "fog [*brouillard*]" of the Chimera's breath as she "departs, describing circles on her way," while the Sphinx sinks (*s'enfonçant*) into the sand on which his enigmatic alphabets remain scribed, to become integrated with both *ainigma* and earth (183). The profound night of Act VII, then, is both grown from a Sphinx-planted ground and atmosphered as Chimeric fog. In a motif that we will return to in chapter 5 and the "long night" of Woolf's "Time

Passes," (profound) night is not the nothingness of oblivion but rather a time and place of indeterminate generation.[22] In the Sphinx's riddle, night is the time when the enigmatic being grows a third leg, or otherwise metamorphoses from a being with two legs into a being with three legs. In night, in other words, the being of the riddle passes into unimaginability and becomes something oblique to the concepts of both human and animal.

"Is it the glow of dawn, or only an effect of moonlight [*Est-ce la clarté de l'aube, ou bien un reflet de la lune*]?" Anthony asks on waking into Act VII. In the absence of the divine ordering voice, there is another vibration of—simultaneously sourceless and many-sourced—light that seems to confound the very separation between day and night itself. And Act VII's co(n)founded light in turn makes possible the shaping of the earth's dust into something other than humanistic being, birthing monstrous creatures—alternative, formerly unrealized arrangements of matter and form, dust and flesh—that now fill the night of Anthony's visions. Among them are the gods, goddesses, and idols of ancient religions; fantastic beings known only through mythical accounts (Unicorn, Griffin, Sadhuzag, Martichoras, and Catobeplas); beings composed of grotesque combinations of animal and human body parts; and creatures of unnatural bodily formations (such as the Blemmyes, headless beings with faces "imprinted upon [their] breasts" and brains in their stomachs, and the Astomi, creatures of air who are "a little more than Dreams, yet not quite beings" and subsist wholly on "breezes and perfumes") (183). Like John Hunter's monstrosity—as contagious, propagative deviation from the uniformity of nature—Flaubertian temptation is the terror of what forms and worlds might emerge if *adamah* were illumined differently, through the womb of another night.

III.

Another name for this otherwise creation, the improvisatory shaping of *adamah* in and through the indeterminacy of Thebaidian light, is the grotesque. Grotesque aesthetics suffuse *La tentation*, which originates in part from Flaubert's obsession with Pieter Bruegel the Elder's 1594 painting of *The Temptation of Saint Anthony* and Jacques Callot's 1617 and 1635 engravings of the same subject. The visual grotesques inspired by Saint Anthony's temptations give us a nearly comprehensive philosophical and aesthetic lineage of the grotesque itself: these include treatments by Martin Schongauer (ca. 1470), Michaelangelo (1487–88), Hieronymus Bosch (ca. 1501), Bruegel the Elder (1594), Bruegel the Younger (ca. 1625), Joos van Craesbeeck (1650), Paul Cézanne (1887), along with the surrealist works of Max Ernst (1945),

Salvador Dalí (1946), and Leonora Carrington (1947).[23] I want to turn briefly to the grotesque as an alternative (and potentially blasphemous) modality of creation in which flesh is inseparable from earth, and untransfigured matter inheres in every finished form. Mikhail Bakhtin, in his landmark survey of the work of François Rabelais, *Rabelais and His World*, emphasizes the "unfinished body" as the site of grotesque creation and growth. Bakhtin's canonical account draws an opposition between grotesque, pre-Enlightenment (often medieval and pagan folk tradition-based) figurations of the body—as fragmented, unfinished, hybridized, and involved in a process of eternal renewal and recomposition with the world from which it springs—and the "new bodily canon" that emerges from the Enlightenment vision of a mechanistic and calculable nature. No longer the body's unruly material and excessive remainder, flesh is bounded into singular, completed bodies that relate according to determinate laws.

The banishment of contradictory, grotesque bodies to the margins of thought was literal in the case of Renaissance art, which often presented individualized, ideal forms, such as the body of Christ, at the center of paintings, with the margins populated by the grotesque figures of human-animal and human-vegetable hybrids common to both pagan ideals of nature worship and multispecies relationality.[24] Bakhtin writes, "The grotesque body . . . is a body in the act of becoming. It is never finished, never completed; it is continually built, created, and builds and creates another body. Moreover, the body swallows the world and is itself swallowed by the world. . . . This is why the essential role belongs to those parts of the grotesque body in which it outgrows its own self, transgressing its own body, in which it conceives a new second body: the bowels and the phallus."[25] Grotesque bodies grow like plants, in the accretive (bad) infinitude of *epiphora* rather than toward telos or the fulfillment of concept. Rather than fixing chaos into form, creation creates with chaos's materials, reorigining life in and as the primordial growth of zoophytic swarms and the improvisatory play of dust and light.[26] In this way, Anthony's opening monologue is also an opening to grotesque growth: once it is possible to modify the "immutable order of the universe [*l'ordre immuable*]" by "putting into play [*puis de les mettre en jeu*]" the laws by which things relate, creation can be liberated from the reproduction of concept, and parts from the organic totality of the body.

In part and in provisional whole, the grotesque (human or inhuman) body is an ongoing relation to its world, made of communions between body and external universe, along with the touching together of the body's own surfaces and depths: "The grotesque ignores the impenetrable surface that closes and limits the body as a separate and completed phenomenon. The grotesque image displays not only the outward but also the inner features of the body: blood, bowels, heart, and other organs. The outward and inward features are often merged into one."[27] The new bodily canon is in many ways an enclosure of profusive growth within concept, giving "an entirely finished, completed, strictly limited body, which is shown from the outside as something individual. That which protrudes, bulges, sprouts, or branches off (when a body transgresses its limits and a new one begins) is eliminated, hidden, or moderated. All orifices of the body are closed. The basis of the image is the individual, strictly limited mass, the impenetrable façade. The opaque surface and the body's 'valleys' acquire an essential meaning as the border of a closed individuality that does not merge with other bodies and with the world."[28]

We can read Oannes's appearance as passage from boundless (self-)relationality to new bodily canon, from the fractious interassemblages of bodies and body parts to the regulations of form that, for Flaubert, become equivalent with both origins and the primordial act of creation. Origin is iterative rather than ontological, however, and decreation, as a (non-regressive) movement toward chaos, subtends all creation. In "Flaubert and the Graphic Arts," Jean Seznec refers to the procession of monsters—and Anthony's beholding of the "birth of life" and "beginning of motion"—in Act VII as Flaubert's "microbian epic," which "had been revealed to him by the reading of Darwin: [Flaubert] had had a glimpse at it himself under the microscope, when he was studying zoophytes with his friend, Pouchet the physiologist."[29] Darwinism, according to Seznec, provides a means of imagining existence prior to form, where everything is still Astomic: "a little more than Dreams, yet not quite beings [*un peu plus que des rêves, pas des êtres tout à fait*]" (183). Flaubert, with Anthony, looks to "the border of the imperceptible world," or what Odilon Redon (artist and illustrator of three separate projects on *La tentation* from 1888 to 1896) calls "the life that is intermediary between animal and plant."[30] Intermediary life is relational life, we might say, made in and from the light-confounded dust swirling between eye and lens, dream and being.

Grotesque creation, which is both prior to and insistent beyond Oannean law / canon, emerges from the methexic collaborations among the preindividual parts of bodies. "Fingers, fish, and wings," according to Oannes, were once blended, and now might blend again, amidst another logic of touching. Nisnas, Blemmyes, Astomi, and other mythical, inhumanist shapings of *adamah* are what grow in the aqueous night of Oanneanism, after Oannes has dissolved once again into the "deep pools left by the Deluge" (127). Consider the Sciapods of Act VII, who are creatures with inverted bodies—their heads are at their feet, their bodies are tied to the earth by their hair, and they absorb light "through the thickness of our heels"—that invert creation itself into an intimacy with earth from which boundless resemblances propagate. They speak of themselves and then become multiplicate, an engendering forest, a swarm of loosed resemblances:

> Their lifted thighs,—resembling [*ressemblant*] the trunks of trees,—multiply [*se multiplient*].
>
> And a forest appears. Great apes clamber through it on all fours:—these are men with the heads of dogs. (184)

The forest appears as both Tree of Life and "monstrous timberwork" of trunks. Materialized likenesses give another, epiphoric alogic of growth: from earth-tied hair and lifted thighs to trees, and from trees to forest, along with the interstitial passages between trees through which the human-dog hybrid Cynocephali clamber. What name for this mode of creation occurring through the infinitude of resemblances and resembling things? *Convenientia*, for Foucault, whose description in *The Order of Things* is itself inflected by Sciapodian creation: "And then, from this contact, by exchange, there arise new resemblances."[31] Surrealism, grotesquerie, plant desire, and in our analysis, *epiphora* and inconceptuality, also name something of such a logos-indifferent modality of gathering and creating.

For the forest that arises from the Sciapods also creates in excess of the Sciapods, through the forestial grammars and forestial thought that inhere in each expressive leaf and branch.[32] Anthony "inhales the freshness of the green leaves," and now "there is a movement [*s'agitent*] among them, a clashing of branches [*s'entre-choquent*], and suddenly there appears a huge black stag, with the head of a bull, having between his ears a thicket of white horns" (185). The Sadhuzag materialized here (a being with seventy-four antlers, "hollow like flutes," that produce both soulful melodies and forest-shattering howls) is followed by the Martichoras (a human-lion hybrid), the Catobeplas (a buffalo with a pig's head), and the Griffin (a "lion with a vulture's beak, and white wings, red paws, and blue neck"). Each forest being coincides with its own enigmatic self-expression, a language of glyphic communions rather than legible signs.

The Griffin, for example, creates with the earth gesturally. He first burrows into the earth to reveal to Anthony his own lineage in the slumbering ancient kings and precious metals buried in its undergrounds [*tombeaux*, *cavernes*, and in Act V, *grottes*], but instead unearths only the inhuman noise/*psófos* gathered in and as its soils. The Griffin speaks to Anthony in Anthony's language, and speaks into the earth in another: he "crows like a cock [*criant comme un coq*]" into it, planting noise into the earth, from which more noise now grows:

"A thousand voices answer him. The forest trembles. And all manner of frightful creatures arise [*Et toutes sortes de bêtes effroyables surgissent*]:—The Tragelaphus, half deer, half ox; the Myrmecoles, lion before and ant behind, whose genitals are set reversely. . . . Mosquitos begin to hum, toads commence to leap; serpents hiss. Lightnings flicker. Hail falls" (187; my ellipsis). Envoiced forest, envoiced soils and sky, all confound the singularity and unitonality of creation. The earth's surface is soon flooded with creatures of "marvellous anatomies," "multiplying by their own contact," who "move about Anthony with a surging motion as though the ground were the deck of a ship [*comme si le sol était le pont d'un navire*]" (187–88). The burrowed-into earth gives only the *epiphora* of an otherwise creation, a counter-geology of gathered noise. And once again, the ground/soil/*sol* does not solve being's enigmas with the answer of the humanist subject and the laws by which matter/*adamah* is hardened into form. Instead, the ground is made fluid, set into motion with the sea. Both Act VII and Anthony here return to the oceanism once repressed by Oannes—"formless world . . . opaque atmosphere, in the deeps of dark waters, when fingers, fins, and wings were blended"—as the "Beasts of the Sea" now tempt Anthony to "accompany us to our immensities, whither as yet no one has descended" (189).

Anthony obliges, yielding self to earth. For the earth tempts Anthony to become earth—as simultaneous plant, water, sound, light, and dust—to "develop myself like the plants,—flow like water,—vibrate like sound—shine like light, squatting upon all forms—penetrate each atom—descend to the very bottom of matter,—be matter itself!" (190). Yet "Day at last appears" to give form and time back to night. "Is it the glow of dawn, or only an effect of moonlight?" Anthony pondered at the beginning of Act VII and now seems to be given answer when he beholds the luminous face of Christ in "the very disk of the Sun [*et dans le disque même du soleil, rayonne la face de Jésus-Christ*]" (190). The appearance of Day is (en)closure, allowing Anthony to resume his prayers [*se remet en prières*] and for what is created in temptation to surrender itself in and to the light of meaning. Yet we have been warned earlier by Hilarion-Lucifer[33] about the deceptiveness of truth, especially as the light of creation comes to be confounded in the gathering of resemblances into another modality of the true. After the procession of the "false gods" of other religions and myth systems has ended in Act V, Hilarion tells Anthony, still mourning for "the souls that have been lost through these false gods": "Dost thou not think that they . . . sometimes . . . bear much resemblance to the true [*comme des ressemblances avec le vrai*]?" (141; ellipses in original).[34] Even the resemblances are conditional, only modes of oblique seeming of a truth that is itself only semblance.

The truth-bearing sun is the paradigm of all meaning, according to Derrida's Aristotelian analysis, precisely because the sun is beyond all sensible and metaphorical apprehension. All metaphor emerges from trying to give name and knowledge to what is beyond all knowing: the turning of the sun. Metaphor is thus heliotrope, sunflower, *tournesol*. In both Aristotelian poetics and the closing of *La tentation*, the sun appears as irresolute *ainigma*: it is both itself and not itself, undeterminable presence and absence, its essence inseparable from the infinitude of what resembles and what illumines with it. "For if the sun is not entirely natural, what can remain in nature that is natural?" asks Derrida. "This object which is the most natural in nature has in itself the capacity to go out of itself; it joins with artificial light, it suffers eclipse and ellipse, has always itself been other."[35] Even the sun's light is neither itself nor its own truth: it vibrates with the dust that it shapes into being, joining with the light of all that resembles (through) it. Sunlight would then be the enigmatic source of all meaning as well as all contagion. Enclosed between opening *pierres* and closing *prières*, divine truth and the heresy of semblance, *La tentation*'s final movement might be toward the becoming-*ainigma* of Anthony himself: integrated into the "matter itself" of the earth, Anthony becomes a patch of earth touched by light, the originary contagion embedded in every act of creation.

IV.

> During the seventeenth century, Anthony was venerated as a saint empowered to ward off the plague in the region of Lorraine, [Jacques] Callot's primary place of residence; this area had been ravaged by several outbreaks of the Black Death.
>
> —Jutta-Annette Bruhn, "The Magic Lantern Slide after The Temptation of Saint Anthony by Jacques Callot" (268)

In what is probably Joyce's most explicit reference to *La tentation*, the disc of the sun rises into the night of "Circe," yet now in the absurdist form of a singing bar of light-giving soap:

> *(He points to the south, then to the east. A cake of new clean lemon soap arises, diffusing light and perfume.)*
>
> **THE SOAP**
>
> We're a capital couple are Bloom and I.
> He brightens the earth. I polish the sky.

(*The freckled face of Sweny, the druggist, appears in the disc of the soapsun.*) (*U* 15.338–41)

The singing soapsun makes comedy of the purifying and truth-bearing functions of the sun, offering a simoniacal communion between the discourses of advertising and theology that permeates nearly every episode of *Ulysses*.[36] By scrubbing away the dirt that inheres in and as all flesh—and by separating purified earth from purified sky—the light of the soapsun is meant to protect against contagion. Bloom, like Anthony, is earth touched by cleansing light—one half of a *symbolon*, or "capital couple"—and similarly caught between the normative demands of individuating purity and the temptation to descend and to unselve with earth and its beings.[37] The soapsun is one among a series of variants on Homeric *moly*, the herb that Hermes gives Odysseus to protect him from Circe's transformative sorcery. In a letter to close friend and confidant Frank Budgen, Joyce refers to *moly* as the "invisible influence (prayer, chance, agility, *presence of mind*, power of recuperation) which saves in case of accident."[38] Other variants on *moly* include a potato that Bloom carries with him—a gift from his late mother and referred to as "Preservative against Plague and Pestilence" (15.1952)—Bloom's wife Molly (for homophonic as well as symbolic reasons), and Bloom's imagined encounter with Rudy, his stillborn son who died eleven years prior, and whose appearance as a shepherd boy and "changeling" helps to secure Bloom within a stable paternal lineage (15.4957).

Rudy's appearance at the close of "Circe," however, is as mystifying as the sunface of Christ (and its parodic counterpart in the druggist Sweny's). Rudy emerges simultaneously as a lineaged self—though one that is also blended into a resemblance with Stephen Dedalus and with the night itself—and as a metamorphic changeling outside of all law and lineage. The appearance, in other words, is a riddle that Bloom solves by giving the name of Rudy, and that we might tend to interpret as the episode's gift of *moly*. The meaning of Joycean *moly*, however, is itself another *ainigma* that continues to mystify readers of *Ulysses* over a century on.[39] In Joyce's own words, "*Moly* is a nut to crack" as well as "a plant with many leaves," and in Homer's original description in the *Odyssey*, *moly* is "dangerous for a mortal man to pluck from the soil, but not for the deathless gods. All lies within their power."[40] As both protective herb and interpretive hermeneutic, *moly* belongs to the earth rather than to mortal men, and our extractive handling of it might only lead to a confused knowledge and a deeper confounding of self with earth. In this sense, the enigmatic nature of *moly* might lie in its simultaneous capacity to protect from infection *and* to immerse us in the transformative contagion of the world and of its resembling things. Rudy-as-*moly* gives both recovery of self as well as an irresolution of riddle, a nut still to be cracked. *Moly*, finally, would then also figure as Derridean-Platonic *pharmakon*: a substance or an interpretive practice (writing, most particularly) that is at once remedy and poison.[41]

The recuperative power of *moly*, Joyce muses in the aforementioned letter to Frank Budgen, "would also cover immunity from syphilis (σύ φιλις = swinelove?)."[42] *Su philis* (σύ φιλις) as "swine-love" is an inaccurate, yet still seductive, etymology that locates the idiom of contagion in "Circe" in the thought of forbidden resemblances and intimacies across species. (Interestingly, syphilis itself was known, in the early twentieth century, as the "Great Imitator," supposedly by Sir William Osler, "due to clinical presentations in its late stages simulating 'almost every disease known to man.'"[43]) Unlike Odysseus, Bloom is not immune from swine-philia—a love for pigs (and a love for consuming them) that culminates in his becoming swine via Bella/o Cohen's sorcery—and from the myriad forms of transformative infection that emerge from swine-love. With Bloom as one of its primary mediums, "Circe" traffics unabashedly in the contagion of likenesses, which cannot be held apart or held to concept by the laws of light, self-preservation, or *symbolon*. Bodies—beginning with Bloom's and including those of his resurfaced ancestors—become growing grounds for the propagation of likenesses across human, animal, vegetal, and lithic/telluric orders. In this way, "Circe" grasps toward something like the disorder of *convenientia*, the blended syntax of another creation, in which everything in the world touches together and resemblances multiply by contact. A contagion, finally, that is untreatable by either Homeric/Joycean *moly* or by Sweny's pharmacy.

Circean contagion begins by undoing the coupling of mind and body, which is perhaps the first form of immunity against the contagion of the world and its resemblances. Syphilitic infection is explicitly linked in Joyce's (and "Circe's") imagination with the condition of "locomotor ataxia," defined as an inability to control the movements of one's body brought on by syphilis. In a conversation between Stephen's friend, Lynch, and Florry, one of the women in the brothel:

> LYNCH: (*Laughs.*) And to such delights has Metchnikoff inoculated anthropoid apes.
>
> FLORRY: (*Nods.*) Locomotor Ataxy (*U* 15.2590–92)

The reference is to Russian zoologist and immunologist Élie Metchnikoff's experiments on infecting apes with syphilis to study its effects on humans. "Circe" opens the possibility that a loss of self(-preservation) experienced as locomotor ataxia might be something other than pathological privation. Swine-philic ataxia, instead, might be another way to think the relationality of bodies across the determinations of species, and without the mediation of mind. The logos-bound mind cannot "grasp . . . the relationship" of two copresent or juxtaposed "realities," according to Breton; for Spinoza, the mind "can only wonder at" the motions, speeds, and relations of the body to which it finds itself attached (*E* II, P2 Schol).[44] Joyce also claimed to Budgen that "the rhythm [of 'Circe'] is the rhythm of locomotor ataxia": an indeterminate rhythm, made amidst the improvisatory play of unselved bodies and parts.

Ataxia's (ir)rhythms also ground the composition of "Circe," which itself is made from fragments of other parts of *Ulysses*, now thrown together into new configurations of matter and form, to be wondered at. As Richard Brown argues in "'Everything' in 'Circe,'" "We might discover that there is nothing in 'Circe' that is not a recapitulation of something occurring in the earlier parts of the book or even that there is nothing in the earlier parts of the book that does not get, in some way, re-cycled or re-cannibalised here."[45] Similarly, Hélène Cixous claims that "Circe" "replays all parts of *Ulysses* in one scene" and "decomposes each and everyone into his several selves."[46] "Circe" is the site of the book's enfolding—or perhaps invagination—of itself into itself, and place for the creation of "everything." The everything of "Circe," additionally, branches from the everything of *La tentation*'s relationally constituted universe: if the world is "a whole [*ensemble*] of which all parts mutually influence one another, like the organs of one body," then it might be possible "to modify what appears to be the immutable order of the universe [*l'ordre immuable*]" (17). Metaphor in "Circe" becomes an agent of profanation, medium for the material-affective touching together of part to part, everything to everything, across orders of being. And yet, as we will explore in the next sections, Circean metaphoricity also holds in itself a tension between metaphor as a liberatory contagion among beings, lineages and everythings; and metaphor as a means of (imperialist, misogynistic, ableist) dehumanization wielded against those excluded from the universality of Man.

For what is a metaphor when it is given in stage directions, as the performative enfleshment of an impossible event in and across bodies and parts in (self-)relation, rather than the offer of name and logos to what was formerly outside of meaning? When a metaphor's expressivity is no longer confined to metaphysical abstraction—by laws of *symbolon*, light of soapsun—metaphors come to infect reality, shrouding what is called truth amidst a piling up of untransfigured resemblances. In extension and in thought, Bloom's body becomes ataxic, something other than a possessed self, as soon as it touches itself to nonself, in the form of two nonhuman body parts: the pig's and sheep's foot that he purchases and carries with him in parcels at the beginning of the episode. Bloom emerges from the pork butcher's with "a parcel" "in each hand . . .

one containing a lukewarm pig's crubeen, the other a cold sheep's trotter, sprinkled with wholepepper. He gasps, standing upright. Then bending to one side he presses a parcel against his ribs and groans" (15.157–61). Bloom's motions detach from the command of mind, as his body shifts from jerked upright to side-bent form, with a parcel pressed to his ribs. The parcels are extensions of his own body—his new hybrid paw-hands—that mediate material contact with his environment and with himself. Yet they are also symbolic figurations for his becoming-pig via Bella / o Cohen and his imagined encounter with Rudy, who carries with him a "lambkin" in the last line of the episode and is himself identified as both shepherd and as lamb (15.4967). The functions of metaphor as both enfleshment and as abstraction / name-giving come to meld, indeterminately, into a condition we might call material-metaphoric. Material-metaphor is made from the contagious passage of body into body, part to part, in which resemblances among (multispecied) beings achieve articulation in and as reality.

Bloom's body is not only spatially but also temporally unselved, as its touching together with nonhuman parts also leads to the inhumanization of his own familial and paternal lineage. Bloom's (deceased) father and grandfather materialize in "Circe" as similarly animal-appendaged bodies that disrupt the idea of paternal succession as an abstract passage of the name across generations. In surrealist fashion, the stage directions attach an incomprehensible profusion of animal / material-metaphors to the bodies of Bloom's progenitors, making them into grotesque, multispecied aggregations, or the Flaubertian "frightful creatures" of "marvellous anatomies" in Act VII of *La tentation*. Fittingly, Joyce claimed "Circe" as "an animal episode, full of animal allusions, animal mannerisms."[47] The metaphor-bound progenitors are also points where humanistic fictions of lineage entangle with their inhumanist geohistories. We saw something of this confounded patrilinealism in *La tentation*, during the Griffin's unearthing of the inhuman noise of the earth in place of the slumbering kings and the ancestral origins of human civilization that he promised to reveal to Anthony. It inheres, too, in the unimaginable temporality of Odradek's unthreading and in the inhuman rustle of his laughter: expressivities that insist beyond lineage, family, house, and other modes of imposing determinate time and end (*Ziel*) upon being. Contrary to Stephen Dedalus's conception of paternity as a disembodied, "mystical," and "apostolic" relation only accessible through language and its attendant fictions, "Circe" gives lineage as carnalized swine-philia (as well as biophilia, and even geophilia) in which paterfamilias is an interspecied assemblage rather than a metaphysical "name" (9.838, 844).

V.

> The episode of Circe has changed me too into an animal.
>
> —James Joyce, letter to Carlo Linati (September 6, 1920)[48]

We can think of Bakhtin's pre-Enlightenment grotesque world as a realm of non-noncontradiction: beings can be both themselves and not (them)selves simultaneously, and bodies are both the self-enclosed individuals of novelistic character and gatherings of multiplicate parts that commune ataxically with their external worlds. The "formless world" from which Oannes emerges is formless not because forms have not come into existence, but rather because the relations among existing forms ("fingers, fins, wings") have not yet been hardened into concept. As "the first consciousness of Chaos," Oannes allows conceptual meaning to take place, furrowing an immanently relational field into a determinate formal order. "Circe," in this way, is Oannean dreamwork, grotesque thought of the joining of bodies and parts prior (and ongoingly indifferent) to the awakened laws of cultivation and composition.

Eating, too, is a gathering, and an opening of self to the material-affective contagion of other bodies in the world. The transformative capacity of swine-love in "Circe" is occasioned both by Bella/o's sorcery as well as the seemingly mundane act of entering a butcher shop. Both are acts of affinity for (becoming-)swine as well as passages toward the grotesque, everywhere-relationality of a blended earth. "He that kills an animal shall become that animal," as Manes, founder and prophet of Manicheanism, counsels Anthony in *La tentation* (58). A similar thought passes across Bloom's mind in the "Lestrygonians" episode: "Don't eat a beefsteak. If you do the eyes of that cow will pursue you through all eternity" (*U* 8.535–36). (And both read as potential transgressions against the divine injunction to Peter from the Apostles: "What God has made clean, you are not to call profane."[49]) Yet both Bloom and Anthony reject the Manichean principle of seeking spiritual purity through hermetic withdrawal from all relations and metaphysical removal from the matter of the earth. For Manes, unimplicated beautiful soul, pure asceticism is liberation of the light enclosed within the body. Anything less leads to the corruption of both body and spirit. Manes tempts Anthony toward deeper hermeticism and detachment of spirit from the earth, claiming that, "the proper aim [*but*] of every creature is the deliverance of the ray of celestial light imprisoned within matter [*enfermé dans la matière*]. . . . Food absorbs the celestial light. . . . Therefore, abstain! fast!" (58).

Anthony's fasting, however, is what allows the outside to gather in and as his body, while Manes's fasting is meant to empty the body of everything but the pure light—no longer commingled with the dust of creation—enclosed [*enfermé*] within it. The consumptive habits of Bloom—true devotee of the Hegelian animal religions—might seem the near-polar opposite of Anthony's. At the same time, however, what Bloom seeks in nonfasting tends to coincide with Anthony's overarching desire to (de)compose self with and through the materials of the earth, as opposed to seeking self-protection and purity through radical asceticism. Bloom is introduced, in "Calypso," through an almost ritualistic worship of carnivorism:[50] "Mr Leopold Bloom ate with relish the inner organs of beasts and fowls. He liked thick giblet soup, nutty gizzards, a stuffed roast heart, liver slices fried with crustcrumbs, fried hencod's roes. Most of all he liked grilled mutton kidneys which gave to his palate a fine tang of faintly scented urine" (*U* 4.1–5). Bloom's consumption, as both epiphoric accretion and limitless jouissance, calls to mind the insatiability of the Rabelaisian giants Pantagruel and Gargantua (and perhaps even Messere Gaster, the stomach-deity from Book IV of the *Pantagruel*). Pantagruel's chronicles begin at infancy, when "at each one of his feeds, he golloped down the milk of four hundred and six cows," and one morning "seized [one of the cows] under the hocks and bit off and ate her udders and half her belly including her liver and kidneys."[51] This passage into animal consumption, or what Hegel calls the "guilt of animal religions,"[52] is one of the central features of Bakhtinian grotesque realism, which emphasizes "the lower stratum of the body, the life of the belly and the reproductive organs," as the site that joins death, negation, and destructivity with rebirth and regeneration.[53] Read in light of his Pantagruelian legacy, Bloom emerges as a grotesque gathering of consumed beings and body parts, to which the pig's and sheep's feet are added in "Circe."[54]

But Bloom does not eat the crubeen and trotter, as his Pantagruelian inclinations from "Calypso" might suggest. Rather, he holds them in each hand, melding them with his own body as his hands hybridize into paws—or more precisely, into hand-paw-parcel assemblages—that remain so hybridized until he decides to feed the meat to a hungry dog (Garryowen) trailing him. Bloom presses a "parcel" (15.160) against his ribs to catch his breath, and after a near encounter with an oncoming trolley, "brushes a mudflake from his cheek with a parcelled hand" (15.198). Soon Bloom "begins to bestow his parcels in his filled pockets but desists" (15.291–92), as if he is materially, fleshly incorporate with them and cannot let them disjoin from his hands. Parceled Bloom proceeds with a "stiff walk," a motion which makes him contemplate the need to take up his floor exercises (performed "on the hands down") and which makes him recall "that awful cramp in Lad Lane" from some time ago. He attributes the latter to "something poisonous I ate. Emblem of luck. Why? Probably lost cattle. Mark of the beast. (*he closes his eyes an instant*) Bit light in the head. Monthly or effect of the other. Brainfogfag. That tired feeling. Too much for me now. Ow!" (*U* 15.207–11). The cramp he recalls is the result of his eating "lost cattle," or in contemporary usage, "beef that has been illegally slaughtered or horse meat that has been substituted for beef."[55] In this case, he literally does not know of what his body is composed, or what parts of what other bodies—what grotesque amalgamations of beings—he has internalized to make his body both ataxically uncontrollable as well as oriented toward earth, on the hands down. ("Mark of the beast," then, is both a metaphor for divine condemnation and the impression that the ingested beast has made on Bloom's body.) If *symbolon* is the metaphysical joining of what is thrown together, the body and its dissociate limbs now thrash in space arrhythmically, an alogic throwing outward of what has been thrown together in and as logos.

Syphilitic ataxia, the product of profane swine-love: Bloom has taken beings into his body and in so doing has himself become a multispecied gathering, as *La tentation*'s Manes once prophesied. The crubeen and trotter are material-metaphoric extensions of self, physical appendages as well as imaginative morphoses of body. For example, during a vision of his (deceased) father, Rudolph, Bloom "*hides the crubeen and trotter behind his back and, crestfallen, feels warm and cold feetmeat*" (15.256–57; italics original). "Feetmeat" is a literal, fleshly description of what he is holding as well as a contagious figuration of the preceding milieu, in which Bloom, "*shocked, on weak hams, . . . halts*" (15.241; italics original). In this way, the pig flesh he holds occasions other, alike enfleshments across his body: Bloom's legs now pass into halted "hams" in the stage directions. Earlier in the day (also in the "Calypso" episode), Bloom follows a female customer out of a different pork butcher's shop to get a glimpse of "her moving hams" (*U* 4.172). The mirrored reversal of this phrase in "Circe," which now refers to Bloom's own, "halt[ed]" legs, has the effect of turning Bloom's consumptive gaze back on his own body, now become a gathering of the body parts that he has devoured in the past, both physically (animal parts) and visually (the woman as animal/meat,[56] a recurrent motif in "Circe").[57]

Even as Bloom's is a body composed of/alongside animal bodies, or a body to which animal parts have attached themselves, we can still speak of "Bloom," as we did of Saint Anthony, as an entity in the singular, even if Bloom is no longer (or never was) a self-consistent being. Between enclosures of stone each bordering an abyss, Anthony is both a selved being and a glowing patch of ground, a medium through which the temptations—alternative shapings of *adamah* gathered across space and time—achieve world. Circean Bloom, similarly, is knotted as a self situated in space and time—the husband, father, friend, and citizen whose wanderings we have been following up to now—and as a multispecied gathering of bodies and parts, a piling up of enfleshments across the space and the time of a body become both unselved and geohistoricized.

For the ataxia dispossessing Bloom's body also comes to infect Bloom's (paternal) lineage, which itself becomes swine- and zoo-philic through the inhumanistic mediation of the stage directions. The reanimated bodies of Bloom's father and grandfather, never encountered outside of "Circe," become improvisatory assemblages of enfleshed parts, touching across formerly irremediable temporal and spatial distances. Soon after leaving the porkbutcher's shop, Bloom encounters in a vision his father, Rudolph, who appears to Bloom not with parcels but with "vulture talons" in place of hands in the stage directions. "What you making down this place?" he scolds. "Have you no soul? (*with feeble vulture talons he feels the silent face of Bloom*) Are you not my son Leopold, the grandson of Leopold? Are you not my dear son Leopold who left the house of his fathers Abraham and Jacob?" (15.253, 259–62). Talons mediate "stooped," "horned"-spectacled, vulturous Rudolph's relations to his environment and his contact with Bloom across the chasm between death and life (15.248, 250).

Lineage is both a filial line as well as an inhumanist history, inseparable from the multispecied parts that have joined themselves to the body: an inorganic vitalism of disembodied parts passing across generations. The stage directions transmute resemblances across species (human hand and vulture talon) into reality, making Rudolph into something like a contradictory gathering of body, animal, and metaphor. Rudolph's touching, additionally, seems to meld the humanist and inhumanist varieties of Heideggerian *berühren* ("to touch, to be affected by"). A stone's touching of earth and a lizard's touching of stone, Heidegger argues, are "above all not the same as that touch which we experience when we rest our hand upon the head of another human being."[58] In "Circe" and its staging, however, a metaphysical filial line is mediated through an inhumanist touching between talon and "silent face." Or perhaps a contagion of interspecied resemblances (among vulture, pig, sheep, human) passes into humanistic-metaphysical lineage and thought, offering another modality of creation and making something monstrous of the fiction of patrilineal descent. Descent itself, however, implies directionality, and there is no clear direction to contagion's movement, whether from Bloom to Rudolph or Rudolph to Bloom (or from "Circe" itself to both Bloom and Rudolph). Perhaps it is not simply filial succession that has been contaminated, but (temporal and spatial) directionality itself.

VI.

For what descends, or what falls, in descent is not simply reproducible form and stable conceptual identity, but also the inhumanist geohistories that have been sedimented alongside the lineages of humanist earth-writing. The contamination of Bloom's filial line extends to (or perhaps extends from) Bloom's grandfather, Virag, who subsequently appears to Bloom as another gathering ground for inhuman, material-metaphoric body parts. In Stephen Dedalus's aforementioned musings in "Scylla and Charybdis," by contrast, patrilineal descent is pure abstraction, "a legal fiction," or a "mystical estate, an apostolic succession" which is passed on through language and the symbolic order rather than through the body (*U* 9.844, 838). The relation to the father is "founded upon the void," "upon incertitude, upon unlikelihood," and does not have the immediate certainty of "love of the mother" (*amor matris*) (9.842–43). In its Lacanian valences, fatherhood is a structural position rather than a physical one: the father is always conflated into a symbolic function, as lawgiver, as the *nom-du-père* that is also the law-of-the-father.[59] For Stephen, "fatherhood, in the sense of conscious begetting, is unknown to man. It is a mystical estate, an apostolic succession, from only begetter to only begotten. . . . *Amor matris*, subjective and objective genitive, may be the only true thing in life. Paternity may be a legal fiction" (9.837–39, 842–44). In "apostolic succession," a filial line is rooted in spiritual connectivity—the transmission of logos through the agreements of human discourse, the touching of human hand to human head that is metaphysically removed from earthly relations—as opposed to the "true," embodied relations of nonmetaphysical touchings.

The absent center or locus of the paternal name gives rise to a chain of metaphorical substitutions which attempt to designate it, and it is through these attempts on the part of the subject that the connection to the father is achieved, amidst the unconscious realm opened up by the play of language. As Jean-Michel Rabaté writes concerning the motif of Stephen's search for a (symbolic) father, "To be a father, symbolically, does not imply merely a real paternity; on the contrary, it takes death, absence, and radical otherness into account. . . . A symbolic father is not simply the father of a son, as such can be left to the real father's function; a symbolic father is, as it were, the father of a father—a grandfather in a way—who fades away to become increasingly identified with a pure name."[60] The *nom-du-père* is not (or is no longer) simply the unattainable or unspeakable "pure name" of the symbolic, spiritualized father. As symbolic father and progenitor, Virag—despite his status as "pure name" in Bloom's imaginary—is given as a grotesquely enfleshed body: *le-corps-du-père*, perhaps. Circean paternity / genealogy, in this way, is an inhumanist assemblaging, an unknowable geohistory unto itself.

As a monstrously exaggerated version of his son's and grandson's, Virag's body is a further transit for the passage of imaginative resemblances into world. Protruding beyond parceled hands and vulture talons, Virag's body is variously and contradictorily staged / materialized through a series of enfleshed likenesses, a throwing together of unruly parts. For example, "*his weasel teeth bared yellow*" (15.2339), "*his yellow parrotbeak gabbles nasally*" (15.2414), "*his glowworm's nose running backwards over the letters which he claws*" (15.2435–36), "*he leans out on tortured forepaws*" (15.2574), "*he sticks out a flickering phosphorescent scorpion tongue, his hand on his fork*" (15.2600–601). Further, Virag—fragmented bestiary—is also given through a dissonant series of inhuman expressions: supplementing his "howls" with a "mooncalf nozzle," Virag also "*prompts in a pig's whisper*" (15.2412), "*gobbles gluttonously with turkey wattles*" (15.2433–34), "*crows derisively*" (15.2404), and "*with gibbering baboon's cries . . . jerks his hips in the cynical spasm*" (15.2602–3) (all italics in original). Like the Sphinx and Jabberwock before him (an alternative filiality), Virag is an irresolute riddle composed of contradictory embodiments, or symbolically inconsistent parts that touch together in indifference to a whole. What has both a "glowworm's nose" and a "mooncalf nozzle," "tortured forepaws" and a "hand," "weasel teeth" and a "parrotbeak," for example? Virag's being, in this way, unfolds through the alogics of Aristotelian *epipherein*, an excess or piling up of names onto the thing under designation that renders it unpossessable by conceptual or other forms of capture.

The expressive fragments gathered in and as Virag are perhaps only thinkable through the otherwise modes of joining given in surrealist and grotesque thought: the Bretonian "juxtaposition of two realities" that the work of reason ("man's power") cannot connect, as well as the grotesque amalgamation of "antagonistic parts" within a simultaneously individual and multiplicate body.[61] These parts also touch in and across time. Each given (inhuman) expressivity materializes through a verb—"his yellow parrotbeak gabbles nasally," "he leans out on tortured forepaws," "gobbles gluttonously with turkey wattles," for example—as the stage directions continue a monstrous, uncontrollably profusive scription of what might variously be called a grotesque being, an unsolvable riddle, an inhuman geology, a surrealist artwork, and a multispecied metaphorics.[62]

VII.

Is there something like an ethics of mattering in the stage directions in "Circe" and in the episode's inhumanization of paternal lineage?[63] And what ethicality, if any, to be made from the routing of Man's temporal and spatial relations through the multispeciesism of metaphors? These questions arise, for me, from another, inescapable dimension of the metaphorical labor that transits through Virag: the use of animal/multispecies metaphor as misogynistic insult. As an intensification of Bloom's voyeuristic consumption in the pork butcher's shop (and elsewhere), Virag's first actions upon descending the chimney flue are to describe and to visually consume the female bodies around him through a mixture of faux biological language and taxonomic orderings, regarding them primarily as "mammals" and "insects." "Her beam is broad," Virag remarks of one of the women. "She is coated with quite a considerable layer of fat. Obviously mammal in weight of bosom you remark that she in front well to the fore two protuberances of very respectable dimensions" (*U* 15.2355–58). He continues, "Such fleshy parts are the products of careful nurture. When coopfattened their livers reach an elephantine size. . . . That suits your book, eh? Fleshhotpots of Egypt to hanker after" (15.2361–62, 2365–66). Virag speaks into being a metaphorics of Man, made from the intertwined misogynistic, racist, imperialist, and ableist logics by which beings are determined and distributed along a human/nonhuman axis.

In *Animacies: Biopolitics, Racial Mattering, and Queer Affect*, Mel Chen employs the term "animacy hierarchy" (borrowed from linguistics) to name the deterministic conceptualities by which beings are made to occupy specific, ordered positions within the categories of human/nonhuman, life/nonlife, or being/nonbeing. In its most basic definition, "animacy" is "a quality of agency, awareness, mobility, and liveness," and an animacy hierarchy is that which "conceptually arranges human life, disabled life, animal life, plant life, and forms of nonliving material in orders of value and priority," or by degrees of animacy.[64] (Heideggerian distinctions between human, animal, plant, and stone seem to operate along similar lines, as do the gradations on the Aristotelian *scala naturae*.) Insults, metaphorical and otherwise, operate by using language or other forms of expression to reduce a being to a lower level within the prevailing animacy hierarchy of a given sociopolitical context. Animacy hierarchies, then, are formed when particular social, political, and cultural judgments and valuations of being become raised into an implicit universality. Racist, misogynistic, transphobic, and ableist insults symptomize a "fragile humanity" that needs to normatively secure itself through the transit of what Chen calls a "racialized animality" enacted by a dominant animacy hierarchy.[65]

Expanding on the social and political consequences of such modes of valuation, and the metaphorics that arise from them, Sunaura Taylor argues that animal metaphors in particular are often deployed to enforce determinant and normative embodiments (an experience to which the author, who has arthrogryposis, has been subjected from a young age). However, what is implicit in these metaphorical comparisons is that it is insulting and hurtful to be compared to an animal only because of how poorly and exploitatively animals are treated in society. Taylor writes,

> What does it mean to be compared to an animal? To be called a "Monkey Woman"? Is there any way to consider these metaphors beyond the blatant racism, classism, and ableism these comparisons espouse? I find myself wondering why animals exist as such negative points of reference for us, animals who themselves are victims of unthinkable oppressions and stereotypes. . . . But how do we treat animals? We treat animals as things to be used for our benefit—we watch them in zoos, gawking at their different bodies, their different ways of moving and acting. We label and number their body parts for purposes like meat, milk, and leather. In short, we treat them as if their bodies exist solely for us. As a freak, as a patient, I do not deny that I'm like an animal. Instead, I want to be aware of the mistreatment that those labeled "animal" (human and nonhuman) experience. I am an animal.[66]

There is a distinction to be made between exclusionary, oppressive dehumanization and the transformative, collaborative inhumanism toward which Taylor's identifications tend. Rather than signaling an acceptance of one's dehumanization, identifying with an animal may instead be an interrogation of the normative structures and animacy hierarchies that have historically oppressed and continue to oppress Black, Indigenous, and people of color, colonized and diasporic peoples, women, queer and transgender people, poor people, and disabled people. As Taylor claims, "Perhaps we need to ask how we can assert both our humanity and our animality. How do those of us who have been negatively compared to nonhuman animals assert our value as human beings without either implying human superiority or denying our very own animality?"[67] And what, we might add, would an identification, an animacy, and a metaphorics open to radical, inconceptual relation with nonhuman and inhuman beings look like?

Virag's particular animacy hierarchy—rooted in the conjunction of misogyny, imperialist exoticization, and speciesism—arises, Oannes-like, to fix the chaos of relational beings and bodies into preexisting (gendered, raced, and specied) form and type: coopfattened, protuberant, mammalian, fleshhotpots of Egypt, for example. Like the pre-Oannean world, however, animacies are fluid arrangements of matter. Animacies, as Chen notes and as Taylor implies, can thus be retroped into alternative embodiments and relationalities, new joinings of bodies and parts. Internal to Circean metaphoricity, then, is the tension between these two modes: metaphor as the methexic, inhumanist contagion of bodies, beings, and lineages; and metaphor as a dehumanizing conceptuality wielded by Virag and by all oppressors.

A metaphorics of dehumanization, stemming perhaps from Virag's own contagious use of animal metaphor as insult, persists throughout the episode's most famous scene. This is when sorcerer-dominatrix Bella/o Cohen transforms Bloom first into a woman and then into a pig, calling to mind the swine into which Circe transforms Odysseus's men.[68] While these transformations may seem to offer transgressive potentials for rearticulating dominant animacy hierarchies, Bella/o's sorcery (and Bloom's recovery from it) still operates on an intertwined misogyny-speciesism axis similar to the one deployed in Virag's metaphors:

> **BELLO**
>
> Down! *(he taps her on the shoulder with his fan)* Incline feet forward! Slide left foot one pace back! You will fall. You are falling. On the hands down!
>
> **BLOOM**
>
> *(her eyes upturned in the sign of admiration, closing, yaps)* Truffles!
>
> *(With a piercing epileptic cry she sinks on all fours, grunting, shuffling, rooting at his feet: then lies, shamming dead, with eyes shut tight, trembling eyelids, bowed upon the ground in the attitude of most excellent master.)* (*U* 15.2846–55)

This transformation is at once conceptual and material: Bloom's masochistic surrender to Bello also alters the fleshly composition of his body. Bloom's sinking is partially determined by the pig and sheep parts gathered into him at the beginning of the episode to mark his ataxic bodily dispossession. As he then contemplated his floor exercises ("On the hands down") amid the contagious enfleshment between his body and other bodies, now Bella/o fulfills Bloom's earlier desire for descent by forcing him toward a new intimacy with ground, earth, and animal being.[69]

Bella/o's sorcery, however, works against the regenerative possibilities—for a new animacy hierarchy, for alternative arrangements of bodies and parts, for another modality of lineage and of descent—made from the vibrant resemblances among beings. In Bella/o's animate imaginary, and in Bloom's masochistic collaboration with it, becoming-woman is an equal abjection to becoming-pig. Adjacent to both, as Hélène Cixous argues, is Bloom's becoming maid to Bella/o, a position of twinned sexual and economic servitude, as the "domestic of a multiplied master."[70] What subtends these varied dynamics of becoming and transformation in the midst of shifting relations of power, perhaps, are the logics of consumption that open "Circe": in a reversal of his insatiable, Pantagruelian desire for female and animal bodies, Bloom now becomes the dual/multiplied object of Bella/o's consumption.

Yet Bloom's is still a willed dehumanization, regulated by the terms of the masochist's contract, in which the submissive specifies and decides on the humiliations and punishments to be experienced at the hands of the dominant, who is in this way paradoxically kept under the control of the masochist.[71] As one of Bloom's namesakes, Leopold von Sacher-Masoch specifies in a contract dictated to his wife, Wanda von Sacher-Masoch, shortly after their marriage in 1873 (three years after the publication of Sacher-Masoch's *Venus in Furs*, a central inspiration for "Circe"):

> You shall renounce your identity completely.
>
> You shall submit totally to my will.
>
> In my hands you are a blind instrument that carries out all my orders without discussion.
>
> ....
>
> You shall be neither a son nor a brother nor a friend; you shall be no more than my slave groveling in the dust.[72]

The loss of identity, lineage, bodily possession, and the culminating immersion in dust is mediated by the terms of the contract, in which the masochist self-surrenders to achieve an otherwise unthinkable pleasure. In the words of Severin, protagonist of *Venus in Furs*:

> "I want your power over me to become law; then my life will rest in your hands and I shall have no protection whatsoever against you. Ah, what delight to depend entirely on your whims, to be constantly at your beck and call! And then—what bliss!—when the goddess shows clemency, the slave will have permission to kiss the lips on which his life and death depend." I knelt and leaned my burning forehead on her knees.[73]

Bloom's submission can thus be thought paradoxically as a means of self-recovery (as well as an opening of the body to Severinian jouissance, beginning with the touch of human forehead to human knee) in which he allows himself to be commanded into abjected but legible forms by Bella/o (pig, horse, cow). (Masochism as yet another iteration of *moly*, perhaps?) Bella/o transmutes the fluid, shifting resemblances among different beings and orders of being into imporous concepts, with fixed coordinates along an animacy hierarchy that unilaterally maintains the subordination of animal, female, and "epileptic" being. While Bloom still directs the terms of his dispossession, he does so only through the arrangements and hierarchies that Bella/o (and earlier, Virag) has set in place. Recovered Bloom, in other words, keeps intact the existing structures of mastery—consumer/consumed, human/nonhuman, and to borrow from Cixous, buggerer/buggered (*enculée*)—along with the valuations of kinds and orders of livingness, through which his contractual submission is routed.

The contract expires, however, and the (atemporal/Circean) interval of masochism seems to allow Bloom a means of repossessing his lost body after its ataxic wanderings. Repossession, however, comes at the cost of dominating Bella/o, as if reaffirming both the power that the masochist has been secretly wielding during submission as well as the existing animacy hierarchies upon which these forms of power are predicated. Bloom now brandishes zoometaphors as a means of dehumanizing Bella/o, within the same configuration of animacy enacted by both Bella/o and Virag: "*Passee*," he tells Bella/o. "Mutton dressed as lamb. Long in the tooth and superfluous hair. . . . Your eyes are as vapid as the glasseyes of your stuffed fox (15.3112). He continues: "Clean your nailless middle finger first, your bully's cold spunk is dripping from your cockscomb. Take a handful of hay and wipe yourself" (15.3123). Bloom individualizes himself by participating in the same (contagious) imperialist-misogynist-speciesist logics of domination expressed through his grandfather. In other words, Bloom self-possesses his body by degrading another through determinate and restrictive modes of metaphorical expression, an instance of what it might mean to become (multiplied) "master" over metaphor. A vicious circle, inviting new imaginative oppressors to ascend the animacy hierarchy through the sociosymbolic dehumanization of others. This is precisely what happens when Blazes Boylan—the character whom Bloom suspects of having an affair with his wife, Molly, earlier in the day—comes to being in "Circe," literally cuckolds Bloom by hanging his hat on Bloom's newly "antlered head," and, to Bloom's own voyeuristic satisfaction, begins to pleasure Molly (15.3764).

Circean metaphoricity/sorcery, then, carries within it this split between two types of contagion. On the one hand, the contagion brought on by ataxic surrender of self-possession to an indeterminate, swine-/zoo-/geophilic relationality. And on the other hand, the contagion of humanist-metaphysical mastery, as the power to become Man-like by wielding an idiom of dehumanization against all that might too closely resemble one's self. What emerges from a lineage that is simultaneously apostolic-conceptual (the inheritance of the name that is also the inheritance of a humanist conceptuality that cannot abide the blended syntax of a resembling earth) and inhumanist-ataxic (the descent of a monstrous counter-geology, made of confounded gatherings of beings, bodies, parts, and expressivities irreducible to origin's logics)? The answer is itself another riddle, emerging in the final scene of "Circe," wherein Bloom imaginatively encounters the being/nonbeing of his lineage in the form of his son, Rudy, eleven years after his death as a stillborn baby.

VIII.

As Bloom looks into Stephen's face at the end of "Circe," the stage directions describe Bloom's vision of Rudy as an eleven-year-old "fairy boy" or "changeling":

> *(Silent, thoughtful, alert he stands on guard, his fingers at his lips in the attitude of secret master. Against the dark wall a figure appears slowly, a fairy boy of eleven, a changeling, kidnapped, dressed in an Eton suit with glass shoes and a little bronze helmet, holding a book in his hand. He reads from left to right inaudibly, smiling, kissing the page.)*
>
> **BLOOM**
>
> *(wonderstruck, calls inaudibly)* Rudy!
>
> **RUDY**
>
> *(gazes, unseeing, into Bloom's eyes and goes on reading, kissing, smiling. He has a delicate mauve face. On his suit he has diamond and ruby buttons. In his free left hand he holds a slim ivory cane with a violet bowknot. A white lambkin peeps out of his waistcoat pocket.)*

Rudy offers another possibility of lineage: not the grotesque material-metaphoricity of Rudolph's and Virag's multispecied bodies but rather a lineage culminating in an angelic child that realigns the symbolic terms of fatherhood with Dedalian "apostolic succession." With "his fingers to his lips in the attitude of secret master," Bloom assumes the role of symbolic father to Rudy, passing onto him the "mystical estate" of fatherhood shrouded in ritual, and mediated through an abstracted language and silent reading, containing rather than creating temptations. Fatherhood, further, is devoid of any material-affective contact: the climax of the scene is Bloom's ("inaudible") utterance of the name, not an embrace or another bodily relation that could be mediated by talons, paws, beaks, hands, or any other manifested appendage. What is joined joins according to the logic of Aristotelian-Heideggerian *symbolon*: a determinate holding together between beings upon the metaphysical ground of logos, a relation ontologically prior to enfleshment. Not, finally, the traumatic *überleben* of Odradek—an ongoing livingness exceeding the time of family, lineage, and species—but rather the spiritualization of lineage through the radical absence/absenting of the beings raised into concept.

Metaphor, too, becomes respiritualized: the single lambkin that Rudy holds—placid, individualized zoon—protects from the material-metaphoric contagion of bodies by fractious parts in relation. As conceptual (nonpartialized, nonparceled) metaphor for Rudy's innocence, the lambkin is made whole and makes whole, in contrast to the sheep and pig feet once gathered into Bloom.[74] With a fatherhood mediated through the name and the concept rather than the epiphoric aggregations of Rudolph and Virag, Bloom's body recovers from its fragmentation into uncontrollable parts/"parcels," giving to Bloom the self-possession that masochistic surrender ultimately could not provide. Bloom, "silent, thoughtful, alert" now "stands on guard, his fingers at his lips in the attitude of secret master," in contrast to both his "grunting" and "shuffling" at Bello's feet (as he remained "bowed upon the ground in the attitude of most excellent master"), and his ataxic dispossession at the opening of "Circe." By possessing Rudy symbolically, through the mediation of the fleshless name, Bloom in turn comes into possession of what is beyond possession: the radically other, Rudy, that exists only in the space of the conditional, and his own lineage as a mystical thread uniting presence and absence. Insofar as any possibility of "*presence of mind*, power of recuperation" and the existence of any savior from "accident" exists in "Circe," we might find it here, in Rudy/*moly*.[75]

Accordingly, the first sentence of the next episode, "Eumaeus," gives language back over to recuperated humanist conceptuality/"presence of mind," and character back to nongrotesque, nonmultiplicate individuation: "Preparatory to anything else Mr Bloom brushed off the greater bulk of the shavings and handed Stephen the hat and ashplant and bucked him up generally in orthodox Samaritan fashion which he very badly needed" (*U* 16.1–3). The "preparatory" relinearizes Circean atemporality—it is also the first hour of a new day, since "Circe" takes place between 11:00 p.m. and midnight, "Eumaeus" from midnight to 1:00 a.m.—offering a metaphysical ground for the stable progression of events, scaled back to the level of individuate human character. The material-contagious remnants of Nighttown, the shavings covering Stephen (resonant vibrations of/with Thebaidian golden dust), are "brushed off" from his shoulders.[76] "Mr Bloom brushed off . . . and handed": a Eumaean corrective of "he brushes a mudflake from his cheek with a parcelled hand" in "Circe." Bloom recovers his hand: no longer an amalgam of ataxic parts, his body is now a joined unity (or recuperated *symbolon*), in which "hand" is a verb, an expression of the body's determinate agency in time rather than a name for a parceled, multispecied (melded human, swine, sheep) organ dissensually thrown together with the body. Enfleshed metaphor and its contagions, too, no longer spread upon the inoculated, unresembling ground of the "preparatory": to "buck up" here is to strengthen, to raise up—into upright posture as well as humanist conceptuality—rather than a material-metaphoric growth, in the form of antlers, for example. Rudy's appearance, like the Christ-Sun to Anthony, seems to fulfill what was promised by the purifying soap-light of Sweny: alexiteric *moly*, protection of being and lineage from the unmaking intimacies of resemblance and from the viral communions of interspecied parts.

But Rudy is also given as both *ainigma* as well as Anthonean/surrealistic temptation, occurring in the mist of Circean night rather than the grounded, Eumaean time of the preparatory. Before encountering Rudy, Bloom looks down at the "face and form" of Stephen, asleep in the carriage, and earlier, watches as Stephen "turns on his left side, sighing, doubling himself together" (15.4934). We might thus read Rudy as a manifestation of Bloom's desire to be symbolic father to Stephen, a further act of sorcery in which Stephen is transformed or "doubled" into envisaged Rudy: multiplied child or changeling of a multiplied master/father. (Additionally, Stephen's face is itself mediated through another resemblance/lineage: "Face reminds me of his poor mother," Bloom thinks [15.4949].) Both apostolic fatherhoods, in turn, are mediated by an even deeper enigmatic relation: Bloom "communes with the night" before looking down on Stephen's face and form (15.4949). Joining himself to what is beyond possession—the Flaubertian *nuit profonde* in which resemblances multiply and alternative arrangements of *adamah* arise—Bloom also opens himself, like Saint Anthony before him, to self-dispossession amidst the temptations of metamorphic night. Night gives a figure that both vibrates among multiple lineages and belongs to no lineage: a being that is both "fairy boy" and "changeling," an otherworldly presence as well as an infinitely resembling presence, substituted for another in the night, to be wonderstruck at.

The utterance of "Rudy" may then be Bloom's answer to an irresolute riddle: the giving of a (singular) name to what is a fluctuating profusion of beings—fairy, changeling, schoolboy, "silent" Masonic master, lambkin and lambkin-holding shepherd—as a means of fixing the temptation into the determinate symbolic terms of lineage and linear time. (Aristotle: "The very nature of the riddle [*ainigma*] is this, to describe a fact in an impossible combination of names."[77]) In addition to a doubled Stephen, Rudy may also be another Virag, another riddle—or set of contradictory and otherwise embodiments—that the name of Man (or boy) can only provisionally determine.

La tentation also concludes amidst both *ainigma* and the passing of night into (Eumaean) day:

> *Day at last appears;—and, like tabernacle curtains uplifted, clouds of gold* [nuage d'or] *uprolling on broad volutes unveil the sky.*
>
> *Even in the midst thereof* [au milieu], *and in the very disk of the sun* [dans le disque-même du soleil], *beams the face of Jesus Christ.*
>
> *Anthony makes the sign of the cross, and resumes his devotions* [se remet en prières]. (191)

Like the emergence of Rudy / *moly*—simultaneous remedy and poison, self-recovery and unselving riddle—the face of Christ-Sun ("invisible influence," "power of recuperation") returns selfhood to Anthony while also further enigmatizing creation itself. The appearance of pure light protects from the temptations of a mingled creation, a confounded sky and earth, imagination and reality. But we can also see in the divine, individuating face of Christ an amalgamation with the disk of the sun as well as with the golden dust (from *poudre d'or* to *nuage d'or*) of the opening, itself a co-vibration of light with earth, seen between (*au milieu*) stones. In the opening stage directions was "a golden dust [*poudre d'or*] so fine as to become confounded with the vibrations of the light," an originary contagion of dust and light that has now perhaps become etherealized in and as the sky itself. The closing image is both a purity and a contagious multitude, a riddle, an (earth-imbued) sky unveiled that could bring either self-possession and the raising of *adamah* into perfectible form, or the descent of further temptations and the improvisation of a new relation between dust and light, another (un)ground of creation.

In Geoffrey Harpham's analysis, the grotesque accepts all interpretations, as surrealism accepts all worlds within itself. Metaphors, too, "speak in riddles [*metaphorai . . . ainittontai*],"[78] according to Aristotle in the *Poetics*, and riddles are contradictory, disindividuating gatherings of names (and in "Circe," enfleshments) unbound from concept. Body, name, lineage, animateness, species, and concept are forms imposed upon the improvisatory play of material-metaphoric bodies and parts in extension and in thought. But forms themselves are historical as well as geological constructs, ongoingly composed by the shapings of matter and flesh that constitute them. In this vein, Theodor

Adorno notes that all forms are "sedimented contents. In them survives what is otherwise forgotten and is no longer capable of speaking directly."[79] Perhaps what is sedimented then speaks, or expresses, in riddle. We do not know what Anthony prays for when he "resumes his prayers [*se remet en prières*]": does he pray for protection from the contagion of the world's resembling things, and from the mingling of true with false light by which contagion is born? Or might his prayers be a continuity of his preceding desire to yield self to profounder immersion in and with the first materials of the earth? Anthony wanted earlier "to shine like light, to squat [*me blottir*] upon all forms," exchanging the security of a humanism made in stone for the potential to become(-with) both forms themselves and the light that touches all forms. In each of these possibilities (and perhaps in an infinity of others as well), Anthony prays to be given form by earth, whether as a bounded self or a boundless geology or as both simultaneously. Do the earth's sediments and its vibrant dust then speak through the medium of Anthony and Anthony's enigmatic prayers? To be an earth-form—like Anthony, like Bloom, like *adamah*, and like all beings—is perhaps to be a riddle the earth glyphs upon itself; it is to be at once a self and a container for the infinitude of resemblance, the everywhere-touching of dust and light.

In this sense, it matters little whether Anthony prays for self-preservation or for radical unselving, since all prayer is a means of participation in the earth's unfolding of itself through the materials, beings, and relations by which its *ainigma* is spoken. To detach from the expressive materiality of the earth, to pray for ascent into the pure light that resembles nothing and that nothing resembles, is still to conceive of oneself in relation to earth. For Kafka, to and with whom we turn in the next chapter, the "true language of prayer [*Gebetsprache*]" is "humility [*Demut*]," "at once adoration and the firmest of unions." Humility, additionally, "provides everyone, even him who despairs in solitude, with the strongest relationship to his fellow man."[80] Prayer is a sociality—as well as an ecology, even a planetarity—achieved through the acceptance of one's earth-being, a lowliness that allows all to join all. This connection is intensified in the English word "humility"/French *humilité*, which has its origin in the Latin *humus* (soil): humility is a striving toward an earth-relationality or soil-being through which all other relations among beings are made possible. Kafka's later short fiction intimates that faithfulness to the earth and to what is given by earth, at the expense of the consistency of self, may be both the only incorruptible act of worship and the ground of faith itself. Read this way, what Anthony strives toward in resumed prayers—and what Bloom seeks in metamorphic *moly*—is perhaps faith in another planetarity to come: a way of being collectively improvised, enigmatically said by earth.

Chapter 4

Soils

Kafka's Inhuman Fidelities

> In response to grandiose plans to settle Palestine or build a Jewish state, it was in the summer after writing down [the *Zürau Aphorisms* (1917)] that Kafka went to the *Pomologishes Institut* to learn gardening. We can only speculate as to what attracted Kafka to tending the earth; maybe it was the way soil, contrary to its political reputation, freely abandons its consistency, the way soil yields itself to alien growths.
>
> —Paul North, *The Yield: Kafka's Atheological Reformation*

> Humility provides everyone, even him who despairs in solitude, with the strongest relationship to his fellow man, and this immediately, though, of course, only in the case of complete and permanent humility. It can do this because it is the true language of prayer, at once adoration and the firmest of unions. The relationship to one's fellow man is the relationship of prayer, the relationship to oneself is the relationship of striving; it is from prayer that one draws the strength for one's striving.[1]
>
> —Franz Kafka, *Zürau Aphorisms* 106

Humility is acceptance of one's own belonging to others—a worship (*Anbetung*) of what one is joined with in firmest union (*festeste Verbindung*). A worship, too, of the joyful relations that both constitute a being and supersede the possession of a singular self. Saint Anthony intuited something like this in the desire to abandon self to the blended matter of the earth, as did Bloom in his surrender of individuated being to the gatherings given in swine- and geophilia. But while the *Gebetsprache* (prayer-language) of humility provides "the strongest relationship [*Verhältnis*] to one's fellow man [*Mitmenschen*]," earth-faith in late Kafka seems more the province of nonhuman rather than human beings. Complete and permanent humility (*völliger und dauernder Demut*) may only be possible through another ontology of both self and of ground in which a being raised above the earth and abstracted from earth relations is too strongly a self and can no longer speak the social language of prayer. (In contrast to the "serf-mindedness" of *Demut*, *Hochmut* [pride] is literally "high-mindedness.") For the dog-philosopher of *Forschungen eines Hundes* (Investigations of a Dog), any being removed from ground—most particularly the dogs that inexplicably float in the air, known as *Lufthunde* (soaring dogs)—is incapable of philosophy, reflection, observation, or belief (294). The dog-philosopher and the burrower of *Der Bau* (The Burrow) both open the possibility that the fullness of faith and the perfection of humility could only ever be realized through strivings toward the "the nourishing earth [*der nährenden Erde*]" of both dog- and burrow philosophy. There may be an inhumanism central to faith itself, as both enigmatic narrations/narrators wager: faith is an intensive attunement to what is given by earth, to and past the point of complete self-abandonment.

Inhumanist earth-faith—as devotional striving with the soils of the earth—is also a means toward an even more immersive (inhuman) modernist planetarity. What we might call the planetarity of late Kafka—along with the planetarities of late Woolf and of eco-metamodernist poetry in the chapters to follow—proceeds from the acts of striving and acts of faith by which nonhuman beings compose socialities with earth, in profounder indifference to a humanist imaginary. Walter Benjamin suggests something of this indifference in "Franz Kafka: On the Tenth Anniversary of His Death": "It is possible to read Kafka's animal stories for quite a while without realizing that they are not about human beings at all. When one encounters the name of the creature—monkey, dog, mole—one looks up in fright and realizes that one is already far away from the continent of man. But it is always Kafka; he divests human gesture of its traditional supports, and then has a subject for reflection without end."[2] How to think inhuman strivings with earth while minimizing (in the impossibility of completely removing) the mediations of human character and humanist scale, along with other traditional supports? Both Kafka's and Woolf's late works seek—within inescapable material, aesthetic, and specied/embodied limits—new grammars of relation among beings and earths (and materials, and soils, and skies) without recourse to the possessionism of human languages.

As both enigmatic phenomenon and instrument of faith, sound becomes central to each of the following chapters' envisioned planetarities. The incongruity of the word "envision" in this context is intentional, and through it I mean to draw attention to the lack of a word in English to express an attempt to *hear* toward the future, or to imagine[3] a future through hearing rather than vision (as given in the word "envision," as well as "imagine," and even "theorize" [from the ancient Greek θεωρία (*theōria*): "looking or watching"]).[4] Sound's (material and epistemological) excesses unmake sense, but in so doing, they also make possible different futures and otherwise relations to earth. Alicean nonsense, as we explored, begins with noise gathered and noise propagating in the undergrounds of logos, in the earth's sediments. Burrow philosophy (what we might call *Bauschaft*) also begins with attempted attunement to the noise of the underground, or what might be called both nonsense and the raw materials of a magical language to come. The burrower hears an earth sound and, in relationally striving alongside it, invents a listening apparatus that is also the cocreation of another world, even a cosmos, of meaning. Refusing the "poison" (*Gift*) of abstract individuation, Kafka's dog-philosopher seeks a faith made in and of earth—rooted in the ancient canine practice of giving nourishment to a nourishing earth—that forms the ground of an ontological sociality amongst all dogs. Starlings make new sonic collaborations with trees, and from these arise new sound objects expressible only as fragments of entangled language and noise that Woolf knows as "syllables." The undulations of corals, in Alexis Pauline Gumbs's *Dub: Finding Ceremony*, are both a form of singing (collaborative co-vibration with ocean) as well as the expressive unfolding of a vast, unarchivable intergenerational and interspecies memory.

Human(ist) consciousness, and all of the ableist/sanist machinery of sapience, sense, logos, and abstract conceptuality through which it knows, is inadequate as a means to think these and other acts of inhumanist striving (*Streben*). While "the relationship to oneself is the relationship of striving," relationality itself is only possible through a humility that is also the acceptance of one's belonging to earth. Perhaps one only strives toward what one does not or cannot possess, whether a self or any of the relations out of which a self is constituted. Striving would then be a self-indifferent fidelity to ground, and in our context, a fidelity to what is given in and through underground burrow, canine *Grund*-religion, starling-tree assemblage and bird-syllabled forest, post-extinction ocean ceremony, and other inhuman (meta)modernist sound- and noise worlds.

Spinozist *conatus*—as both an orientation toward the everywhere-being of God/Nature and as a grammar of relation—may help us attune to the paradoxical quality of striving as simultaneously directed inward ("the relationship to oneself") and radiating outward into the sociality of prayer ("the relationship to one's fellow man"). In the *Ethics*, "the striving [*conatus*] by which each thing strives to persevere in its being is nothing but the actual essence of the thing" (*E* III, P7). As discussed in the introduction, a being's essence *is* the relational expressivity by which being and world blend: an essence is not an identity but an ongoing interpretation of the divinity or divine Substance that we know as both God and Nature (held as one and the same in the *Ethics*). Conative striving—like the growth of plants, and like all inhumanistic modes of growth—is without direction or telos, without desire for a form to approximate or a singular self to attain. We can think of *conatus* instead as a desiring with or a desiring alongside, an own-being that is only possible through improvisatory communion.

The ateleology—and in our terms, the earth-oriented humility—of conative striving also helps us to more precisely envision an inhuman grammar of concept- and meaning formation, or what we have been calling an *inconceptuality*. Recalling the introduction and chapter 1, the inconceptual is that which unfolds indifferently to concepts and to the possessionism of language. The inconceptual offers a means to think what (else) happens in worlds, when worlds are no longer humanistically determined. European Enlightenment-humanist conceptualities, as we have discussed, begin from a claim to possession, as given in the origin of the German *Begriff* (concept) in the *greifen* of "to grasp" or "to seize." What Nietzsche calls *Urerlebniss*, the occluded, preconceptual material origin or lived experience, comes to be etherealized as concept over the course of a language's expansive progression. As a mechanism of possession, the concept works by effacing, or letting fall away (*fallenlassen*), differences among the phenomena and experiences that it sublates.

Such possessiveness is internal to language itself, according to Kafka's aforementioned *Zürau Aphorism* 57: "Language can be used only very obliquely [*andeutungsweise*] of things outside the physical world [*der sinnlichen Welt*], not even metaphorically, since all it knows to do—according to the nature of the physical world—is to treat of ownership [*Besitz*] and its relations."[5] Yet what would it mean, as Paul North poses in *The Yield*, to have a language rooted in abandonment rather than possession, a language capable of expressing the nature of being as a "belonging-to" another, or as "self-dispossession" rather than a mode of ownership over one's inhabited world?[6] "Language, in the limited realm of human experience," North elaborates, "claims property and specifies its relations; for everything else, it 'hints.' In fact, language is the quintessential medium of possession, of itself, of objects, of the self, of the future and of the past. . . . There is a covert claim to possession in every kind of language that isn't hinting."[7]

For Kafka, a "magical language" rooted in hinting toward—rather than attempting to name—the unpossessable is what allows us to speak of the supersensible, of God and Heaven, provided that the (Odradekian) obliqueness of parable, riddle, or prophecy is not positivized into a determinate solution or moral instruction. North continues, "All possessive language hints at a magical language that would not possess but abandon. And the magical language implies, at the same time, that these highest most difficult to possess things are its special province and indeed its chattel."[8] The "tiny linguistic gestures" of Kafka's *Das Schloß*, for example, provide a mode of hinting at the unpossessable authority, the eponymous castle, which continually asserts its right to ownership over the inhabitants of the town.[9] This magical language is perhaps what allows us to think the content of striving in its obliqueness to and incommensurability with concept. The earth may be as unpossessable as the highest heavens, however, and as the dog-philosopher demonstrates, acts of faith can be directed to both ground and sky. In his scientific understanding (a "*Wissenschaft*" made identical with a "*Hundeschaft*"), there are "two chief methods of procuring food: namely the actual preparation of the ground, and secondly the auxiliary perfecting processes of incantation, dance, and song," the latter functioning both as an appeal to the sky's agents and as a (sonic-)ritualistic means of enchanting the earth itself to summon food from its sky (303). The burrower's faith, similarly, is a practice of self-unmaking devotion to the sound syllabary given him by the earth.

Such magical languages and strivings are grounded even more deeply in devotional abandonment to the unpossessable earth, rather than aimed at the super-sensible. The inhumanist faiths of both stories, finally, are given through the modalities of weakness, indeterminacy, and surrender: faith for both narrators is a set of beliefs, practices, and comportments that do not cohere into something as strong and stable as a law, a religion, a self, or a solution. Humility, as soil- and earth-belonging, is acceptance of one's own weakness as well as the weakness of the bonds by which one's world is held together. In weakness opens another kind of gathering among beings and grounds, a diffuse relationality that stays unrouted through logos and its laws of joining.

I.

Weakness is also a hermeneutics—and even an epistemology—unto itself. "Interpretation," according to philosopher of religion Paul Vattimo, is "a process of weakening, a process in which the weight of objective structures is reduced."[10] A weak epistemology of world allows for the flourishing of wayward articulative and interpretive relations, which resist what Paul Saint-Amour calls "bad totalities," or the "strong," determinate theoretical forms such as "doctrine, coherentism, triumphalism, and sovereign self-understanding."[11] Weakness can thus function as a refusal to adopt the onto-epistemological mastery of the sovereign subject in favor of what Saint-Amour calls the "countervailing models of the subject as distributed, precarious, dependent on and even co-constituted with other beings, objects, environments."[12] Adjacent to weakness, perhaps, is abandonment, which similarly involves, as North argues, the rejection of forms of epistemological strength—such as the Nietzschean will to power and the "act [*Tat*]"—to clear space for "alternatives to power": "the non-act—secession, surrender, the yield."[13]

To give some sense of what a weakly composed world might look like, let us turn to an incomplete short story in Kafka's diary (May 27, 1914), a transcription of a vision descending on him before sleep. A mysterious, ownerless white horse appears in the midst of "the city of A," wanders into a trucking yard and through A's empty, twilit roadways as it gradually accommodates to the "life of the streets": "Its slow pace could frighten no one, it never strayed [*verliess*] out of the roadway or from its own side of the street; when it was obliged to stop for a vehicle coming out of a cross street, it stopped; had the most careful driver been leading it by the halter it could not have behaved more perfectly."[14] A policeman takes and keeps hold of the horse, "thinking that the owner would soon be along after the runaway animal [*entlaufenen Tier*]," and the story abruptly concludes here, with the horse suspended between two modalities of law: the policeman and its unknown (or perhaps imaginary) owner.[15] In the next entry, Kafka writes,

> It has meaning [*Sinn*], but is weak [*matt*]; its blood flows too thin, too far from the heart. There are still some pretty scenes in my head but I will stop regardless. Yesterday the white horse appeared [*erschien*] to me for the first time before I fell asleep; I have an impression of its first stepping [*getreten*] out of my head, which was turned to the wall, jumping across me and down from the bed and then disappearing [*verloren*]. This last, unfortunately, was not refuted [*widerlegt*] by writing the story.[16]

The horse "appears [*erscheint*]" twice: first, as it emerges from Kafka's head as he falls asleep and second, in the first line of the story he composes: "The first appearance [*erschien*] of the white horse was in the middle of the autumn afternoon in a small but not very busy street in the city of A."[17] The story is given as a body, with blood and a beating heart, a material-affective medium through which the dreamed horse can imprint traces and persist in its wandering. According to Walter H. Sokel, Kafka's ideal story is one that joins inner and outer worlds, subjective poetic vision and linguistic/social expression, in "perfect correspondence."[18] Writing/storying is the fleshly substantiation of vision, allowing imagination and experience to coincide in a single story-body.

But the (less-than-ideal) story's is also a "weak," anemic body, barely able to sustain its own being, much less give flesh to the visions stepping out of the author's head or even to negate the disappearance or "forgetting [*verloren*]" of these unfleshed visions. However, the weakly envisioned story and horse also hint toward what happens otherwise in worlds, what goes ungrasped by the strong forms of concept, reason, law, and the promise of a mythical unity between inner and outer being. A weak story, a weak world, is one that cannot possess its contents, or what passes (or "steps [*getreten*]") through it, and thus one in which its gathered beings relate indifferently to structures of meaning and claims of ownership. A horse exceeds the determinations of the law while never straying from the city's architectural rhythms: like Odradek, who is "perfectly finished [*abgeschlossen*] in his own way" and "causes no harm to anyone [*schadet . . . niemandem*]," the white horse "behaves perfectly [*fehlerfrei*]" and "frightens [*erschrecken*] no one."[19] And in further resonance with Odradek, the fugitive (*entlaufenen*) horse insists beyond the structures of ownership, in indifference even to its own negation or disappearance.

Similar weak articulative practices and nonstructures constitute the inhuman worlds of Kafka's later fiction, in particular "The Metamorphosis," "Josephine, the Singer," "A Report to an Academy" (along with "The Burrow" and "Investigations of a Dog"). Each unfolds amidst a dispossession from strong forms of subjectivity (a family, a people, a species, a home, and finally, abstract knowledge itself), leaving their protagonists to improvise being in the absence of determinate logos, and for burrower and dog-philosopher, amidst the gathered noise of the earth. Humility/soil-being is also a recognition that one does not and cannot possess what gathers in and as one's world. Such a world is only expressible through a magical language that would dwell with—rather than sublate into concept—the particulars given by earth and sought in its soils: a language without a grammar of ownership.

II.

In this sense, we can think of both *Bauschaft* and *Hundeschaft* as weak epistemologies as well as weak metaphorics: they are ways of knowing and ways of relating (and joining, and touching) to what one knows, that either remain unaware of or else consciously refuse determinate possession. The strivings of *Bauschaft* and *Hundeschaft,* additionally, unfold not in direct opposition but rather in indifference to humanistic conceptuality: neither protagonist interacts with a human being, and even though the dog-philosopher shares a world with humans, human being and thought have no significant place within a canine epistemology. Even Kafka's Rotpeter, the languaging ape from "A Report to an Academy" (*"Ein Bericht für eine Akademie,"* 1917), who spends the majority of his life in captivity by human beings, sees European humanity as a means of mindless domination rather than a higher order of being toward which to aspire. Writing from a state of physical and epistemological capture, Rotpeter forgoes the strong, sovereign—and to him, unimaginable—form of freedom (*Freiheit*) for the seemingly more recessive *Ausweg* ("a way out"). By mastering the discursive codes and practices of European Enlightenment humanism, Rotpeter is afforded a way out of the physical bonds of captivity, though this *Ausweg* comes at the cost of an even greater epistemological unfreedom.

To Rotpeter's *Ausweg* we can contrast the burrower's obsessive formation of an interior, his ongoing faithfulness in depth as a means of protection: the creation of *Eingang/Eingangsbau*, or a way in. We can also contrast to Rotpeter's adopted humanism the burrower's obliviousness to the humanistic language of the report, as both a document and as a narrative genre. Even though the burrower can theorize beyond his own body, or imagine himself as a multiplicity of selves, he still cannot abstract himself (or any of his selves) from his earth relations. As magical language or language of prayer/striving, burrow expression can only exist in obliqueness to humanistic form, particularly in its inability to abstract earth phenomena and earth relations into possessable concepts. Burrow language, as we will explore, is ongoingly remade alongside the burrower's shifting comportments toward phenomena and reconfigured at every moment around the varied interpretive possibilities given by earth. Such a language negates, erases, and contradicts itself freely and indifferently amidst a metamorphic rather than narrative time,[20] gives multiple, irreconcilable embodiments to the imagined sources of phenomena, and privileges faithfulness to the particulars given by earth over the self-consistency of the perceiving/hearing being.

Burrow language, finally, is accretive, epiphoric in what it gathers and in what gathers in it: it is not a vessel for the passage of meaning—from one context to another or from inside-burrow to outside-burrow, for example—but rather locus for the piling up of the untransfigurable phenomena found amidst the earth's soils. In Henry Sussman's terms, "the subterranean rodent . . . fills his 'Bau' with the proliferating ruminations that he can neither master nor subordinate."[21] To philosophize in a burrow is to be filled with ruminations that do not totalize into meaning, because here (and perhaps everywhere) all thinking is inseparable from the soil in which it is thought. For Spinoza, as we discussed in chapter 1, thinking is the material arrangement of sentient relations, or the "ordering" of images within the mind as they have been (contingently) impressed on the body by the world: "[Memory] is nothing other than a certain connection [*concatenation*] of ideas involving the nature of things which are outside the human body—a connection which is in the mind according to the order and connection of the affections of the human body" (*E* II, P 18 Schol).

The burrower strives toward an essence, toward a more precise understanding of self, that only exists in relation to the material environments through which he strives. Stanley Corngold suggests something of the material-affectivity of burrow thought in *The Necessity of Form*, where he claims that "the burrow also exhibits qualities of a sensate body. . . . Creature and burrow, being inseparable, are equally sensate."[22] In this way, the burrower exists in the weakness of indetermination, with no means of abstracting self from environs and no nonburrow logos through which to mediate relations. To burrow, then, is to stay amidst a vulnerability to the intrusive thoughts, traces, and phenomena that pile up in and as mind.

Der Bau begins amidst an irresolute oscillation between belief and doubt in the protections offered by the burrow, an irresolution that becomes internal to faith itself, for the burrower: "I have constructed my burrow and it seems [*scheint*] to be very successful [*Ich habe den Bau eingerichtet und er scheint wohlgelungen*]" (4). The entire story will be the atemporal unfolding of this proposition, with the weak seeming of appearance—as in the *erschien* of the white horse—functioning not as privation but rather as an opening to modes of (weak, diffuse, nonmetaphysical) joining that elude possessionist languages and grammars, along with interpretation itself.[23]

There is a joy—inexpressible or unsayable (*unsagbare*)—in burrowing that we miss when we reduce the burrower's strivings to the determinate, so-called instinctive needs to hunt prey and protect from predators. The delimiting of (inhuman) striving to instinctual behavior also relies on a binary of human intellection/animal instinct, which has been critiqued at least since Darwin's *Origin*, in which he claims that instinct has no singular definition, that no single trait of instinct is "universal," and that "a little dose . . . of judgment or reason, often comes into play, even in animals very low in the scale of nature" (225).[24] The time devoted to burrowing, which includes both the material digging as well as the impossibly narrated expression of it, opens the burrower to a series of vulnerabilities that provide the dramatic substance of the story. But burrowing also provides certain pleasures and satisfactions (*Arbeitsfreude*) which a reductive model of animal instinct cannot account for. To belong to rather than to possess—to be vulnerable to the phenomena given in and by earth—is itself a form of joy for the burrower:

> My burrow takes up too much of my thoughts. I fled from the entrance fast enough, but soon I am back at it again. I seek out a good hiding place and keep watch on the entrance of my house—this time from outside—for whole days and nights. Call it foolish if you like; it gives me inexpressible joy [*unsagbare Freude*] and reassures me. At such times it is as if I were not so much looking at my house as at myself sleeping, and had the pleasure [*Glück*] of being in a profound slumber and simultaneously of keeping vigilant guard over myself. (334; translation modified)

To think in a burrow is also to think *as* a burrow. In striving toward a more precise sense of self or own-being in the burrow, the burrower finds himself (joyfully) multiplied into several selves dispersed within and without the burrow: a self that watches and a self that sleeps, already an imagined relation between at least two beings. Through contradictory embodiments, a self is continually reformulated according to the different capacities it exerts in different parts of the burrow: a self that sleeps, a self that watches, but also a self that digs, a self that eats and stockpiles, and soon, a self that listens. What Ted Geier in *Kafka's Nonhuman Form* calls a "coexistent ecology," and the "intuition of an inarticulate coexistence" among the burrower's (real and imaginary) encounters with other underground beings, might already have emerged here amidst the burrower's relational selves.[25]

Yet burrow thought and burrow language must accommodate not only a multiplicity of selves but also the presence of "invaders [*Feind*]." One of the burrower's impulses, when contemplating the burrow from outside, is to turn burrow language into a means of imaginative slaughter of these enemies. When the burrower temporarily exits the burrow, he feels "new powers [*neue Kräfte*] awakening in my body for which there was no room, as it were, in the burrow, not even in the Castle Keep [*Burgplatz*], though it had been ten times as big" (333). One of these awakened powers may be the power of (momentarily) overcoming vulnerability with the narrative performance of bloodshed:

> If he were actually to arrive now, if in his obscene lust he were to discover the entrance and set about working at it, lifting the moss; if he were actually to succeed, if he were actually to wriggle his way in in my stead, until only his hindquarters still showed; if all this were actually to happen, so that at last, casting all prudence to the winds, I might in my blind rage leap on him, maul him, tear the flesh from his bones, destroy him, drink his blood, and fling his corpse among the rest of my spoil. . . . But nobody comes and I remain dependent only on myself [*ich bleibe auf mich allein angewiesen*]. (337; my ellipsis, translation modified)

The somatic-material thought of this passage first enfleshes the imaginary, conditionally given *Feind*, and then disenfleshes it again, adding it to the burrower's growing, epiphoric heap of corpses. But it is the final sentence given here that seems to suggest a disposition aside from the murderous possessiveness of the preceding thoughts. We might read in this description of nonarrival a desire for sociality, for at least one other being to depend on (*angewiesen*), to unfold oneself with, or to make world or make burrow with. Additionally, the burrower here seems to be confronting the possibility of his own dispossession from the burrow or, rather, the possibility that no one owns a burrow because a burrow, like the earth itself, cannot be possessed either in extension or in thought.

The burrower's claims to possession and performances of strength, then, happen amidst the weakness of seeming that we saw in the first sentence: "I have constructed my burrow, and it seems to be very successful." As Kári Driscoll notes in *Towards a Poetics of Animality*, "The narrator's sense of powerlessness and vulnerability is thus compounded by a gnawing suspicion that he is not truly at home in his burrow, but indeed rather intruding on another creature's territory. The dichotomy between inside and outside which the narrator has worked so hard to maintain is thus undermined from within by indigenous creatures that render the narrator himself an outsider in this burrow."[26] These oscillations—between empowerment and vulnerability, possessing and being possessed, faith and doubt, inside and outside—however, are internal to the workings of burrow language and its grammars of joining beings with earths. As we have explored, burrow language joins in the absence of the strong forms of meaning and totality given in concepts, abstractions, and other modes of ordering (earth-)phenomena in relation to a singular self. Burrow language limns other relations to the ground apart from (humanistic/agrilogistic) furrowing and forms of language and thought that do not privilege the self-consistency of the (singular) subject at the expense of thinking its/their relations with earth. In the words of *Zürau Aphorism* 72, "The same person has perceptions that, for all their differences, have the same object, which leads one to infer that there are different subjects contained within one and the same person."[27] *Der Bau*'s external world—the earth and its phenomena—may be a self-consistent one, while the burrower is a fractured, multiple, and weak set of subjects or subjectivities striving to articulate his own knottedness through an inescapably possessionist language.

Humility is faith in the content of what is given by earth rather than in the consistency of a self. For the burrower, faith comes to manifest as self-unmaking devotion to an omnipresent sound. An (earth/burrow) sound is both singular and multiple at the same time, encounterable only by a multiply selved being, or a being made many-selved by the sound's expression. Simultaneously internal and external to the burrow, the invasive/indigenous sound undoes logics of possessive hearing or understanding, and instead calls for alternative comportments by which to imagine and to (self-)create alongside it. In the burrower's strivings to form a relation, a common notion, with what is beyond possession, we can track the emergence of a magical language made of hints, grammared in and as the unpossessability of sounds and other phenomena. At first, when the burrower is awakened by "an almost inaudible whistling noise [*an sich kaum hörbares Zischen*]," he comprehends without hesitation: "I recognized what it was immediately; the small fry [*tätiges Volk*], whom I had allowed far too much latitude, had burrowed a new channel somewhere during my absence, this channel must have chanced to intersect an older one, the air was caught there, and that produced the whistling noise [*das zischende Geräusch*]" (343). But it becomes impossible to locate the source of the whistling, which seems not to have a single source but perhaps "two centers" (345), or may even be emanating from "all the passages" (346). Like the white horse's weak flickering in and out of being in Kafka's diary entry, the sound draws its power from its indeterminacy, a weak expressivity that paradoxically allows it to be omnipresent—passing, like the horse, through the interstices of its textual world—and uncontainable by the strong determinative form of cause and effect.

In attempting to discover the source of the noise, the burrower explains,

> In such cases as the present it is usually the technical problem that attracts me, for example, from the noise, which my ear can distinguish in all its finest shades, so that it has a perfectly clear outline to me, I deduce its cause, and now I am on fire to discover whether that conclusion is valid. And with good reason, for as long as that is not established I cannot feel safe, even if it were merely a matter of discovering where a grain of sand that had fallen from one of the walls had rolled to. (344)

While able to "deduce its cause" in one moment, in subsequent moments the burrower's constant investigations reveal nothing about the source of the noise or the phenomena or fellow being(s) responsible for producing it. Moreover, his ear cannot really, as he claims, "distinguish [the noise] in all its finest shades," since he alternately refers to the noise as an undifferentiated sound punctuated by regular pauses: "the same thin note, with regular pauses, now a sort of whistling [*Zischen*], but again like a kind of piping [*Pfeifen*]" (344), "exactly the same noise wherever I may hear it" (345), and as a differentiated set of sounds: "Already I have almost fancied sometimes, when I have listened carefully, that I could distinguish, if very indistinctly, differences of tone" (345). And toward the end of the story, "the noise seems to have become louder, not much louder, of course—here it is a matter of the subtlest shades—but all the same sufficiently louder for the ear to recognize it clearly" (351). At first, he claims that "I did not hear [the noise] at all when I first arrived, although it must certainly have been there" (343) and then apparently reverses course by counter-claiming, "I cannot suddenly begin to hear now a thing that I have never heard before though it was always there" (347). Over time, the burrower hears the noise "more distinctly, for my ear has grown keener through practice, though in reality it is exactly the same noise wherever I may hear it" (347), and yet, "my sensitiveness to disturbances in the burrow has perhaps become greater with the years, yet my hearing has by no means grown keener" (347).

We can think this logically fraught series of (apparent) negations, traces, erasures, and contradictions as the foundational elements of an inhumanist or magical language. Each articulation of the sound gives rise to new selves, crafted according to the particular composition of the burrower's world in each moment of time and in relation to the metabolic activity of the earth/burrow and its dwellers (the small fry, a swarm of unknown creatures, another burrowing animal, the pipings of winds, for example). As Henry Sussman writes in his deconstructive analysis of the story, "The [narrating] voice of the animal is . . . also the voice of construction [of the burrow], the voice of the rhetorical constructs employed in this particular production."[28] In its weak omnipresence, the sound becomes attached to different bodies, different relational possibilities, as the burrower more vividly comports himself toward the burrow and the phenomena it gives.

When he thinks the noise is coming from the wind blowing through the channels bored into the walls by the small fry, he understands his own hearing capacity in a certain way, suggesting that the noise *had* always been there but that he was unable to hear it before settling back into the burrow after his voyage outside. Upon moving to the explanation that the noise comes from the small fry's burrowing (and not the wind blowing through tunnels they have bored), however, he begins to think that the noise might be coming from an "unknown animal," or a "swarm of unknown . . . tiny burrowers" (347). Now he understands his hearing capacity another way, claiming that he cannot "suddenly begin to hear now a thing that I have never heard before though it was always there" and that his hearing "has by no means grown keener" (347). Finally, when he imagines that the sound is coming from a single great beast digging its own burrow adjacent to his own, he imagines that the sound is emanating from a single, mobile, "encircling" source, which is the "muzzle" of the beast as it draws in breath (354). In imagining the sound as an indication of this terrifying beast, the sound becomes "louder," which runs counter to the punctual regularity of the sound that he had earlier emphasized:

> The noise seems to have become louder, not much louder, of course—here it is always a matter of the subtlest shades—but at the same sufficiently louder for the ear to recognize it clearly. And this growing-louder [*Stärkerwerden*] seems a coming-nearer [*ein Näherkommen*]; still more distinctly than you hear the increasing loudness of the noise, you can literally [*förmlich*] see the step that brings it closer to you. You leap back from the wall, you try to grasp at once all the possible consequences brought by this discovery. (351; translation modified)

As the sound becomes attached to different causal sources and perceptive experiences, the burrower continually reformulates his understanding of his own powers and capacities in relation to the fragile collectivity between inner self and external world generated in each moment. The ongoing knotting of self and sound is another gathering mode, giving provisional response to the question of what else happens—or what else grows from the soils of the earth—aside from what can be grasped by humanistic thought. What other earth-futures become possible when universalist abstraction and the name of Man is no longer the primary solution to the riddles of both sound and being? Without the strong form/invention of directional time, what seem to be contradictions in both logic and narrative might be more aptly considered traces of a being's time-indifferent striving: toward more and more precise articulations of self, toward more energetic involvements of mental and bodily capacities with one's thought and extended world.

An inhumanist sensorium, then, is a means of joining body to world in the absence of determinate concepts and logics of possession. In the words of Karl Marx's *Economic and Philosophic Manuscripts of 1844*, the senses are "directly in their practice theoreticians," though their theoretical content can only manifest with the abolition of private property. Private property, for Marx, has "made us so stupid and one-sided that an object is only *ours* when we have it—when it exists for us as capital, or when it is directly possessed. . . . In place of *all* physical and mental senses there has therefore come the sheer estrangement of *all* these senses, the sense of *having*."[29] Marx seeks in the liberation of the senses the affirmation of human being, in which the object of perception becomes a "social, human object—an object emanating from man for man."[30] The burrower retains a similar faith and commitment to a nonpossessive relation to phenomena, but one that is unrouted through Man entirely.[31] In both cases, being is an ontological sociality among selves, phenomena, and other beings. Abstraction from communal relations, as burrow thought attests to and as the dog-philosopher of *Forschungen eines Hundes* concludes, impoverishes both philosophical understanding and being itself.

III.

The inhumanly heard sound(s) of *Der Bau* is almost the definition of acousmatic sound avant la lettre,[32] as Mladen Dolar, Brian Kane, Kata Gellen, and others have argued: it is sound with a hidden, uncertain, and perhaps undiscoverable source.[33] In Michel Chion's analysis of early sound cinema in *The Voice in Cinema*, the *acousmêtre* is the unknown origin of sound or a voice in the film that takes on otherworldly and near-supernatural powers.[34] This all-powerful acousmatic presence has been deployed throughout the history of sound cinema almost from its inception, from the voice of the criminal mastermind Dr. Mabuse in *The Testament of Dr. Mabuse* to Norman Bates's mother in *Psycho* to the cybernetic voices of Alpha 60 in *Alphaville* and HAL 9000 in *2001: A Space Odyssey* to the mystical, mythical, and phantasmatic power of the "white voice" in Boots Riley's *Sorry to Bother You*. The drama in these and other experimental films revolves around the search for the origin of the sound—a drive to attach sound to body, a process that Chion calls "disacousmatization"—after which the animating conflict is either resolved or further deferred. But as Mladen Dolar observes, there is something acousmatic in every sound, even when the body is discovered and disacousmatization has occurred. This is because the "source" of voice or sound "stems from an undisclosed and structurally concealed interior," leaving indeterminate the relation between voicing/resonant body and emanated sound.[35] The lag, gap, or *ainigma* that emerges between the sound and its source—the place from which fantasy emerges, according to Dolar—is never quite filled, just as meaning is never achieved through the grafting of content onto the signifier and deterministic solution does not dispel the ongoingness of what is gathered in the riddle.[36] The connection between sound and source always retains such a mystical dimension to it, one that is analogous or at least akin to the gap between phenomenon and concept, in which the difference-creating powers of phenomena recompose perceiving subjects into unbound, nonsingular selves.

To form a relation with the burrow sound, to hold it in common with the body, however, the burrower must remain beholden to what is given in sound. As mentioned in the introduction, the noise world of the burrow is one that undoes both self and the self-tone (*Ich-ton*) by which one mediates relations to one's surroundings. A sound without location is also an everywhere-sound: it provides the shifting modalities through which it solicits interpretation, while remaining itself opaque to all interpretation.[37] To join one's body to such a sound—to attempt to harmonize *Ich-ton* with demonically indeterminate *Bau-ton*—is to join with what is beyond meaning and possessability. Such listening, as insistent attunement to unmodulable noise, is enough to conjure now another underground being: a second burrower, monstrous in its extensive dimensions and irreducible to the epistemological modes through which the narrator has crafted his burrow. Into the auditory imagination of the burrower emerges the "great beast [*großen Tieres*]" (and later, "strange beast [*Fremde Tier*]") that is "dangerous [*gefährlich*] beyond all one's powers of conception [*Vorstellbarkeit*]" (353, 358). The burrow has served throughout the story as a metaphorics and an epistemology, a means of knowing and conceptualizing self through its fluctuating relations with earth, as well as a storehouse of images and bodies (the Castle Keep) to form oneself in relation to. The intrusion of the locationless sound is thus also the intrusion of the radical outside, embodied in an unimaginable being that undoes all of the burrower's powers of thought and extension. "The noise can be heard everywhere, and always at the same strength, and moreover uniformly, both by day and night": the unimaginable beast exists only in and as sound—as ubiquitous and panoptic *acousmêtre*, simultaneously everywhere and nowhere—that cannot be held to concept (354).

Narration stops here, incompletely, held in suspension at the point of striving toward an impossible joining, toward what cannot be taken into the understanding: an indeterminable force made in and of earth/earth sounds but inexpressible in burrow or any other language. Unlike language, however, the burrower's earth-faith is incorruptible. Such faith has been and continues to be a devotion to phenomena—a fidelity to sound and to earth, to the content of one's strivings—in the (ontological) absence of the sound's/beast's source and at the expense of the singular consistency of self. Fidelity to the unknown and unknowable: if our faith were faith in something we understood enough to positively or negatively conceptualize, it would be a corruptible or lesser faith than the true. According to Paul North, faith is a turning—not toward a negative or a "nothing"—but toward the unknown in a manner that emancipates what Kafka calls "the indestructible [*das Unzerstörbare*] within oneself," the enigmatic interior proper to us yet hidden from us at the same time, and that we cannot conceive of ourselves without.[38] *Conatus* is perhaps a striving toward an indestructible core, the articulation of an essence that is only possible through more vivid and precise unfoldings of one's being in and with world.

In this sense, faithfulness to sourceless phenomena/sound, and fidelity to the unknown and unknowable object to the point of complete self-surrender, may also be the ground of all humility and the originary grammar of a prayer language. Truly to participate in the dynamic unfolding and materialization of the earth would be to believe so deeply in phenomena and in one's intensive and extensive embodiment with them that the self is unthinkable in abstraction from them, and must fracture into multiplicate selves if a common being with the earth is to be felt and thought. (Humility, as we recall from *Zürau Aphorism* 106, is "at once adoration and the firmest of unions.") For Kafka, however, this struggle is accompanied by the world's unfolding itself with and within you, and by the emancipation of the indestructible within your inner being, as noted in the *Zürau Aphorisms*.[39] To strive is to become gathered in and as soil, to partake in the earth's own improvisatory self-expression. The burrower never dies, despite his persistent fears of death at the paws of a monstrous creature sounding in and sounded by the burrow. Accordingly, the last line of the story, "But all remained unchanged [*Aber alles blieb unverändert*]—" breaks off in mid-sentence and amidst the persistence of the burrower's fidelity to what is made in sound and in earth (359).[40]

The last of Kafka's *Zürau Aphorisms*, 109, reads, "It isn't necessary that you leave home. Sit at your desk and listen [*horsche*]. Don't even listen, just wait [*warte*]. Don't wait, be still and alone. The whole world will offer itself to you to be unmasked [*Entlarvung*], it can do no other, it will writhe before you in ecstasy [*verzückt wird sie sich vor Dir winden*]."[41] A stilling that is constituted by a series of erasures: don't leave home, but sit and listen; don't even listen, just wait; don't wait, but be still and alone. In nonthinking and nonbeing, a state that is more passive than even waiting—since waiting still presupposes directional time as well as the fulfillment of arrival—a thorougher engagement with the earth and its generative unfolding becomes possible. Listening, waiting, sitting, stilling are modes of existence rooted in abandon, oriented around what the body can attune itself to rather than what it can possess. The burrower's final desire—like Saint Anthony's wish to "divide myself everywhere,—be in everything . . . descend to the bottom of matter"—is to become so involved with the soils of the earth that he is literally imperceptible, unable to be heard or otherwise sensed amidst his self-expressive devotions to earth:

> So long as I still knew nothing about [the beast], it simply cannot have heard me, for at that time I kept very quiet, nothing could be more quiet than my return to the burrow; afterwards, when I dug the experimental trenches, perhaps it could have heard me, though my style of digging makes very little noise; but if it had heard me I must have noticed some sign of it, the beast must at least have stopped its work every now and then to listen. But all remain unchanged. (359)

Burrowing assumes a faith directed not away from the world and toward the super-sensible but one that reaches into the earth for unmasking or revelation, striving toward a self-unfolding that is also the expressivity of the earth and the phenomena it has given. An inhuman faith—which may simply be a name for faith itself—requires a language and a conceptuality rooted in abandon, not just as the refusal to possess but rather as a devotional comportment: to be abandoned to every given everywhere. We can thus think the inhuman prayer language/*Gebetsprache* of *Der Bau* as an orientation toward earth that strives toward new socialities, that partakes in the "inexpressible joy [*unsagbare Freude*]" of an unselved belonging to soil.

IV.

> Had you not been lying on the ground among the animals, you wouldn't have been able to see the sky and the stars wouldn't have been set free. Perhaps you wouldn't have survived the terror of standing upright.[42]
>
> —Kafka, letter to Felice Bauer, March 3, 1915

> Whence does the earth procure this food? [*Woher nimmt die Erde diese Nahrung*?]
>
> —Kafka, "Investigations of a Dog" ["Forschungen eines Hundes"], 288

Another irresolute riddle, an asking after an unknowable relation between phenomenon and cause/source, sets into motion these investigations of a dog. What is more magical, and less comprehensible, than the generativity of the earth, the growth of its soils? Inhuman modernism, as we have been discussing, arises largely in response to the problem of finding a thought supple and light enough to hold multispecied beings and grounds in common, without recourse to concepts and other modes of grasping. We can think these problems alongside Kafka's dog-philosopher, or even dog-modernist, for whom canine being and canine ontology are conceivable only in relation with the nourishing ground. Like *Der Bau*, *Forschungen eines Hundes* tracks the varying comportments of a nonhuman narrator in his strivings to hold what is beyond possession: an impossibly embodied sound in the former and, in the latter, a form of metaphysical thought that would enable knowledge of what is external to the body's relations. But *Hundeschaft*, in this respect, forays further and deeper than the burrower's *Bauschaft*: the dog-philosopher's abandonment of the power of abstraction is the result of a conscious and self-reflexive decision to let fall away what does not sustain the sociality of canine being in its faithfulness to earth. Abstract knowing comes to be defined as "poison [*Gift*]," as that which removes one from the commons made in and of "dogdom [*Hundeschaft*]" and dog-earth relations, sundering being from the bodies through which it is incomprehensibly expressed (291).

The onto-canine thought of the story begins amidst the insistence, or unchangeability, of the dog's being through the course of his potentially life- and species-altering investigations: "How much my life has changed, and yet how it has remained unchanged at bottom [*Wie sich mein Leben verändert hat und wie es sich doch nicht verändert hat im Grunde*]!" (278). Like the burrower's investigations, which similarly leave the core of the (devotional, earth-made) self "unchanged [*unverändert*]," the dog-philosopher's research into the sources and causes of phenomena have taken him to the point of abstraction from the earth and from the communities made through relations to earth. *Grund* is multivalent, simultaneously the sustaining commons of earth/*Erde* and the metaphysical ground of reason, understanding, cause, and concept. The dog-philosopher's temptation is toward the *Grund* of *logos*—as his methodo(g)logical emphasis on *begründen* ("to produce reasons") and *Begründung* ("reason, justification") would seem to suggest—as a means of solving investigation's first *ainigma*: the origin of nourishment.

But there is also a thought and a time that precedes investigation, and this is the ancient canine wisdom—parable, riddle, or perhaps aphorism—from a mythical and primordial dog mother: "Water the ground as much as you can [*Mache alles naß, soviel du kannst*]" (287). Arche-thought and arche-writing/arche-wetting, this "simple rule with which the mother weans her young ones from her teats and sends them out into the world" is "the essence of all knowledge [*der Extrakt aller Wissenschaft*]":

> And in this sentence is not almost everything contained? What has scientific inquiry [*Forschung*], ever since our first fathers inaugurated it, of decisive importance to add to this? Mere details, mere details, and how uncertain they are: but this rule will remain as long as we are dogs. It concerns our main staple of food: true, we have also other resources, but only at a pinch, and if the year is not too bad we could live on this main staple of our food; this food we find on the earth, but the earth needs our water to nourish it and only at that price provides us with our food, the emergence of which, however, and this should not be forgotten, can also be hastened by certain spells, songs, and ritual movements. But in my opinion that is all; there is nothing else that is fundamental [*grundsätzlich*] to be said on the question. (287)

Has the dog perhaps mistaken one ground for another? What if the two primary foundations of canine wisdom—"Water the ground as much as you can" (literally: "Make everything wet, as much as you can") and "From whence does the earth procure this food?"—were really two different ways of composing a relation to the ground and its soils? The dog hears "Water the ground as much as you can" and assumes a causal link between making water and the earth's generation of food, which in turn leads to his question about how the earth generates sustenance for the beings who water it. However, there is no thought of reciprocity inherent to the primordial dog-mother's statement, no sense that one should give to the earth in order to receive something else in return. To make wet is an enjoining not toward a specific action but rather toward a devotional orientation or comportment to earth, a giving without thought of recompense. Water the earth because this is a way of holding your being in common with the earth's, rather than "water the earth so that you may possess what grows from it." A new thought of earth opens in humility and in abandonment, a magical relation that is beyond the grasp of the conceptual edifice through which the dog-philosopher initially attempts to rationalize the earth's productivity. And it is this primordial, possessionless relation to which the dog returns to by the end of the story, refusing the gesture of "raising [*erheben*]" his paw on the "ladder" of scientific knowledge that threatens to lift and to abstract him away from the communions to be made with ground (316).

Philosophy, canine or otherwise, may only be possible under or on ground. To raise or to be raised, like the *Lufthunde* (soaring dogs),[43] the dogs that somehow remain suspended in the air contrary to all known physical laws (they are actually being carried by their human owners) is also to be removed from both the "nourishing earth" and the "dog community":

> But far more strange to my mind is the senselessness [*Unsinnigkeit*], the dumb senselessness of these existences. Nothing justifies [*begründet*] their being: they hover in the air, and that is all, and life goes on its usual way. . . . Why do they hover up there, letting their legs, the pride of dogs, fall into desuetude, preserving a detachment from the nourishing earth [*der nährenden Erde*], reaping without having sowed, and said to be particularly well nourished [*genährt*] at the cost of the dog community [*Hundeschaft*]? (294; translation modified)

The *Lufthunde* take without thought the earth's nourishment. Their dumb or silent detachment from earth, from community, and from their own bodies also marks an inability to philosophize:

> They are perpetually talking, partly of their philosophical reflections [*philosophischen Überlegungen*], with which, seeing that they have completely renounced bodily exertion, they can continuously occupy themselves, partly of the observations which they have made from their exalted stations; and although, as is very understandable considering their lazy existence, they are not much distinguished for intellectual power [*Geisteskraft*], and their philosophy is as worthless as their observations [*Beobachtungen*], and science [*Wissenschaft*] can make hardly any use of their utterances, and besides is not reduced to draw assistance from such wretched sources. (295)

The ground is the watered earth, but it is also what gives sociality as well as—or as a form of—nourishment. Observation (*Beobachtungen*) and scientific knowledge (*Wissenschaft*) grow from ground, and from the gatherings that ground makes possible. Kafka suggested something of this inhumanist ground-philosophy years earlier in a letter to Felice Bauer (the epigraph to this section) in which he gives this interpretation to one of her dreams: "Had you not been lying on the ground among the animals, you wouldn't have been able to see the sky and the stars wouldn't have been set free. Perhaps you wouldn't have survived the terror of standing upright."[44] In refusal to leave the ground—and refusal to become Man—also emerges another relation to both ground and cosmos in which the surrender to what is given by earth and in community with other beings is the highest form of liberation.

Hundeschaft, similarly, is a philosophy of unraised or unsublated gathering, or *Beisammensein* (together-being), out of which the "laws and institutions" of dogdom arise: "We are drawn to each other and nothing can prevent us from satisfying that communal impulse [*Drängen*]; all our laws and institutions [*Gesetze und Einrichtungen*], the few that I still know and the many that I have forgotten, go back to this longing for the greatest bliss [*dem größten Glück*] we are capable of, the warm comfort of being together [*dem warmen Beisammensein*]" (279; translation modified). To hold together is to philosophize. To hold together is also to rethink the metaphysicality of *Grund* or to offer a differently, immanently made *Begründung* (as rational foundation) from which alternative conceptualities, metaphorics, and modes of making knowledge might grow. Like Aristotelian metaphorics, dog *Begründung* also seems to operate by "giving name to nameless things," by finding a place for strange phenomena (the *Lufthunde*, most particularly) within existing systems of knowledge.[45] "For all the senseless phenomena [*unsinnigen Erscheinungen*] of our existence, and the most senseless most of all, are susceptible to investigation [*lassen sich nämlich begründen*]. Not completely of course—that is the diabolical jest—but sufficiently to spare one painful questions": in its alethic mode, investigation seeks a name, a place, or a ground for what is nonsensical or outside of comprehension (295). Or more precisely, the most nonsensical phenomena (*unsinnigen Erscheinungen*) "can be justified" or "grounded [*lassen sich nämlich begründen*]" (295).

Investigation, however, gives material, sensate, and social—rather than metaphysical—*Grund* for the nameless things of the world. All being and knowing, all laws and institutions, are constituted through ground and its soils, and through the together-being of *Beisammensein*. Dogs, it is emphasized, "all live together literally [*förmlich*] in a heap [*Haufen*], all of us, different as we are from one another on account of numberless and profound modifications which have arisen in the course of time. All in one heap!" (279; translation modified). Knowledge and thought emerge epiphorically, through the piling up of many beings into a together-being that insists through—and relates across—evolutionary and historical difference. Only what piles up together, in a heap, on the ground, can be given place in thought, and all philosophy worthy of the name proceeds from here. In Jean-Luc Nancy's words, "Philosophy is responsible for what its name means: *philein*, to love in friendship, not to aim to appropriate for oneself. Not only to not appropriate *sophia* but to be in *sophia* (wisdom, knowledge) of *philein*, of friendship. What is friendship? It is care, proximity and non-solitude. Not only between humans, but between all beings."[46]

The dog-philosopher, however, can only fully realize the philosophy of *Beisammensein*, the ancient canine *sophia*, by first passing through the abstract universalism toward which he is tempted throughout the story. To this end, he boldly declares at one point, "All knowledge, the totality of all questions and all answers, is contained in the dog [*Alles Wissen, die Gesamtheit aller Fragen und aller Antworten ist in den Hunden enthalten*]. If one could but realize this knowledge, if one could but bring it into the light of day [*in den hellen Tag*], if we dogs would but own that we know infinitely more than we admit to ourselves!" (289–90). However, investigation's modality as both Enlightenment and as *aletheia* fails, since it cannot disconceal what it cannot think with or about.[47] Most particularly, it cannot give name or place to those beings and phenomena that are constituted through abstracted, individualistic detachment rather than the joyful nonsolitude of the heap. Nonheaped beings, in this sense, can only be given in terms of their incongruity with ground. *Lufthunde* are unthinkable within an ontology of *Beisammensein*, but even more so are those other sky beings, the upright humans who have lifted both themselves and these philosophically challenged dogs from the ground.

The dog-philosopher does not think with or about human beings, according to most critical accounts of the story, because of a narrative conceit in which the story's dogs cannot physically perceive humans. If the dog-philosopher could experience the determining presence of human agency, as these accounts emphasize, then investigation's mysteries would cease. As John Winkelman writes in an early essay on "Investigations," "If the fictional dogs were able to see, hear, feel or smell their human masters, the mysteries which torment them would be cleared up immediately; as it is, the existence of human beings results in the phenomenon of food whose origin must forever be the subject of fruitless speculation or at any rate tormenting uncertainty."[48] Similarly, in "Kafka's Hybrids," Margot Norris argues that dog-philosophy is a satirical reversal of anthropocentric attitudes: "This very occlusion, this failure to give the human recognition, functions as the mirror of human anthropocentrism."[49] And more generally, "in the mirror of Kafka's story, we recognize ourselves as failing to recognize that we are merely burrowers enfolded in an organic universe teeming with myriad life forms that are each to themselves the centers of their universe."[50]

In contrast to these accounts, let us attempt to think (with) dog-philosophy as a striving toward and alongside the earth's soils, not in epistemological failure but rather in indifference to all sky-beings and their concepts. Like the Sphinx's, the canine *ainigma*—"From whence does the earth procure its food?"—is not solved with the name of Man and through the work of logos. For humans are neither the origin of food nor the answer to all mysteries. Dog-philosophy asks after the multiple, indeterminate agencies by which the earth gives nourishing phenomena, not after a singular location (the human hand, or the human mind, for example). In fact, dog-philosophy bemoans the tendency of traditional canine knowledge to simplify this larger question into a self-evident, conceptual tautology:

> Our scientific knowledge, which generally makes for an extreme specialization, is remarkably simple in one province. I mean where it teaches that the earth engenders [*hervorgebracht*] our food, and then, after having laid down this hypothesis, gives the methods by which the different foods may be achieved in their best kinds and greatest abundance. Now it is of course true that the earth brings forth all food, of that there can be no doubt; but as simple as people generally imagine it to be the matter is not; and their belief that it is simple prevents further inquiry. (302–3)

If the nature of Aristotelian *ainigma* is "to describe a fact in an impossible combination of names," we might think the *ainigma* of canine earth as something like "an impossible combination of causes" by which nourishment is given to beings.[51]

What does the dog know of the earth and its nourishment, and what has investigation revealed of these mysteries? Food, the dog-philosopher discovers, has two primary sources: it either appears on the ground, or it descends from the sky. Science tells him that there are "two chief methods of procuring food: namely the actual preparation of the ground, and secondly the auxiliary perfecting processes of incantation, dance, and song" (303). Through experiments, he takes this insight a step further by determining that the watering and scratching of the earth causes food to appear on the ground and that song and incantation causes food to emerge from the sky. These ritual or devotional performances seek a deterministic relation to earth and sky, where ground is prepared and songs are sung in exchange for nourishment. But these are forms of a lesser faith than what the primordial wisdom enjoins: "Water the earth as much as you can."

Still, the primordial is a wisdom that can only be attained through a tarrying with investigation, to and past the point of the exhaustion of investigation's truth-bearing capacities. The aim of investigation has earlier been metaphorized as the collective chewing of a bone to reach its marrow—the inner world of things that dog-philosophy strives to reach—wherein the highest philosophical ideal, or "ultimate science [*einer allerletzten Wissenschaft*]" (316), is to abandon the marrow, along with any form of consumption and any comportment toward phenomena that would separate being from the joy-giving sociality of the heap (291).

One does not reach the ultimate science by ascending toward it, however, since knowledge resides in the ground and in the heap rather than in the sky and its sky-beings. Scratching the soil and singing into the sky both bring food, as the dog-philosopher has discovered, but the inversion of singing into the ground to summon food from above merely leads to further perplexity, since "sometimes the food did not appear" but then "would appear" later, "as if my strange performance had caused some confusion at first, but had shown itself later to possess advantages, so that in my case the usual barking and leaping could be dispensed with" (304). What we readers know, or can surmise, is that the human companion is at first confused at the dog's actions of singing into the ground, then eventually decides to leave him the food. Yet it is not entirely accurate to claim that the dog-philosopher is unaware that someone or something is bringing him the food. He clearly registers the presence of another entity, though one only expressible in the subjunctive: "it was exactly as if my performance had caused some confusion at first [*so als wäre man zuerst beirrt gewesen durch meine sonderbare Aufführung*]." These actions function as an appeal to another being or entity or causal force, even though the dog-philosopher is unable to locate this external agency within a stable, situated, and present body. The entity bringing food is a sky-being, but even the agencies of sky-beings and the sky itself, in canine cosmology, are located within the earth/ *Grund*. The science of food holds that "the earth draws one kind of food out of itself and calls down [*herabruft*] another kind from the skies [*der Höhe*]": even in appealing to the sky, devotions are still ultimately directed to the ground by which the sky and its beings are made cosmologically legible and from which all causes arise. "It is the soil that the incantations must be whispered to, the soil that must be danced to" (304). In this sense, dog-philosophy / *Hundeschaft* may also be a prayer language (*Gebetsprache*), a devotion to ground at the expense of one's own self-possession, the surrender of being to the mysteries of soil. Watering the earth and its soils is an act of faith and a comportment of humility to the source of all being and all causes, rather than a self-preservative transaction. To belong to (rather than to possess) soil is also to partake in the mystical processes of generativity and nourishment through which soil "yields itself to alien growths," in Paul North's terms. A being yields oneself to what yields in return, for North. In our modulation of these terms, one grows with what grows in soil, perhaps, and this is how one partakes in the earth's ongoing improvisation of its own being.

To this end, fasting—the penultimate stage of the investigations—begins as an attempt to determine the causality of soil and to solve the riddle of food and earth-generation at last. Like Anthony's fasting for the sake of spiritual and imaginative purification in *La tentation*, the dog-philosopher's fasting is a means of reaching beyond oneself, toward the *Außer-sich-sein* (to be outside or beside oneself) from which universalist concepts and other modes of abstract understanding are made (314). Fasting also brings the dog-philosopher to the point of death, since it destroys both the body *and* the sociality by which he has been nurtured as much as by food: "It seemed to me as if I were separated from all my fellows, not by a quite short stretch, but by an infinite distance, and as if I would die less of hunger than of neglect" (311).

Investigation began with the sonic spectacle, witnessed in puppyhood, of a troupe of musical dogs (*Musikhunden*) performing on the street, contradicting nature's laws by walking on their hind legs and producing a disorienting "blast of music which seemed to come from all sides, from the heights, from the deeps, from everywhere [*von allen Seiten, von der Höhe, von der Tiefe, von überall*], surrounding the listener, overwhelming him, crushing him" (282). And now investigation reaches its conclusion in the company of a singing hound, whose song also spatially and sensorily disorients, taking the starving, dying dog-philosopher momentarily beyond the limits of self in a way that fasting alone never could:

> I thought I saw that the hound was already singing without knowing it, nay, more, that the melody, separated [*getrennt*] from him, was floating on the air in accordance with its own laws [*nach eigenem Gesetz durch die Lüfte schwebte*] and, as though he had no part in it, was moving toward me, toward me alone. Today, of course, I deny the validity of all such perceptions and ascribe them to my overexcitation at that time, but even if it was an error it had nevertheless a sort of grandeur, and is the sole, even if delusive, reality that I have carried over into this world from my period of fasting, and shows at least how far we can go when we are beyond ourselves [*Außer-sich-sein*]. And I was actually quite beyond myself [*völlig außer mir*]. (314)

In acousmatic detachment from the body that produces it, the song "floats"—like the isolated *Lufthunde*, and like causality itself, which investigation has decoupled from determinate sources—and threatens to take the dog-philosopher out of his body, to somewhere beyond dogdom.

In *Kafka and Noise*, Kata Gellen's term for these momentary (sonic) perceptions of the beyond on the part of nonhuman narrators is "species transcendence," which "names [the] moment of recognition—when an individual perceives something from a realm beyond, consciously or not, with or without attempting to explain it."[52] And yet the solution to the riddle of investigation, in Gellen's otherwise compelling account, remains Man, as if the human were the only "beyond" or outside-self that a nonhuman being/species might aspire to or transcend toward.[53] The dog-philosopher has already perceived and self-consciously rejected transcendence's logics after seeing what they make of beings and what they make of philosophy: to leave the ground and to become individualized, like *Lufthunde* and other sky-beings, is to forgo the communal strivings with ground through which canine being and thought are constituted.

Giving oneself over to the floating and all-enveloping sounds of the air, letting the body float with the disembodied and sourceless melody toward the abstract conceptuality of the *Außer-sich-sein*: this is the dog's last temptation. (In an inversion of Saint Anthony's, the temptation is to become sky rather than to descend to the bottom of matter and to be earthed, ensoiled.) Investigation turns briefly here to the "science of music [*Musikwissenschaft*]" and its connections to the "science of food [*Nahrungswissenschaft*]" and back to the "theory of incantation, by which food is called down," before the dog-philosopher eventually gives up on solving this mystery due to his own "scientific incapacity," along with his "limited powers of thought," "bad memory," and "inability to keep my scientific aim continuously before my eyes" (314, 315). As the closing sentences read,

> If I wanted to brag I might say that it was this very instinct [*Instinkt*] that invalidated my scientific capacities, for it would surely be a very extraordinary thing if one who shows a tolerable degree of intelligence in dealing with the ordinary business of life, which certainly cannot be called simple, and moreover one whose findings have been checked and verified, where that was possible, by individual scientists if not by science itself, should *a priori* be incapable of raising his paw even to the first rung of the ladder of science [*die Pfote auch nur zur ersten Stufe der Wissenschaft zu erheben*]. It was this instinct that made me—and perhaps for the sake of science itself, but a different science from that of today, an ultimate science [*einer allerletzten Wissenschaft*]—prize freedom [*Freiheit*] higher than anything else. Freedom! Certainly such freedom as is possible today is a wretched business. But nevertheless freedom, nevertheless a possession [*Besitz*]. (316; translation modified)

Abandoning (or letting fall away) the universality of concepts—the sublation of particulars into an abstract sky-logos by which beings are deterministically joined to grounds—is humility. Rather than ascend the ladder of science, the dog-philosopher instead embraces the sustaining communal relations and bodily comportments toward *Grund* that he also names as "instinct." But this is not the unmediated, unreflective instinct usually assigned to animal behavior in which a nonhuman animal's actions are completely calculable and predictable based on its adherence to fixed laws that govern the reproduction of a species.[54] Akin to the satisfactions and pleasures (*Arbeitsfreude*) that the burrower finds in burrowing/building, canine instinct here is driven by specific modes of intellection and even the possibility of freedom, to be attained through strivings toward more precisely thought and felt relations between self and earth. In such a weak formulation of world, the (instinctive, experiential, aphoristic) knowledges that hold canines in common are too fragile, too provisional and unscientific, to be wielded as determination.

Philosophizing dogs, then, live and world in weak modalities of suspension in which being is held indeterminately between two grammars of striving: toward ground and sky, community and singularity, food and song, *sophia* and science, for example. Like the runaway horse in Kafka's diary entry, "held . . . in the middle of the road" by a policeman awaiting its owner, and like the burrower suspended at the point of an impossible joining to the burrow's omnipresent and locationless *acousmêtre*, dogs philosophize in the interstices between forms of determinate thought. To refuse the ladder of science is also to affirm suspension and, more particularly, to affirm suspension through a material-affective metaphorics: abstractions (science, knowledge, freedom) are given meaning only in relation to the body (the paw touching the ladder of science), and knowledge is produced contingently, through what touches or refuses to touch together. Another, counter-Heideggerian term for this, in dog-philosophy, is *freundliche Berührungen*: the "friendly touching" or "friendly contiguity" among dogs in which opens a path toward an "ecstasy [*Entzücken*]" inadmissible to knowledge (290). Such touchings, such contingent as-structures and affectivities, are what inform the dog-philosopher's prophecy of an "ultimate science [*einer allerletzten Wissenschaft*]" to come, one that allows for the "freedom" of relating to the earth's phenomena in the absence of the deterministic rationalities of *Nahrungswissenchaft* and *Musikwissenschaft*. This is perhaps the only way that freedom can be defined as something that one can possess: the final "possession [*Besitz*]" is the freedom from the logic of possession itself (316).

The dog-philosopher has earlier labored through a metaphorics / as-structure of suspension, as the path toward an "ultimate science" of joyful abandon. Knowledge is the marrow of a bone chewed in common by all dogs but accessible only through a disposition of "monstrous" individualism:

> But it is not merely flesh and blood that we have in common, but knowledge [*Wissen*] also, and not only knowledge, but the key to it as well. . . . The hardest bones, containing the richest marrow, can be conquered only by a united crunching of all the teeth of all dogs. That of course is only a metaphor [*Bild*], and exaggerated; if all teeth were but ready they would not need even to bite, the bones would crack themselves and the marrow would be freely accessible to the feeblest of dogs. If I remain faithful to this metaphor [*Bleibe ich innerhalb dieses Bildes*], then the goal of my aims, my questions, my inquiries, appears monstrous [*Ungeheuerliches*], it is true. For I want to compel all dogs thus to assemble together, I want the bones to crack open under the pressure of their collective preparedness, and then I want to dismiss them to the ordinary life that they love, while all by myself, quite alone [*weit und breit allein*], I lap up the marrow. That sounds monstrous [*ungeheuerlich*], almost as if I wanted to feed on the marrow, not merely of a bone, but of the whole canine race [*vom Mark der Hundeschaft*] itself. But it is only a metaphor [*Bild*]. The marrow that I am discussing here is no food; on the contrary, it is a poison [*Gift*]. (291; translation modified)

Knowledge is a communal pursuit, a (figurative) crunching of exteriors to reach the inner being of things. But only an individual can possess the marrow that is abstract knowledge, "not merely of a bone" but of all *Hundeschaft*. This capacity for abstraction—of things, beings, and most of all, the canine species—is another name for the transcendence that would sunder a knowledging dog from other dogs: a poison that destroys both body and community, insofar as canine being is only possible in and through the sociality of *Beisammensein*. Such (a) possession fits the inferior logic of *Wissenschaft* and the "ladder of Science" in contrast with the abandonment and possession-indifference of the ultimate science. The dog-philosopher, finally, can only strive toward this ultimate science by refusing to place his paw on the step that would uplift or raise (*erheben*) an individualized dog from the philosociality of the heap.

To pursue abstract knowledge to the exclusion of the community is to become "monstrous [*ungeheuerlich*]," the same condition into which Gregor Samsa awakes in *The Metamorphosis*, as *ungeheueren Ungeziefer* ("monstrous vermin"). Gregor's monstrosity, too, is linked to his separation from unadministered social and communal ties. Unlike Gregor, however, the dog-philosopher's monstrosity is figured not as abjected animality but rather as raised individuation. This is a trait he earlier associates with the "all sorts of [noncanine] burrowers" of his world, as he observes "how little inclined they are, compared with us dogs, to stick together, how silently and unfamiliarly and with what a curious hostility they pass each other by, how only the basest of interests can bind them together for a little in ostensible union, and how often these very interests give rise to hatred and conflict" (279). The ultimate science may, after all, merely be a restatement of the ancient wisdom: "Water the ground as much as you can." Only relate to an unpossessable earth through the concepts and knowledges made by (inhumanist) touching (*freundliche Berührungen*) rather than (humanistic) possession. Let the *ainigma* of the origin of spiritual and phenomenal nourishment—*Woher nimmt die Erde diese Nahrung?*—insist through any and all determinations, whether those furrowed by canines or anyone else.

The riddle of the earth returns us to the story's beginning: "How much my life has changed, and yet how it has remained unchanged at bottom [*Wie sich mein Leben verändert hat und wie es sich doch nicht verändert hat im Grunde*]!" (278). To be "unchanged at bottom [*im Grunde*]," perhaps, is to comport oneself to earth/*Grund* in humility and in persistent indifference to possessive conceptuality and the steps of science leading to it. Such comportment is also a faithfulness to the earth and its phenomena as well as to what Kafka calls "the indestructible within oneself," the essence or *conatus* of being that emerges through the ongoing, devotional yielding of self to earth.[55] The ground, in

Zürau Aphorism 24, exists only in relation to what is gathered on it: "Grasp [*begreifen*] the good fortune [*Das Glück*] that the ground [*der Boden*] on which you stand cannot be any bigger than the two feet covering it [*bedecken*]."[56] There is no such thing as an abstract ground, as given here, since ground/*Boden* is constituted through one's embodied engagements with it, the (material-metaphoric) feet that cover it and that bind being to ground. To place the foot/paw on the first step of the ladder of science is also to transform ground into something other than and less than ground, as that which is no longer held in common with being and community, no longer joyfully gathered or piled upon, or made wet.

To stay with ground is also to adopt another language of joining—as magical language or language of prayer—that both *Der Bau* and *Forschungen* labor toward: *begreifen* here is to "grasp" the "good fortune" or "joy" (*Glück*) of a new, common relation to ground rather than to conceptually possess it. The dog-philosopher similarly concludes his account through a modality of joy brought by abandonment: "I confess all of this frankly, even with a certain joy [*einer gewissen Freude*]" (345; translation modified). True *Hundeschaft* is a recognition that being, knowledging, worlding, common-making are only possible through faithful strivings with and through ground. The burrower, in contemplating the aggregative relations that he has made between his multiplicate selves and his environed world, also comes to feel what he calls an "inexpressible joy [*unsagbare Freude*]" (334; translation modified). Both *Hundeschaft* and *Bauschaft*, then, are faiths as well as metaphorics, modes of joining selves with earths in the unsayable joy of being's abandon. In the words of the aphorism immediately following (#25), "How is it possible to rejoice in the world [*über die Welt freuen*] except by fleeing [*flüchtet*] to it?"[57]

Fleeing/flying to the world is an act of rejoicing in what is and in what one belongs to: where else could one flee toward (or even flee from) except the world? The inexpressible joy of being in relation with the earth emerges through self-surrender, through the humility of soil-being, which opens not a new world or a world beyond this one, but rather another comportment toward one's belonged-to world. We see something of this soil-comportment, too, in Kafka's interpretation of Bauer's dream: "Had you not been lying on the ground among the animals, you wouldn't have been able to see the sky and the stars wouldn't have been set free. Perhaps you wouldn't have survived the terror of standing upright." Gathered on and with the ground (and in a pile with the other animals, perhaps), another relation between sky and ground becomes possible. The *Lufthunde* live in the sky, but they still do not see the sky or what fills it, aside from their own floating bodies. They also do not see, know, or even think about how the sky might be filled otherwise. Ground-/

soil-beings see the sky together, and in their together-seeing, the cosmos are liberated, become something other than reflections of the humanisms and humanistic futures projected into the sky and stars. In a motif that we will return to in chapter 6, the Adam of Milton's *Paradise Lost* answers Eve's riddle of the stars—"But wherefore all night long shine these, for whom / This glorious sight, when sleep hath shut all eyes?"—with the agonistic determinations of a sky- and star-conceptuality ceaselessly at war with darkness:

> These have their course to finish, round the Earth,
> By morrow evening, and from land to land
> In order, though to nations yet unborn,
> Minist'ring light prepared, they set and rise;
> Lest total darkness should by night regain
> Her old possession, and extinguish life
> In nature and all things[58]

Together-seeing is liberation from the "Minist'ring light" of a deterministic cosmos, the improvisation of another sky alongside fellow soil-beings and beings of the dark. In late Woolf, an inhumanized sky is one in which no particular mode of seeing has determinative power over another. Neither is it the "total darkness" of oblivion and the extinguishing of "life / In nature and all things," even as Woolf writes amidst the increasingly real possibility of human extinction in the mid-twentieth century and at one of the potential starting points of what is now called Anthropocene. Woolf's sky is instead a gathering place for the earth's own expressions—noise, song, sounding, coloring, flight, and vibrationality of the earth and its beings—that fill oblivion and insist through the extinctions to come.

From our place on the ground, with the other animals, let us now see what we can of both the still-glittering starlight Eve once saw, and the strange futures given in and as an inhuman sky.

Chapter 5

Sky

The Inhuman Comedy of Woolf's *Between the Acts*

> "Dare we, I asked myself, limit life to ourselves? May we not hold that there is a spirit that inspires, pervades . . ." (The swallows were sweeping round him. They seemed cognizant of his meaning. Then they swept out of sight.)
>
> —Woolf, *Between the Acts*

> "The earth opens up, and we laugh."
> "[*La terre*] *s'ouvre, nous rions.*"
>
> —Édouard Glissant, *The Ripening* [*La lézarde*]

The swallows' flight gives response to the *ainigma* posed by the Reverend Streatfield: we may hold that inhuman being makes worlds still from humanism's ruins, striving in excess of—and in obliqueness to—human consciousness and the directionality of its time. The swallows' is a seeming cognizance that is ultimately indifferent to logos and its modes of holding: sweeping and encircling in place of determining, scribing worlds too fragile for sublation into form. As with the white horse in Kafka's diary entry, "seeming" is weak appearance or weak presence, only thinkable through the oblique modalities of hinting, abandon, gesture, and striving. At the core of the reverend's remarks—occasioned by Miss La Trobe's pageant tracing the history of Britain from its prehistory to its present, the central event of *Between the Acts* (hereafter *BTA*)—is the question of how to think or to comport oneself to what follows in extinction's wake. How, in other words, to imagine what is beyond the time of the human (or "out of sight"), through conceptualities that have only ever been humanistic? "Am I treading, like angels," the reverend adds, "where as a fool I should absent myself?" Spiritual transcendence or self-directed disappearance: both are strong forms of faith in the persistence of a humanism or human sovereignty unrouted through the material-affectivity of the worlds that it is ongoingly ending. Woolf's late works seek a metaphorics weak enough to think decreation—or the posts of both humanism and extinction—and to imagine (with) the inhumanisms still to come in ecocide's aftermath.

Between pre- and post-(human-)extinction is a vanishing, decaying world, as mourned in both Woolf's final, unfinished novel and in diary entries written alongside it. The "world seen without a self" of Bernard's philosophizing in *The Waves* (1931) wakes into the rise of fascism in Europe and the onset of a second world war that may well bring what Woolf calls the "complete ruin, not only of civilisation, in Europe, but of our last lap" (August 17, 1938).[1] "That's what makes a view so sad," Mrs. Swithin muses in *Between the Acts*—through a simultaneously familial, humanistic, and geoplanetary "we"—on the distant fields that "will be there . . . when we're not" and, implicitly, the unknown selves that will see them (53). The world seen without a self is also thought of the strange futurities—given in the weak modality of seeming—emerging from inhuman modernism's planetary analytic and resonating in the geo-imperial earth-writing now called Anthropocene.

Tracing the disjunctive temporalities and futurities brought by the Anthropocene—as a metonym for the variegated discourses that have arisen in response to anthro-/sociogenic climate change—Elizabeth DeLoughrey writes,

> Engaging with the Anthropocene means that we must simultaneously consider the deep geological time of the planet—in comparison with previous epochs—as well as the futurity of the human as a species. Futurity is marked not just by concerns about human survival, or ameliorating species extinctions, but also by the fact that one cannot locate a stratigraphic marker for this epoch until a geologically significant period of time—such as tens of thousands of years—has passed. Thus, the Anthropocene is both forward-looking and a future retrospective, characterized by "anticipatory logics" and anticipatory mourning.[2]

Read as proto-Anthropocenic thought, *Between the Acts* is simultaneously a proleptic elegy for the extinctions to come and an opening to the comedic unimaginability of a ground and a sky that persist beyond humanistic possession. Comedy is also an ecological modality: the surrender of determination to forces, beings, or temporalities other than the human.

Woolf theorizes the inhuman dimensions of comedy and laughter in a 1905 essay, "The Value of Laughter." Woolfian laughter is weakness, the province of "children and silly women," and "the voice of folly and frivolity inspired neither by knowledge nor by emotion."[3] Aligned with both animals and the larger domain of the non-Man, laughter "gives no message, conveys no information; it is an inarticulate utterance like the bark of a dog or the bleat of sheep, and it is beneath the dignity of a race that has made itself a language to express itself thus."[4] At the same time, however, Woolf notes that "laughter is the one sound, inarticulate though it be, that no animal can produce."[5] Comedy and laughter are both mediated through nonhuman beings *and* withdrawn from them, a theoretical formation that operates as well in both Mark McGurl's and Henri Bergson's accounts of the inhumanism of comedy. Of Bergson's *Laughter* (*Le rire,* 1900), McGurl writes, "Comedy can only be human, but it is also true that it arises only in the relation of humanity to something other."[6] Further, even as Bergson locates laughter strictly within human social communities, he nonetheless inhumanizes both the origin and the propagation of laughter, as uncontrollable "reverberation" without clear beginning or end: "[Laughter] is not an articulate, clear, well-defined sound; it is something which would fain be prolonged by reverberating from one to another, something beginning with a crash, to continue in successive rumblings, like thunder in a mountain."[7]

The Bergsonian comic, too, is only possible through an "absence of feeling," a comportment of indifference toward the objective content of one's surrounding world: "Indifference is [the comic's] natural environment, for laughter has no greater foe than emotion. . . . Try, for a moment, to become interested in everything that is being said and done; act, in imagination, with those who act, and feel with those who feel; in a word, give your sympathy its widest expansion: as though at the touch of a fairy wand you will see the flimsiest of objects assume importance, and a gloomy hue spread over everything. Now step aside, look upon life as a disinterested spectator: many a drama will turn into a comedy."[8] The difference between ecological mourning and ecological comedy may only be a temporal or scalar one: geological time goes on in comic indifference to what humans mourn. Humanity may relate to "something other" through comedy, but this something other also makes comedy of humanity's own relations. In Walter Benjamin's theory of allegory (drawn on extensively by DeLoughrey), "seen from the perspective of death, life is the production of corpses" (and "history the building of ruins," as John McCole adds in his analysis of Benjaminian allegory).[9]

Yet another dialectical turn of this perspective might see corpses, ruins, and extinctions as sites for the ongoing generation of being. In this vein, the Bakhtinian comic holds that the laughter of medieval and Renaissance folk cultures is a means of partaking in the creative regeneration of the earth: metabolic processes in which the laugher is remade in and through what they laugh at. In *Rabelais and His World*, Bakhtin writes,

> The people do not exclude themselves from the wholeness of the world. They, too, are incomplete, they also die and are revived and renewed. This is one of the essential differences of the people's festive laughter from the pure satire of modern times. The satirist whose laughter is negative places himself above the object of his mockery, he is opposed to it. The wholeness of the world's comic aspect is destroyed, and that which appears comic becomes a private reaction. The people's ambivalent laughter, on the other hand, expresses the point of view of the whole world; he who is laughing also belongs to it.[10]

Comedy is fundamentally social for Bakhtin, as it is for Bergson, and offers a means of participation with the earth's creative unfolding of itself.[11] Laughter, especially amidst and as a response to cosmic terror, is a means of belonging to rather than possessing, a surrendering of self-preservation to the comedic ongoingness of the earth and its reverberant elements.[12]

What would make the Bakhtinian comedic vision into a posthuman or an inhuman one might be the unimaginability of the earth's future generation: the Bakhtinian folk laugher sees their body made incorporate with the earth's cycles of regeneration, whereas inhuman comedy emerges as thought of earthly growths or generations indifferent to human being. The posthuman character of comedy arises, as McGurl argues, largely from the gap or incongruity between the vast extensive and intensive dimensions of inhuman worlds ("the inhumanly large and long") and the comparative smallness of humanistic means of thought and representation.[13] Odradek's is "the kind of laughter that has no lungs behind it [*ein Lachen, wie man es ohne Lungen hervorbringen kann*]. It sounds somewhat like the rustling [*Rascheln*] of fallen leaves."[14] When asked, "Where do you live?" Odradek responds, "No fixed abode [*Unbestimmter Wohnsitz*]" and laughs this lungless, inhuman laugh at his own indeterminacy (*Unbestimmter*) and his comedic insistence (*überleben*) beyond temporalities both familial and human.[15] Odradekian laughter, too, is elemental, expressible only through a sonic rhyme (and a weak seeming) between inhuman laughter and the rustling of leaves, a metaphoric joining of inhuman to inhuman, material to vegetal being. Laughter without lungs, earthly generation without humans: these seem to be inversions of the Bergsonian comic. Here, comedy and comedic futures arise through the relation of the *inhuman* to something other than the human.

In Melba Cuddy-Keane's analysis, *Between the Acts* grows from the comic tradition. In line with ancient Greek comedic conventions, both the novel and La Trobe's pageant are structured and punctuated by a "communal chorus" or "choric voice" that rearticulates community by inviting participation from new (human and nonhuman) voices and sounds.[16] "The narrative act of transforming all voices into chorus," according to Cuddy-Keane, is a political and an ecological gesture that also suggests something of the subversive folk forms of festival, dance, and revelry—the *kômos* (revelry) of comedy (*kōmōdía*)—by which existing orders of being are opened to indeterminate futures.[17] Included within the choric voice of *Between the Acts*—and as we will explore in the next section, the "gigantic ear" that hears it—are also the methexic revelries of nonhumans, the making of worlds from sonic communions. Toward the novel's end, for example, a flock of starlings inverts the path previously articulated by the swallows, sweeping directly into a tree, "pelt[ing] it like so many winged stones" until it becomes "a rhapsody, a quivering cacophony, a whizz and vibrant rapture" (209). Lithic birds articulate new assemblages between sky and earth—multispecied gatherings of animal, (stone), and plant—out of which are produced noises simultaneously inhuman and divine. To Miss La Trobe, the pageant director and sole earwitness to this phenomenon, the choric "syllabling" voiced by the starling-tree assemblage is also the possibility of another expressivity: the relational composition of earth noises, or "words of one syllable," that are not yet—and perhaps never to be—assimilable into language. As La Trobe wanders the village fields, graveyard, and tavern post-pageant, she hears voices acousmatically, as syllables—"words without meaning"—decoupled from speaking bodies and instead planted into the earth: "Words of one syllable sank into the mud. . . . The mud became fertile" (212). What growths and futures rise from sound-fertilized mud is neither for La Trobe nor the novel itself to give. We are left, rather, amidst the comedic, revelrous generation of what Woolf calls the "syllabling" of "life, life, life without measure" that is indifferent both to the humanisms of the present and the (post-)humanisms projected into the future.

The creative fertility of worlds growing without (human) selves reads incongruously with Woolf's diary entries from the late 1930s, in which a humanistically unthought world is often conceived as the terrifying limit point of culture's collapse into barbarism. Animal life, in particular, terrifies—in the sense of frightening away from *terra* (earth), as given in the Latin *terreō*—because of its complex alignment with fascism, whereby Woolf messily envisions animals simultaneously as fascism's metaphorical victims and agents. From October 2, 1938: "War broken out already L. thought. Then the statement that all poisonous snakes at the Zoo would be killed, & dangerous animals shot—Vision of London ravaged by cobras & tigers. . . . All this mixture of minute detail; with invocations to God; with Hitler baying & the Germans howling; then the composed & cultured voice breaking in, say about not taking pets."[18] Woolf's final works, however, turn from the irremediable regressivity and violence of a human world become (materially and metaphorically) animal and toward thinking the impersonality of worlds still to be composed beyond the time and consciousness of humans.[19]

Both La Trobe and "Anon," the primordial balladeer in Woolf's eponymous, unfinished essay of 1941, sing of the ending of worlds as well as their ongoingnesses. In the last line of "The Reader" (1941), Woolf's (also unfinished) essay accompanying "Anon": "We are in a world where nothing is concluded."[20] Anon, the "common voice singing out of doors," and sound-gathering La Trobe both environ worlds impersonally, orienting song toward what might emerge in the absence of human consciousness: the birdsong from which human art mimetically arose for Anon, the future growth of planted sounds for La Trobe. The name La Trobe, originating from the Occitan *trobador* (troubadour),[21] suggests La Trobe's close connection to the impersonal generativities of both Anon and the inhumanist worlds ("the bird whispering reed haunted fen") that Anon sought to sing.[22] The world "beneath our consciousness," Woolf writes in "Anon," is "an anonymous world," conceived both as the preconscious primitivity voiced by the singer and, in *Between the Acts*, as fertile ground for the comedic futures growing from humanism's ruins.

The world's anonymity or impersonality is the terror of arche-lithic night in *To the Lighthouse* (hereafter *TTL*). Darkness is aqueous in "Time Passes": a "downpouring" and an unsurvivable "flood" that "confound[s]" both "sea and land," bodies and minds, and the (Oannean) logics of solidity and self-preservation: "there was scarcely anything left of body or mind by which one could say 'This is he' or 'This is she'" (*TTL* 125, 126). Anonymity is a passage to "oblivion" (139), a "long night" (137) made from gathered nights—"Night, however, succeeds to night" (127)—in which every gesture is drowned in an "immense darkness" (125). Every expression of being is made equivalent with its own voiding: "Sometimes a hand was raised as if to clutch something or ward off something, or somebody groaned, or somebody laughed aloud as if sharing a joke with nothingness" (126). In Louise Westling's eco-phenomenological analysis of "Time Passes": "The vast forces of the nonhuman world flow over and erase all the intricate structures of the human community."[23] The eternal night of "Time Passes," additionally, seems to have a precedent in Milton's *Paradise Lost*, in which the adjacent realms of Night and Chaos are given both as the "ancestors of Nature" and the anarchic non-places, outside of God's creation, in which "time and place are lost."[24] "eldest Night / and Chaos, Ancestors of Nature / hold eternal anarchy" in the

hoary deep, a dark
Illimitable ocean without bound,
Without dimension, where length, breadth, and highth
And time and place are lost.[25]

And yet there is "a force working; something not highly conscious" that preserves against decreation; returns time, place, and dimension to Night's illimitable ocean; and impels (human) life forward through the inhuman narrativity of "Time Passes" (*TTL* 139). We might think this as the labor power of the cleaning women, Mrs. McNab and Mrs. Bast, which holds open the possibility of an inhabitable world for the novel's central characters to wake into from oblivion.

Oblivion, for both La Trobe and late Woolf, on the other hand, gives ground for inhumanistic growths and futures. "This is death," La Trobe remarks to herself during a moment in the pageant when the actors' voices are lost amidst wind and rustling leaves, the stage is emptied, and the audience left to contemplate the failure of "illusion." As she leans against a tree, "paralyzed" amidst the muting of the world that she has conjured, a herd of cows fills oblivion (or "continue[s] the emotion") with "the same yearning bellow. The whole world was filled with dumb yearning. It was the primeval voice sounding loud in the ear of the present moment" (*BTA* 140). The cows bellow in the "gap" or "distance" of oblivion, joining primeval and present soundscapes across the span of geohistory.[26] This chorus of the cows suggests the ongoing composition of sounds beyond the time and the hearing of humans. When La Trobe, starling-inspired, imagines a new work to be made—after concluding that the pageant itself has been a collective "failure" that "hadn't made them see"—the setting is "the high ground at midnight; there the rock; and two scarcely perceptible figures. Suddenly the tree was pelted with starlings. She set down her glass. She heard the first words" (210). Night has become the time and place of creation—rather than dissolution—of sounds, expressivities, affects, words.

La Trobe thinks land impersonally: not as "a view," but as "land merely, no land in particular" (210). Anonymity, however, no longer terrifies but generates. An impersonal, indeterminate midnight (night succeeding to night) allows new possibilities of human-indifferent relations to arise. (The night of the Sphinx's riddle is similarly a moment of unimaginable metamorphosis or regeneration, depending on how—and who we are when—we read it.) *Paradise Lost*'s Night, as well, is a place and a time of creation: "the wide womb of uncreated Night / Devoid of sense and motion," as Belial terms it in conversation with Satan.[27] Uncreated themselves, Night and Chaos—as anarchs, matrices, and "ancestors of Nature"—are nonetheless the realms from which creation proceeds. Like Oannean creation in Flaubert's *La tentation*, creation in the Miltonic sense is the giving of form to the confused matter—or the primordial "dark materials" housed in Chaos—from which all potential and actual worlds are made.[28] La Trobe's high ground at midnight is anarchic *atopos* ("no place"), an impersonal womb of creation from which other worlds and other materializations might emerge.[29]

In late Woolf, a passage opens between the impersonal and the interspecies, where the (momentary) cession of human consciousness brings the possibility of thinking worlds otherwise, in the anarchic generativity of Night. *The Waves*, for example, is imagined in Woolf's diary (when the planned title was *The Moths*) as "the idea of some continuous stream, not solely of human thought, but of the ship, the night&c, all flowing together: intersected by the arrival of the bright moths."[30] Punctuating the stream of this "abstract mystical eyeless book" are intervals without selves, where bird-singing marks territorial circles on the shore and the sun's motions gradually create and decreate the solidity of the world and its objects.[31] Woolf's novella *Flush* (1933) is written primarily from the perspective of Elizabeth Barrett's eponymous dog (as formidable an investigator and philosopher as Kafka's *Forschungen*'s) during Barrett's courtship with Robert Browning. Flush's *Umwelt* is environed through "eddies of sound and smell . . . set in circulation" rather than words transmuted into meaning, as given by Barrett's "low voice syllabling innumerable sounds."[32] Language in *Flush*, however, as Karalyn Kendall-Morwick compellingly argues, becomes a means of privation from what Woolf calls Flush's "religion" of "smell," around which his entire *Umwelt* is organized.[33] Flush's is a faith and an epistemology made in and through sensate givens, syllablings, scents and the "infinite gradations that lie between" and go "unrecorded" amidst a noncanine *Umwelt*.[34] "Woolf's narrator," Kendall-Morwick writes, "posits language as a confirmation of the finitude of the human *Umwelt*, turning on its head the characterization of animal being as a state of deprivation. Viewed in this way, language both reveals and compounds a human lack."[35] As in *Forschungen eines Hundes*, Flush's dog-philosophy seeks its own infinitude through the "circulation" of the impersonalized and "innumerable" sensations that go unrecorded and unsublated by human consciousnesses.[36]

Natural history offered Woolf a discourse for thinking the generativity of oblivion, or of what persists in and through humanism's long midnight or *nuit profonde*. Along with H. G. Wells's monumental *The Outline of History* (obsessed over by Mrs. Swithin in *Between the Acts*), the ecological thought of *Between the Acts* was also informed by Gilbert White's *A Natural History of Selborne* (1789), which Woolf read in 1937 and reviewed in 1939 in an essay titled "White's Selborne."[37] What fascinates Woolf most about White's eco-aesthetics is his ability to observe worlds without the impediment of "self-consciousness." As Woolf writes in "White's Selborne,"

> The chief question in dispute—it is indeed the theme of the book—is the migration of swallows. . . . With all his faculties bent on this great question, the image of science at her most innocent and most sincere, he loses that self-consciousness which so often separates us from our fellow-creatures and becomes like a bird seen through a field-glass busy in a distant hedge. This is the moment then, when his eyes are fixed upon the swallow, to watch Gilbert White himself.[38]

And further on:

> His mind, like the bird's crop that the farmer's wife found stuffed with vegetables and cooked for her dinner, has nothing but insects in it and tender green shoots. This innocent, this unconscious happiness is conveyed, not by assertion, but much more effectively by those unsought memories that come of their own accord unsought. . . . Even the strident voice of the cricket, so discordant to some, fills his mind "with a train of summer ideas, of everything that is rural, verdurous and joyful."[39]

White's ecology is surrender to unsought sensations, affects, and memories, gathered in a greened mind. Joyful abandon to the earth and its unpossessable, verdurous phenomena is a means of approximating bird thought ("How else can one rejoice in the world except by flying to it?" as Kafka has it). Woolf's emphasis on the "innocence" and "unconscious happiness" of White's observations, however, also reads as a mourning for earthly comportments that are no longer—or perhaps have never been—possible to hold in the face of the extinctions to come. Flower religion comes to be consumed in the warring, sacrificial, ecocidal worlds made by animal religions; consciousness passes into self-consciousness.

White's unselvings might also be read in terms of what Timothy Morton calls "ecomimesis," a motif in which a particular experience of nature is presented as if unmediated by conscious thought and purified of distinctions (and distance) between subject and object.[40] What is missed in ecomimetic discourse, according to Morton, is the nonidentity of nature, its ultimate inassimilability to language and thought. White's *Natural History*, however, stays attuned to the discordances that emerge between thought and world. Amidst surrender to phenomena, White maintains an awareness of his own embodiment and avoids presumptions to the "sheer rendering of the real" that Morton names as central to ecomimesis.[41] Prefiguring the planetary modernism of *Between the Acts*, White's *Natural History* ends not with the innocent faith in the reproducibility of inhuman worlds in language, but instead with the limits of human thought when confronted with a series of "horrible phaenomena" afflicting "every part of Europe, and even beyond its limits": "alarming meteors," "tremendous thunderstorms," a "peculiar haze, or smokey fog" lasting for weeks, a noontime sun that "looked as blank as a clouded moon," and "heat so intense that butchers' meat could hardly be eaten on the day after it was killed."[42] Their source is the June 8, 1783 volcanic eruption of Laki/Lakagígar in Iceland (continuing until February 7, 1784), one of the most devastating flood lava eruptions in recorded human history.

In its first week, as White's and other contemporary accounts note, Laki brought "sulfuric acid, haze, ashfall, and acid rain . . . causing the Sun to appear blood red and reducing its natural warmth" throughout Iceland and then covering "almost all of Europe" in "thick, dry fog" by June 25, reaching as far south as Lebanon and as far east as Russia, China, and Alaska by the end of June.[43] Laki's "sulfuric aerosol veil . . . hung over the Northern Hemisphere for more than five months" and cooled all of Europe and North America by about −1.3°C for the next two to three years.[44] The volcanic haze, acid, and sulfur caused injuries, sicknesses, and deaths to humans, animals, crops, and vegetation throughout Europe, along with a famine in Iceland (called the "Haze Famine" in Icelandic chronicles) lasting from 1783 to 1786 and leading to the death of about 20 percent of the population.[45] Intuited toward the thought of extinction and the potential ending of worlds, or what he calls the "superstitious kind of dread, with which the minds of men are always impressed by such strange and unusual phaenomena," White's descriptive innocence fails to contain what in nature is impersonal, inhuman, and withdrawn from the observer's gaze.[46]

Instead, White can only express oblivion by reaching back to a prior age's ecological aesthetics; specifically, Milton's *Paradise Lost* and the "disastrous twilight" of the sun's eclipse accompanying Satan's appearance in Book I, from which White quotes:

> as when the sun, new risen,
> Looks through the horizontal, misty air,
> Shorn of his beams; or from behind the moon,
> In dim eclipse, disastrous twilight sheds
> On half the nations, and with fear of change
> Perplexes monarchs.[47]

White's account of Selborne comes to a close here, in the disastrous twilight of the sun's (self-)concealing, in night succeeding to night, and in the confounding of what is known of beings and their relations. Woolf gestures toward a similar perplexion in *Between the Acts*, with the self-concealing sun as metaphor, in a sentence that amalgamates White, Milton, and Percy Shelley:[48] "The future shadowed their present, like the sun coming through the many-veined transparent vine leaf; a criss-cross of lines making no pattern" (114).

While Woolf sees in the indeterminacy of beings and futures the inhuman creativity of "lines making no pattern," such perplexion is "dread" for White. It heralds both the collapse of distinction between the local and the planetary amidst Lakagígaran twilight and the end of joyful collaborations with the sensible givens of nature. White's final letter in *Natural History*, immediately following the account of Lakagígar, concludes with the discordant noises of mechanization: the "repeated" rings and "fierce sparks" emerging from "the bells of an electric machine" at South Lambeth, where the author finishes the work in 1789.[49] Such noisings rhyme ominously with White's earlier descriptions of the "sharp and stridulous" "shrilling of the field-cricket," which (as Woolf emphasizes) "marvellously delights some hearers, filling their minds with a train of summer ideas of everything that is rural, verdurous, and joyous."[50] Woolf, however, wants to continue rather than dread or dispel this inhumanly sounded rhyme. In *Between the Acts*, Woolf seeks a thought and a landscape impersonal enough to hold both types of dissonance—natural and mechanical, orthopteric and electric—in a relation that comedically persists beyond humanistic modes of measure, hearing, and determination. The fragments sounded by a phonograph (another kind of "electric machine") couple with the bellowing of a herd of cows in La Trobe's pageant, syllabling an empty field with measureless, inhuman noise. Is there still a joyous communion to be made with an impersonalized earth and the phenomena planted into it? Yes, Woolf might answer, but not for humans.

I.

In late Woolf, as in late Kafka, sound is the primary imaginary through which nonhuman earth makings are (weakly, seemingly) given.[51] *Between the Acts* is partly constructed around a tension between inhuman, impersonal noise—animal, natural, and machinic—and the promise of harmony, or of sound made meaningful.[52] Underlying Mrs. Swithin's obsession with the geohistorical indifference of land to human histories and futures is a belief in a vast, multispecied comedy spanning the entire earth in which all beings and all voices might one day be harmonized. As she ponders during one of the pageant's intervals, "Sheep, cows, grass, trees, ourselves—all are one. If discordant, producing harmony—if not to us, to a gigantic ear attached to a gigantic head. And thus—she was smiling benignly—the agony of the particular sheep, cow, or human being is necessary; and so—she was beaming seraphically at the gilt vane in the distance—we reach the conclusion that all is harmony, could we hear it. And we shall" (175). Swithin's faith in a world-harmony, at the same time, is expressible only in the conditional tense and in the grotesquely enfleshed form of a "gigantic ear attached to a gigantic head." How might we listen with or listen amidst this simultaneously comedic and ecological fantasy? And what is the earth hearing with its gigantic ear, and what harmony or disharmony is it making of everything methexically gathered into it—"sheep, cows, grass, trees, ourselves"—all the noises, agonies, and expressions planted in its ground?

This gigantic ear might be most easily identifiable as the phonograph or gramophone, a technological prosthesis—employed throughout La Trobe's pageant, including in the scene immediately preceding Mrs. Swithin's musings—that enhances the capacities of the human sensorium, allowing the formerly unheard or unhearable to surface into audition.[53] In early accounts of the phonograph, as classicist and media theorist Shane Butler notes, there is an "unmistakable slippage" between its functions as a listening device and as a medium for the reproduction and transfer of sounds.[54] "This little instrument," as described in a *Harper's Weekly* article from March 1878, "records the utterance of the human voice, and like a faithless confidante repeats every secret confided to it whenever requested to do so."[55] In Butler's analysis, the "'faithless' repetition" of the phonograph actually "depend[s] precisely on *faithful* listening to 'every secret' in earshot."[56] We might also call these faithful comportments to tiny or secret perceptions a mode of impersonal listening (akin to Gabriel White's in the preceding century). In Friedrich Kittler's analysis, such devotional listening is only possible without the mediation of a self-hearing or self-organizing consciousness.[57]

What separates this "little instrument" from Mrs. Swithin's "gigantic ear," however, is that the latter's audition is directed toward not only the reproduction but also the *harmonization* of heard sounds. The little phonograph faithfully returns sounds in the present, while the gigantic ear hears beyond extinction, making worlds out of all that has been gathered or sedimented into it outside of human dictates of meaning and harmony alike. We can think metaphoricity itself as a such a gigantic hearing and harmonizing apparatus—both fleshly and figurative. Metaphoricity, in this sense, takes into itself the discordances and syllablings of the world and produces meaning from them, or in Quintilian's terms, "succeeds in accomplishing the supremely difficult task of providing a name for everything."[58] Classical *metaphora* is a transfer or transaction between contexts, a means of spiritualizing the sensate materials gathered in the metaphor into the effluvia of the abstract concept. But murmuring in *metaphora*'s undergrounds, as we have explored, are the dissensual accumulations of *epipherein*: the giving of names that does not provide an object with meaning but only an accumulated excess of names, materials, enfleshments, sounds. (A thing shrouds itself in its own excesses.) Metaphor may also be the untransfigured preservation of piled-up dissonances, a gigantic ear for hearing the aggregate noisings or *psófos* of the earth in their inconceptual relationality rather than in the harmonized universality of logos.

La Trobe's pageant (and subsequent wanderings) as well as *Between the Acts* itself are gathering places for sounds, or modes of improvising with the inhuman futures of sounds. Woolf forays what we can call a dissonant metaphorics as a means of hearing the dissonant syllablings of nonhuman beings—as both composers and listeners of noise—and the inharmonizable methexes of sound made among species and across geoplanetary scales of time. Metaphors work in and as noise in *Between the Acts.* As forms of phonic irruption into a given semantic order, metaphors also, in Derrida's terms, "set off an errant semantics" that generates something other than a reproduction of the humanist meanings to which they are traditionally beholden.[59] In media and sound theorist Douglas Kahn's analysis, noise is both an empirical phenomenon and an abstract mode of knowledge: noises mask, disrupt, or lead astray the sounds or voices deemed meaningful within a given context. But noise is also its own concept, standing in for everything that is extraneous to the auditory, visual, or discursive communication of a message. Noise signifies even in silence, as in Kahn's example of handwriting, where the particulars of the writer's scrawl can be read as "noise," as phenomena that are "undesirable or extraneous" to a message and its transmission. These phenomena, however, express different or perhaps deviant meanings, such as "a person's style, the result of physiological (sickness) or environmental forces (writing on a bus), and the like."[60]

In its interstitiality—its simultaneous belonging and nonbelonging to meaning, significance, and world—noise also mingles beings, makes aqueous the world's solidity: "the existence of noise," Kahn writes, "implies a mutable world through an unruly intrusion of an other, an other that attracts difference, heterogeneity, and productive confusion."[61] (Kahn subsequently theorizes a "wet modernism" through the avant-garde music, poetry, and visual art of mid-twentieth century New York, in which John Cage, Jackson Pollock, Allan Kaprow, George Brecht, and others experimented with the percussive and melodic sounds, flows, and drippings of water to herald an aesthetics oriented around the "progressive dissolution of disciplinary constraints" and "the flow of art into life and vice versa."[62]) In a motif that we will return to at the end of this chapter, media theorist Liliane Louvel theorizes color, as well, as a form of "noise" within literary and other inscriptive media: color "unsettles the literary text, threatens to collapse it [*s'effondrer*] into an impossibility, an incommensurability."[63] In this sense, we can think noise as the medium through which metaphor's errantry proceeds: metaphor brings into a given context of meaning the unsublated particulars of other worlds, growing and generating otherwise. Dissonant metaphor in *Between the Acts* is a means of attempting to hear what grows from mud-planted noise, in the dark and on the other side of oblivion.

La Trobe, too, imagines self and world sonically: "Ah, but she was not merely a twitcher of individual strings," as she imaginatively retorts to one of the villagers; "she was one who seethes wandering bodies and floating voices in a cauldron, and makes rise up from its amorphous mass a re-created world" (153). La Trobe's stirring is an articulation of world prior to its ordering/hardening into forms. (Multispecied) voices mingle with bodies, neither of them subject to the disacousmatization that would solve the *ainigma* of their relational expression and their rising into other world: the insistent jabbering of *woce* (voice), or the giving of *woce* to jabber, in Alicean terms. The pageant, like the cauldron and like the novel itself, becomes a dwelling for bodies, voices, noises, expressivities that gather discordantly rather than into the harmonic resolution of Swithin's imagined earth composition.[64] Just as noise, however, always risks conscription into a sphere of determinate meaning, so the re-created world is only thinkable (and audible) in fugitive moments stolen from a possessionist language.

Metaphors themselves are forms of noise, as Hans Blumenberg wagers, disharmonies to be either muted or resolved into the conceptual.[65] ("Dissonant metaphor," then, might be a tautology). The mark on the wall in Woolf's eponymous short story is a noise-mingled thing, inhuman in its irresolvability into concept and in its boundless generation of world from the contexts and histories speculated into it. The mark's materiality—as "solidity," as "wood," as sensate emissary of "the impersonal world which is a proof of some existence other than ours"—entangles with the narrator's "swarm" of "thoughts" that do not resolve but only further enigmatize what has occasioned them.[66] The mark insists as *ainigma*, sign of "the mystery of life; The inaccuracy of thought! The ignorance of humanity!" beyond the closing determination of it as a snail by the intrusive human(ist).[67]

If determination is a form of implicit violence at the end of "The Mark on the Wall," *Between the Acts* gives such violence in its material-fleshly dimensions. Giles—whose quasi-fascist hypermasculinity, homophobia, and nationalist warmongering often put him in conflict with Mrs. Swithin's detached musings on existence—encounters a strange phenomenon during the pageant's first intermission:[68] "There, couched in the grass, curled in an olive green ring, was a snake. Dead? No, choked with a toad in its mouth. The snake was unable to swallow; the toad was unable to die. A spasm made the ribs contract; blood oozed. It was birth the wrong way round—a monstrous inversion. So, raising his foot, he stamped on them. The mass crushed and slithered. The white canvas on his tennis shoes was bloodstained and sticky. But it was action. Action relieved him. He strode to the Barn, with blood on his shoes" (99).

In the metaphor's passage from sensuous to spiritual meaning, as Davidson argues, "rivers and bottles did not, as they do now, literally have mouths," but once this likeness of river and bottle openings to a mouth was noticed and its meaning was cemented in language, the mouth of a river is no longer a metaphor but rather an objective designation. Yet there is also something monstrous about a river or a bottle with a mouth; the submerged, noisy underground of this (material-)metaphor that still inheres in every expression of it, in Derrida's analysis. A mouthed river or mouthed bottle gathers multiple orders of being into itself and places them into new, humanistically unthinkable intimacies.[69] As another inappropriate intimacy among beings, the snake-toad aggregation registers as visual noise, or as a noisy metaphor, that expresses its own unorderable sense. To Giles it symbolizes intractable conflict that demands both possession and resolution through the work of the concept. Man once again becomes the determinate solution to *ainigma*, the name that harmonizes inhuman clamor into a univocal I-tone. Giles's shoe sublates nature into meaning, supplanting the innocent expressivity of flower religions with the animal sacrifice that instantiates a regime of warring life.[70] As Christina Alt observes in *Virginia Woolf and the Study of Nature*, there is nothing "unnatural" about this scene, as Giles is witnessing "typical feeding behaviour for snakes," which he then misreads as a demand for outside intervention.[71] Fields are furrowed, noises muted, and beings spiritualized into the concepts imposed upon them. Only in death and in silence can the crushed mass stand in for what Giles wills it to stand in for: the inevitable self-destruction of Europe in the late 1930s and the masculinist violence of "action" as its only mode of resolution.[72]

The task of consciousness, according to Blumenberg, is to resolve the "disharmonies" given by metaphor into "the harmony of the data as data of *one* experience."[73] An openness to indetermination—to "record the atoms as they fall upon the mind in the order in which they fall"—is also a means of letting sound what concepts have muted.[74] In *The Waves*, the fascist leanings of Louis are expressed through an uncompromising desire to reduce all things to order. How else to subdue the noise of a "great beast" "stamping" on the shore—heard first in childhood and reverberating through every subsequent phase of life—than through the colonialist determination of every noisy thought, errant atom, "dance of savages" into the ordered harmony of Man?[75]

Giles, too, is consumed by a desire to "command" metaphor, to make metaphor accord with Man's possessionist logics, its "sanded paths of victory driven through the jungle."[76] As discussed earlier, Giles's aunt Mrs. Swithin reads a protean "we" into the landscape overlooked by the Olivers' estate: "'That's what makes a view so sad. . . . And so beautiful. It'll be there,' she nodded at the strip of gauze laid upon the distant fields, 'when we're not'" (53). But Giles in response sees only the furrowings to be made by instruments of war, signs of conflict only resolvable through action:

> Giles nicked his chair into position with a jerk. Thus only could he show his irritation, his rage with old fogies who sat and looked at views over coffee and cream when the whole of Europe—over there—was bristling like. . . . He had no command of metaphor. Only the ineffective word "hedgehog" illustrated his vision of Europe, bristling with guns, poised with planes. At any moment guns would rake that land into furrows; planes splinter Bolney Minster into smithereens and blast the Folly. (53; ellipsis in the original)

Not subject to strong forms of grasping, metaphor and the errant semantics set off by a weakly, ineffectively conjured animal must be quickly and forcefully sublated into the visions of war that Giles reads into the landscape. Thus is an animal sacrificed to the regime of meaning. The passage from Swithin's to Giles's vision—as well as the progression from "bristling like" to "bristling with guns"—is also a passage from the innocent, White-like abandonment to land and its inhuman futures to the determinate furrowing of Man into the visible earth.

But an imagined hedgehog furrows in its own way: brought (weakly) into a determinate context of meaning, the hedgehog sows into this ground an alternative semantics. In the simplest terms, hedgehogs hog hedges. And hedges are simultaneously borders between the fields, lands, or properties they enclose *and* sites of multispecied entanglement among other hedge-hogging beings, including the soil, trees, worms, birds, and insects that (epiphorically) gather in them. (As Albert, the "village idiot" of *Between the Acts*, sings during the pageant: "I know where the tit nests . . . / In the hedgerow. I know, I know" [86].[77]) Hedges exist in and as the interstices between both possessions and concepts, troubling the forms of determinate enclosure that they are meant to enact. Hedges let grow what grows with them. (A hedge, then, is also itself a metaphoricity.) In *Through the Looking-Glass*, the hedgehogs commanded into croquet balls by the Queen of Hearts also mark the point where signification collapses: the rolling and unrolling of a hedgehog is also the expressive activity by which it exceeds both its concept and its ground, motioning the stopped thought of the noun toward the indeterminate futures given in the verb.[78] (Similarly, Odradek's unspoolings, as he rolls down the family staircase, resound beyond lineage and directional time.) Another relation to land unfolds—one that grows obliquely to both Giles's and Swithin's imaginaries—through Giles's unwitting conjuring of a hedgehog and the (boundless) contexts conjured along with it.

As a form of visual or imaginary noise, the hedgehog's bristling does not go unheard by the gigantic ear that is *Between the Acts*. The novel itself opens into a soundscape sown with inhuman noise: after some chatter among villagers about the local cesspool, the voices stop, and "then there was silence; and a cow coughed" (3). Immediately after, "a bird chuckled outside," a "daylight bird, chuckling over the substance and succulence of the day, over worms, snails, grit, even in sleep" (3). Companions of this chuckling bird in turn wake Mrs. Swithin, who has fallen asleep while reading the *Outline of History*: "She had been waked by the birds. How they sang! attacking the dawn like so many choir boys attacking an iced cake" (8). Immersed in an avian noise world, Swithin's waking thoughts materialize as a metaphor that is only legible through the medium of noise: the audible attack of the birds' singing, which we expect this metaphor to further illuminate (or give a name to), is routed incongruously through the visual attack of the iced cake by the choirboys. Audible noise is transfigured into visual noise, and this transfiguration itself *is* the noise that prevents a harmonious enclosure of the metaphor's terms within the agreements of logos. (The choirboys are known not for singing but for their voracious appetite, making their specification as choirboys excessive, unnatural, and well, noisy: "hungry boys attacking an iced cake" or "choir boys singing out of tune," for example, would make more conventional harmonic sense.) Singing birds and choirboys, along with iced cake, coexist in stalled sublation, an epiphoric heap of noises and noisy beings that disrupt—and sound into metaphor's gigantic ear—but do not spiritualize into meaning.

II.

> "We have destroyed something by our presence," said Bernard, "a world perhaps."
>
> —Virginia Woolf, *The Waves*

After Giles crushes/sublimates the snake-toad aggregation, with the blood of these beings—the material-affective traces of determination—still staining his shoes, he walks to the "Noble Barn," where the pageant-goers congregate during the first intermission. The structure, which "had been built over seven hundred years ago and reminded some people of a Greek temple, others of the middle ages, most people of an age before their own, scarcely anybody of the present moment, was empty" (99). In a Gilbert White-inspired path of thought, Mrs. Swithin later imagines the swallows returning to the barn from Africa year after year, from the time when the barn was "a swamp" (103).[79] Ambivalently historied *and* futured, the barn becomes another of the novel's multispecied dwellings, a nesting or hedging place, that birds and other beings craft into a passage between prehistory and post-(human-)extinction. Woolf continues:

> The Barn was empty. Mice slid in and out of holes or stood upright, nibbling. Swallows were busy with straw in pockets of earth in the rafters. Countless beetles and insects of various sorts burrowed in the dry wood. A stray bitch had made the dark corner where the sacks stood a lying-in ground for her puppies. All these eyes, expanding and narrowing, some adapted to light, others to darkness, looked from different angles and edges. Minute nibblings and rustlings broke the silence. Whiffs of sweetness and richness veined the air. A blue-bottle had settled on the cake and stabbed its yellow rock with its short drill. A butterfly sunned itself sensuously on a sunlit yellow plate. (100)

The endless, comedic generativity of the barn's emptiness, which is here anything but the "oblivion" of "Time Passes," is also a node (from the Latin *nodus*, "knot") joining impersonal and interspecies aesthetics. Emptiness is both the absence of human(ism)s in the barn and the medium through which the barn's nonhuman beings (mice, birds, insects, dogs) create. The barn's dwelling places—the holes, corners, pockets, and burrows made in and of nonhuman strivings—are shaped emptinesses, resignified into the material conditions of a livingness that extends beyond the time of humans. (Odradek, too, crafts worlds from household corners.) "Minute nibblings and rustlings broke the silence": beings and their choric expressivities insist beyond determinations, indifferent to a human consciousness that assumes the muteness of the noise worlds beneath its perceptive thresholds.[80]

With emptiness as its medium, the passage is also an incantation of the miniature or larval forms of the five senses: sight ("all these eyes, expanding and narrowing"), sound ("rustlings"), taste ("nibblings"), smell ("whiffs of sweetness"), and touch ("a blue-bottle settled on the cake stabbed its yellow rock," "a butterfly sunned itself sensuously on a sunlit yellow plate"). The narration strives not to name but rather to sound alongside what materially-affectively communes amidst the empty corners and burrows of the barn. In "A butterfly sunned itself sensuously on a sunlit yellow plate," for example, the aural lushness of coupling "s" and "l" sounds seeks attunement to the tonality of the butterfly's sensuous dwelling. A relation of inhuman touching—contrary to the terms of Heideggerian *berühren*—gives rise to worlds that language must, like the syllabling of Anon, gather its expressive resources from sensation to (obliquely) say. As Woolf writes of the sensate dimensions of language in her preliminary notes to "Anon": "the word heard. Its solidity: its depths" (377).

We might think such touching by means of what Julia Kristeva (following Plato) calls the semiotic *chora*, the pre- or arche-linguistic "vocal or kinetic rhythm" that mediates the body's relation to its outside. Like the nonsense out of which sense is composed, the *chora* "precedes and underlies figuration" while marking the very possibility of semantic signification at all. "Deprived of unity, identity, or deity" and "anterior to sign and syntax," the *chora* has only the maternal body as its ground and its "ordering principle."[81] Michel Chion takes the maternal body's role in subject formation literally, locating the acquisition of hearing and rhythmic orientation at the four-month stage of fetal development. "Sonic sensations" are perceived as "rhythmic pressure," the beating of the mother's heart heard and felt aqueously, through amniotic fluid, prior to the detached "aerial environment" in which hearing is mediated after birth.[82] Read through these terms, the arche-linguistic semiosis of the *chora* may also be a mourning for a lost modality of hearing as musicked touching, as the rhythmic, "co-vibrational" unity of two heartbeats no longer sensible in an air world.

The larval expressions arising in the barn—modes of world making grounded in the sensate relationalities subtending air thought and sky conceptuality—belong, too, to *chora*. To attune to them is also to strive toward new, inhumanist grammars of perception that persist, like semiotic rhythm itself, "indifferent to language" and its determinations.[83] We might then think the barn as prefigural or presymbolic growing ground, as the place of an inhuman *poeisis* that both preexists humans and crafts livable worlds from humanism's ruins. Like La Trobe's cauldron of "wandering voices and floating bodies," the barn is also a gathering of multispecied, choric noises and communions. In *A Foray into the Worlds of Animals and Humans*, as we explored in the introduction, Uexküll theorizes the perceptive relations between a being and its environment as a system of tones in which each encountered object takes on a certain tone—a "sitting tone," a "dwelling tone," an "obstacle tone," a "climbing tone," for example—in relation to the being's tonic note, or "self-tone [*Ich-Ton*]."[84] This inhumanly environed gathering of tones—an "amorphous mass" of sounds touching amidst the barn's emptiness—resounds beyond the hearing, the language, and the time of humans. (A tonal gathering to be perceived, perhaps, only by a gigantic ear somewhere.) Human perception of the barn is given as a collective silencing tone, the muting of dissonant expressivities and noise worlds, or the (momentary) resolution of *ainigma* through the name and the tonality of Man.

Mrs. Sands, the Olivers' cook, is the first human to enter the barn:

> But Mrs. Sands was approaching. . . . She could see the great open door. But butterflies she never saw; mice were only black pellets in kitchen drawers; moths she bundled in her hands and put out of the window. Bitches suggested only servant girls misbehaving. Had there been a cat she would have seen it—any cat, a starved cat with a patch of mange on its rump opened the flood gates of her childless heart. But there was no cat. The Barn was empty. (100–101)

The barn's joyful noise worlds are summarily effaced by Mrs. Sands's self-tone, which also stands in for the privation of a humanistic perception that assumes the nonexistence of what it cannot sublate into concept or raise into an air world. (As in *To the Lighthouse*, it seems to be the task of domestic laborers to mark the place of human self-expression amidst an indifferent nature.) Soon after Mrs. Sands prepares the barn for the pageant-goers to enter, Swithin continues to speak aloud her thoughts on the migration of swallows, this time directed to the "empty air" of the barn: "'They come every year,' said Mrs. Swithin, ignoring the fact that she spoke to the empty air. 'From Africa.' As they had come, she supposed, when the Barn was a swamp" (103). The barn becomes an air world upon the emergence of humans and human voices, their silencing tones. But air, like emptiness, is multivalent, modulable into both an anesthetized concept world and a medium for noisy, inhumanist creation. The flight of the swallows is filled air, a means of rearticulating the barn—as a structure, a space, a chorus, and a future site of ruins—through another temporality that indifferently joins, or geohistoricizes, pre- and post-human worlds into a singular, monstrous interval.

III.

Throughout the novel, inhuman tones have a habit of both filling and modulating the voids left in the wake of human voices. As discussed earlier, the novel's opening conversation among the Haineses and Olivers—itself focused on another void: the village cesspool—cedes to an emptiness modulated by bovine noise: "Then there was silence; and a cow coughed; and that led [Mrs. Haines] to say how odd it was, as a child, she had never feared cows, only horses" (3). As with the tiny expressions in the barn, emptiness both allows the cow cough to surface into perceptibility and becomes the medium for inhuman creation. There is a sonic, if not a musical structure at work here—something like: (1) voice, (2) silence, (3) inhuman noise (modulation), (4) new soundscape—that recurs both in the barn and in the scene of another cow-noised punctum that emerges during La Trobe's pageant. In contrast to the tiny perceptions of the barn and the modulating cow cough, here a herd of bellowing cows poses the opposite problem for perception: how to think or make sense of gigantic noises, or expressions too powerful to be apprehended within humanistic epistemologies or tonal systems? During the Valentine and Flavinda portion of the pageant, a comic playlet on mistaken identity modeled on the conventions of Restoration drama, death arrives in the form of an irremediable silencing tone. As with each of the pageant's scenes, it incorporates a chorus of villagers, who play the part of field laborers here: "Digging and delving (they sang), hedging and ditching, we pass. . . . Summer and winter, autumn and spring return . . . All passes but we, all changes . . . but we remain forever the same" (139). However, their earth song, which is meant to "continue the emotion" across an interval bridging scenes of Valentine and Flavinda's courtship, goes unheard, masked by other expressivities: "(the breeze blew gaps between their words)"; "then the wind rose, and in the rustle of the leaves even the great words became inaudible; and the audience sat staring at the villagers, whose mouths opened, but no sound came" (139, 140). For La Trobe, the absence of choral sound is the negation of life: "Illusion had failed. 'This is death,' she murmured, 'death'" (140).

Subtending the world of individualistic characters and their individualized emotions are two impersonal songs of the ground: the choral refrain of the laborers—articulating what Cuddy-Keane calls "the life of the common people in touch with the soil" and cycles of "repetition and recurrence in opposition to change"—and the territorial refrain, or the yearning tone, of the cows in the field.[85] The laborers' grounding refrain—like the laboring force of Mrs. McNab and Mrs. Bast in *To the Lighthouse*—is meant to secure the future of humanistic expressivities across oblivion. Yet the breeze, too, silences this refrain, along with the other human voices in the audience. (The villagers here become the polar opposite of the all-powerful, unseen but omnipresent and omni-audible *acousmêtre*: "the audience sat staring at the villagers, whose mouths opened, but no sound came.")

Amidst emptiness, however, and through the interstitial gaps made by the breeze, emerges the cows' yearning song:

> Then suddenly, as the illusion petered out, the cows took up the burden. One had lost her calf. In the very nick of time she lifted her great moon-eyed head and bellowed. All the great moon-eyed heads laid themselves back. From cow after cow came the same yearning bellow. The whole world was filled with dumb yearning. It was the primeval voice sounding loud in the ear of the present moment. Then the whole herd caught the infection. Lashing their tails, blobbed like pokers, they tossed their heads high, plunged and bellowed, as if Eros had planted his dart in their flanks and goaded them to fury. The cows annihilated the gap; bridged the distance; filled the emptiness and continued the emotion. (140–41)

In the absence of the "illusion" by which humanity presents itself to itself, the cows articulate another relation to the earth and its futures. In a medium made by the silencing of human voices, the cows modulate extinction into comedy: a contagious, communal song—the *ōidē*, or song, of "comedy" [*kōmōdía*]—of the (re)generation of the earth. The comedy of cow song speaks as well to Woolf's theory of laughter as both "beneath" language—or in our terms, earth-oriented—and given as the message-less, information-less, "inarticulate utterance like the bark of a dog or the bleat of sheep," and, we add, the bellowing of cows.[86]

For cow song "annihilates the gap" between life and death, human and nonhuman being, a gap that is and has only ever been a humanistic construction. Along these lines, Derek Ryan argues that the pageant's cows are best read as a Deleuzian, human-nonhuman assemblage, whose bellowing marks a new, provisional, sonic territory that is beyond the grasp of human subjectivity: the "deterritorialisation of humanity" itself.[87] With a differently articulated territory, I would add, comes a different relation both to the ground and to what is generated from it. Louis in *The Waves* devotes a life to finding a means of domesticating the sound of a "great beast" "stamping" on the shore.[88] The pageant cows, on the other hand, form a comedic, "communal chorus," that noises into being another ground that is beyond the time and hearing of humans. Their world-filling, "dumb yearning" knits together primeval and present temporalities while also carrying the content sedimented into their song—the refrain of the laborers, the sounds of the breeze and the rustle of leaves, the emotions of Valentine and Flavinda, the aesthetic "burden" of illusion—into an indeterminately specied future, a noisy oblivion.

Cow song is also given as "voice," a term generally reserved for human utterances, as well as the element that, according to a strain of sound theory most directly represented in the work of Adriana Cavarero, secures the identity of the individual amidst the nonsensical, inhumanist sounds, soundscapes, and noises of animals, machines, and animate and inanimate nature.[89] Woolf's inhuman counter-ontology of voice (as collective, "primeval voice") resonates with media theorist Dominic Pettman's formulation of the *vox mundi*, or "voice of the world," that imagines the voice as a nontotalizable integration of multiply sourced (human, nonhuman, machinic, digital, cosmic) sounds, "expressed in all manner of creatures, agents, entities, objects, and phenomena."[90] As *vox mundi* (or perhaps *vox terra*), Woolf's "primeval voice sounding loud" is not regression to primitivity, but rather a modulation of the sonic-affective content gathered into it toward new, comedic forms by which to fill oblivion, to continue through the impersonalized night or disastrous twilight of human extinction. I like to think of the comedic ongoingness of this cow song—sounding through the failure of illusion and the approach of death—as a punchline accompanying *To the Lighthouse*'s description of the drowning of every expressivity in the long night's aqueous darkness: "Sometimes a hand was raised as if to clutch something or ward off something, or somebody groaned, or somebody laughed aloud as if sharing a joke with nothingness" (126). By saturating the impersonality of oblivion with the overflow of powerful, world-filling yearning, cow song bridges the gap between death and life, extinction and generation, and tragedy and comedy. Cow song turns the terror of night into the comedy of what persists in and beyond oblivion, where nothingness—as a medium for both unimaginable generation and the sharing of jokes—is to be laughed at and laughed with.

Yet humans remain oblivious to the joke. The primeval voice, the comedic cow song, ultimately becomes muted through another modulation, this time by the pageant's audience, to visual and conceptual apprehension: "Suddenly the cows stopped; lowered their heads, and began browsing. Simultaneously the audience lowered their heads and read the programmes" (141). (In the program, we are told, the producer has described a scene that was omitted from the Valentine and Flavinda portion due to time constraints.) This shift from gigantic, communal, multispecied voice—and the gigantic hearing ear of the landscape—to mute, individualized textual consumption also marks the reemergence of the humanistic silencing tone, effacing yet another inhuman noise-world. La Trobe's aesthetics of impersonality—"she was not merely a twitcher of individual strings, she was one who seethes wandering bodies and floating voices in a cauldron, and makes rise up from its amorphous mass a re-created world"—are crafted with the fragility of these worlds in mind, which can only evade determination for so long, before the clamor of *ainigma* is harmonized into the tonality of Man.

IV.

Whether impersonalized clamor can insist beyond tonal determination is the question posed by the closing scene of the pageant, given in the program as "The present time. Ourselves" (178). La Trobe intends to "expose" the audience, to "douche them, with present-time reality" (179), by having village children (or "imps—elves—demons") encircle them with mirrors:

> Out they leapt, jerked, skipped. Flashing, dazzling, dancing, jumping. Now old Bart . . . he was caught. Now Manresa. Here a nose . . . There a skirt . . . Then trousers only . . . Now perhaps a face. . . . Ourselves? But that's cruel. To snap us as we are, before we've had time to assume . . . And only, too, in parts. . . . That's what's so distorting and upsetting and utterly unfair. (184)

As a collection of fragments themselves—many of the "mirrors," or "looking glasses," held by the children are improvised from various, found reflective objects, such as "tin cans," "bedroom candlesticks," "old jars," and "scraps of scullery glass"—the mirrors in turn fragment individuals from formed subjects into reorderable parts. The audience comes to be dissolved in visual noise, into the cauldrons of what we might call both pre-Oannean chaos ("when fingers, fins, and wings were blended" [126]) and the prefigural, somatic relationality of Kristevan *chora*. This aquatic noise world also becomes a medium for the silenced cows, along with other nearby animals, to resume their sonic collaborations across and with oblivion:

> And Lord! the jangle and the din! The very cows joined in. Walloping, tail lashing, the reticence of nature was undone, and the barriers which should divide Man the Master from the Brute were dissolved. Then the dogs joined in. Excited by the uproar, scurrying and worrying, here they came! Look at them! And the hound, the Afghan hound . . . look at him! (184)

Unpossessable noise—visual, aural, and tactile—becomes a fluid medium for the co-dissolution of master and brute. The pageant's final scene, in other words, is the immersion of an air world into the wet, vibrational clamor of the cauldron.

The all-immersive jangle and din—gathering of noisy particulars, *psófos* excluded from logos—also absorbs into itself Swithin's earlier faith in a univocal, interspecies harmony to come. What now remains of harmony's promise is only a conceptless epiphora of (audiovisual) noise: noise joined to other noise through the medium of noise, together improvising a dissonant future. In the ongoingness of the uproar, "which by this time has passed quite beyond control," the pageant characters appear on stage simultaneously to "declaim . . . some phrase or fragment from their parts":

> *I am not* (said one) *in my perfect mind. . . .* Another, *Reason am I . . . And I? I'm the old top hat. . . . Home is the hunter, home from the hill. . . . Home? Where the miner sweats, and the maiden faith is rudely strumpeted. . . . The owl hoots and ivy mocks tap-tap-tapping on the pane. . . . Lady I love till I die, leave thy chamber and come. . . . Where the worm weaves its winding sheet . . . I'd be a butterfly. I'd be a butterfly. . . . In thy will is our peace. . . . Here, Papa, take your book and read aloud. . . . Hark, hark, the dogs do bark and the beggars.* (185)

The different ages that mark humanity's forward progress through time are compressed into a singular yet multiply noised instant of mingled voices, sounds, bodies, enfleshments, and times. The lyric "I," like falling Alice, speaks only the noise of its own ungrounding and loss of a (self-)possessive consciousness.

It falls to the impersonal "voice" of a megaphone in the bushes to articulate the joined noise of fragments into another grammar of relation, and ultimately, into a new language, to be made of "words of one syllable." The "infernal" voice, or "anonymous bray," enjoins the audience to "*talk in words of one syllable, without larding, stuffing or cant. Let's break the rhythm and forget the rhyme. And calmly consider ourselves*" (187; italics in the original). And selves can only be thought as gathered fragments, as affirmed by the gramophone, which now produces an unknown, yet "traditional tune" that takes the "chaos and cacophony"—of the noisy mechanical processes of the gramophone ("filings magnetized"), of its own sounded notes, and of the mirrored parts of the pageant audience—and orders these into a precarious "measure" that closes the pageant. Yet this is a unity still too unfragmentary, too unchaotic to ground La Trobe's newly syllabled earth language and Woolf's dissonant metaphorics. Once the stage has cleared, the "gift" of the pageant dissipates into the ether, now become a "cloud that melted into the other clouds on the horizon." The water worlds seethed in the cauldron evaporate, or arise into sky-concept, and La Trobe sinks back into dejection: "'A failure,' she groaned, and stooped to put away the records" (209).

The fragments thrown and sounded together in the play's conclusion, as La Trobe seems to intuit, can only reflect and reroute the humanisms out of which they emerged. The concluding voice, or "measure," gathers fragments into a self-enclosed *symbolon* freed from the earth's dissonances, including the tonal minutiae of the barn's inhuman noise worlds, gigantic cow song, swallow flight, and soon, the rhapsody made of starlings and tree. But the voice of humanistic, metaphysically guaranteed *symbolon* emanates most clearly from Reverend Streatfield, whose closing remarks on the pageant opened this chapter. "Scraps, orts and fragments! Surely, we should unite?" he concludes, in an attempt to hold together what is in excess of meaning, or what the malfunctioning gramophone ongoingly insists is "dispersed" (192). Even before the emergence of his voice, however, his status as "symbol" and "representative spokesman"—for the audience, for the perpetuity of humanisms—has already become drowned amidst an indifferent comedy of an inhuman "summer silent world," flowing beyond logos's enclosures: "There he stood their representative spokesman; their symbol; themselves; a butt, a clod, laughed at by looking-glasses; ignored by the cows, condemned by the clouds which continued their majestic rearrangement of the celestial landscape; an irrelevant forked stake in the flow and majesty of the summer silent world" (190–91). The impersonality of (earthly and celestial) landscape—an indifference to or nonrepresentation of persons—opens to the regenerative laughter of and beyond oblivion. Looking-glasses laugh and a gigantic earth ear hears: La Trobe's new language, or re-created world, must seethe mingled fragments, methexic noise, into a form that laughs with and alongside oblivion, shares a joke with the nothingnesses of a summer silent world.

V.

"'A failure,' she groaned, and stooped to put away the records" (209). Yet La Trobe's stooping is what allows her to hear the earth differently, perhaps alongside the gigantic ear of the landscape instead of through the reconfigured sounds returned by the gramophone, that "little instrument" which, like Reverend Streatfield, can voice only the humanisms that have been projected into it. In stooping, however, La Trobe hears the expression of another syllabling and another metaphoricity:

> Then suddenly the starlings attacked the tree behind which she was hidden. In one flock they pelted it like so many winged stones. The whole tree hummed with the whizz they made, as if each bird plucked a wire. A whizz, a buzz rose from the bird-buzzing, bird-vibrant, bird-blackened tree. The tree became a rhapsody, a quivering cacophony, a whizz and vibrant rapture, branches, leaves, birds syllabling discordantly life, life, life, without measure, without stop devouring the tree. Then up! Then off! (209)

The starlings throw themselves into the tree, making an alternative, inhumanist *symbolon* of everything that is discordantly and dissensually gathered into this literal Tree of Life. Thus attacked (also a musical term, as the initiation of a sound) the tree is made co-vibrational, shaken out of the stasis that it occupied during the pageant, when trees—like the *grosses pierres* on the Thebaidian summit—were solidities that "barred the music" and "prevented what was fluid from overflowing" (182). The joining of bird and tree is rhapsody—from the Greek *rhapsōidía*, the sewing together [*rhaptein*] of songs [*ōidē*]—a measureless generation of sewn-together noises. We can hear these both as La Trobe's sought-after "words of one syllable" and what Woolf calls "the word heard. Its solidities: its depths." Such inhuman rhapsody, songs / sounds of sewing and the monstrous *symbolon* into which they are sown, resonate beyond the determinations of logos and—like unspooling, laughing Odradek—beyond humanistic time and measure. The whizz, buzz, hum of the starling-(stone-) tree assemblage are inharmonizable syllables, units of joined noise sounding aggregately into the earth's ear rather than "melt[ing] into the other clouds" (209). Like the choric cow song "sounding loud in the ear of the present moment," starlings and tree co-compose an ongoing sound event that fills both the oblivion of the pageant's "failure" and the post-extinction oblivions to come.

Here, stooped to this ground planted with these inhuman sonic-affective articulations, La Trobe seeks a means of creating with the earth noise reverberating beyond extinction. The starlings fly off, scared away by the passing of a villager, and La Trobe approaches the still starling-vibrant tree: "She crossed the terrace and stopped by the tree where the starlings had gathered. It was here that she had suffered triumph, humiliation, ecstasy, despair—for nothing. Her heels had ground a hole in the grass" (210). La Trobe's new art, perhaps, *is* this hole: the improvised making of an ear—a modest rhyme with the gigantic ear of the landscape—through which to hear and to hear with earth. The hole may also be a plot for the planting of sounds into the earth (or into the undergrounds of *metaphora*), from which new inhumanly grown relations and futures might flower.

The hole composed, La Trobe continues her wanderings through the darkening landscape where, in contrast to Swithin's earlier musings, "there was no longer a view—no Folly, no spire of Bolney Minster. It was land merely, no land in particular. She put down her case and stood looking at the land. Then something rose to the surface" (210). This something—growing from earth or rising from cauldron—is, most immediately, an idea for a future play: "'I should group them here,' she murmured, 'here.' It would be midnight; there would be two figures, half concealed by a rock. The curtain would rise. What would the first words be? The words escaped her" (210). Midnight is no longer oblivion, as in the succession of night into night of *To the Lighthouse*, or the disastrous twilight of the sun's concealment for Milton and for White. Rather, impersonalized midnight is both a time and a place of inhumanly noised and futured creation, a passage toward the interspecied comedy of the earth.

As she enters a pub—to her a site of "shelter; voices; oblivion"—she hears through the chatter: "Words of one syllable sank down into the mud. She drowsed; she nodded. The mud became fertile. Words rose above the intolerably laden dumb oxen plodding through the mud. Words without meaning—wonderful words" (212). From the soil grow nonmeanings, fertile sounds by which to fill oblivion. La Trobe's vision is attempted attunement to the inhuman tones generated from earth: "There was the high ground at midnight; there the rock; and two scarcely perceptible figures. Suddenly the tree was pelted with starlings. She set down her glass. She heard the first words" (212). As with the barn's noise worlds, night ground is a rhapsodic gathering of inharmonizable tones by which beings—starlings, tree, stone, ground—environ worlds in the absence of humanistic (self- and silencing) tonalities. Yet La Trobe can only hear such worlds obliquely, mediated through the indeterminate, "scarcely perceptible" tonality of midnight. We leave La Trobe—rhapsode, troubadour, outdoor voice—wandering amidst the incommunicable syllablings of midnight's "first words," or "the word heard," whose "solidities" and "depths" she is still sounding.

In this way, La Trobe's closing strivings resonate with those of Kafka's burrower and investigator: she is seeking a means to express the *Außer-sich-sein* (beyond oneself) of the dog-philosopher's graspings toward (sky-)conceptuality and the burrower's faithful comportments toward inconceptual earth noise. Such a beyond, in both Woolf and Kafka, eludes representation. La Trobe thus becomes a listening burrower, devotee of ground, whose closing actions include digging a hole with the feet ("heels") on which she stands on the earth, a hole which may also be a means to more precisely hear and relate to the sensate givens of earth.[91] La Trobe surrenders the drive toward a conceptuality that would "make them see" in order to embrace the sonic *ainigmata* grown from the ground at midnight. (Her notebook scrawl at this moment, "I am the slave of my audience," is also a ceding of [self-]possession to a future audience that is both the impersonally perceived landscape and the nonhuman and human beings gathered upon it [211].) For the "first words" La Trobe hears are the discordant syllables made from thrown-together starlings and tree—a joining of sky to earth, night to ground—that modulate extinction into the clamor of an inhuman comedy.

VI.

> But none speaks with a single voice. None with a voice free from the old vibrations.
>
> —Virginia Woolf, *Between the Acts*

This comedy might have begun with Anon, who perhaps hears the same first (bird-)words at Woolf's imagined origin of human history. Woolf's "Anon" (and history itself) begins at the end of *Between the Acts*, amidst the prehistoric sound worlds made of birds and trees:

> "For many centuries after Britain became an island" the historian says, "the untamed forest was king. Its moist and mossy floor was hidden from Heaven's eye by a close drawn curtain woven of innumerable tree tops." On those matted boughs innumerable birds sang, but their song was only heard by a few skin clad hunters in the clearings. (382)

Anon's voice—given both as "the human voice" and the androgynous, unhomed, "common voice singing out of doors"—emerges only after the trees are "felled," after agrilogistic, human history is instantiated by the huntsman's axe. (Similarly, Yeats's Wandering Aengus enters the forest/hazel wood—and carves a "hazel wand"—to transform linear history into the cyclical time of cosmic harvest.[92]) Anon's song is both mourning for the choric expressivities lost when forest world is revealed to Heaven's eye as well as a modulated continuity of bird-tree singing into what is imagined as the beginnings of poetic form. From Anon's first song:

> By a bank as I lay
> Musing myself alone, hey ho!
> A birdes voice
> Did me rejoice
> Singing before the day;
> And me thought in her lay
> She said, winter was past, hey ho! (382)[93]

Anon's song is attempted rhyme between self and bird, sung—and only singable—after a self has already been carved from felled trees. The first line composes an I in separation from a (disentangled) bank, an I which grows even more isolate with the next line's "myself alone." The subsequent progression from "me rejoice" to "me thought" is also a movement from lost musicked touching, or co-vibration, between bird and human voice, to the imposition of a sky-conceptuality upon the bird-given syllables. These joyful syllables are given semantic content and placed into temporality, even as we might still hear the speaker's "hey ho!" not as the grafting of human speech onto bird voice, but rather as a sought resonance between bodies that remain in communion even after the forests have been cleared and opened to the sky. (Anon's joy communes, too, with Gilbert White's yielding of thought and memory to the unpossessable verdure growing in and as his mind.)

And yet, like the enfleshment of Adam "from the whole earth, the place within view of the Lord's eyes," Anon can only come into existence after Heaven's eye sees the unshaped forest floor, an event leading to the differentiation of human voice from the inhumanist *poeisis* of bird-tree noise. Anon, then, is born from the touching of sky and earth/forest, his birth leaving an unhealed "green scar" on the surface of the earth, the place seen, envoiced, and set into temporality by Heaven's eye. In Woolf's words, Anon journeys along "the down and the green scar not yet healed," a green scar that persists into the earth's present, as given more explicitly in an earlier draft of "Anon": "On the down is the green scar still, to mark the road along which the travellers came, past the circle where they were buried; past the stone that marks the meeting place or the burial place" (382, 403). The green scar—as ancient earth-writing, a geography born of the clearing of forests, and an instantiation of human history that might also be an inscription of the Anthropocene itself—also contains the histories of both Anon and the earth, as we can still hear in the impersonal, "common" singing of Anon.[94] As Peter Adkins argues, in *The Modernist Anthropocene*, the "impersonality [of Anon] expresses a broader 'generality' and extends beyond the human world, since Anon stands as both the human speaker and the world that the speaker is expressing."[95] Like cow song, Anon's singing allows "every body" to "share . . . in the emotion," melding communal affect with the expressive voice of the earth itself, a voice that is "not self conscious," that "can borrow," "can repeat," and "repeats over and over again that flowers fade; that death is the end," and "never tire[s] of celebrating red roses and white breasts" (382, 397–98). Anon's green song is the endless, repeated modulation of death into the comedic celebration of fertility and growth, or what Woolf also calls a "nameless vitality" (398). As the stone near the scar marks both burial place and gathering place, so Anon's song is both proleptic mourning for the extinctions to come and celebration of the earth's repetition of its own nameless livingness. As earth song, *vox mundi*, or emissary of naive flower religion, Anon's voice both gives rise to the English literary tradition and becomes silenced by its historical and formal progression toward the mute "individuality" of the playwright (and later, in the post-Elizabethan age, by the "man who writes a book"). No longer capable of gathering audience into the terran commons formed by Anon's first words, playwright and printed word now return to the audience "their own general life individualised in single and separate figures," amidst a fully cleared forest and still scarred landscape in which "Anon is dead" (398).

La Trobe hearing the tree-starling syllables is perhaps also an attempt to hear the earth that Anon heard, its first gifted words. Woolf notes the persistence of Anon and Anon's earth into the present: "By shutting out a chimney or factory we can still see what Anon saw" (382). And as Woolf emphasizes in an earlier draft of "Anon," "Nor is Anon dead in ourselves" (424). La Trobe digs a hole in the earth to hear the earth. Her hearing is directed both backward to what Anon heard and forward to the post-extinction futures beyond humanistic imaginability, toward the oblivions filled by the inhuman *poeises* of new Anons, singing into being new relations to earth. In addition to hearing Anon's first words, La Trobe's backward listening may also be a means of returning to the early words of a twenty-two-year-old Virginia Stephen written during her own communions with earth: "If you lie on the earth somewhere you hear a sound like a vast breath, as though it were the very inspiration of earth itself, & all the living things on her."[96] In this sense, La Trobe herself—as character, composition, and as *doppelkünstler* or *doppelhörer*—may be a medium, or listening device, for Woolf's own earth-(h)earing. From this already impersonalized somewhere—the gigantic breathing mouth of the earth perhaps?—arises a breath-like sound creatively suffused into all living things, grounding all sounds and sounding beings in the tonic of the earth's breath. Somewhere is another voice preexisting and continuing alongside the voice of Anon that breathed both forests and Anon into being, and that sounds still after the clearing of the forests and the death of Anon. (We might thus hear a resonance between the earths of Stephen's diary entry and of Rainer Maria Rilke's apostrophic address in the Ninth Duino Elegy (1923): "Earth, isn't this what you want: to arise within us / invisible?"[97]) Even when La Trobe hears the first words made from gathered starlings and tree and first sung by Anon, she is perhaps only hearing the breath-like sound of the earth from which all sounds are modulated. To modulate is to transfer from one mode (or in music, from one key signature) to another, but to modulate is also to arise from preart-iculate, choric ground into mode itself. In its Spinozist inflection, modulation is the distribution of divine Substance (equivalent with both nature and God or "God's essence") into modes that creatively express Substance through the content of their strivings on and with earth (*E II* D1). Stephen is, in this sense, attempting to hear—through co-vibration, through the touching of bodies—what is prior to modulation itself: the vast breath-like sound of the earth—thinkable only through the obliqueness of simile—before its modal distribution into sounding beings, before the eye of Heaven scarred green the earth.

The earth's sound inspires beyond temporality, its essence never fully hearable in what modally expresses it. The multispecied noise worlds (of barn creatures, cows, sweeping swallows, starlings and bird-blackened tree) sound in and as the comedic continuity of the earth's unhearability, giving more noise to the vast inharmony that first inspired them. In a diary entry from March 31, 1940, Woolf hears earth and its inspirations again, but refuses to modulate what is earth-given into the terms of representationalist thought. Woolf writes,

> I would like to tell myself a nice little wild improbable story to spread my wings after this cramped ant-like morning. . . . The story?—oh about the life of a bird, its cheep cheep—its brandishing of a twig by my window—its sensations. Or about Botten becoming one with the mud—the glory fading—the million tinted flowers sent by the doleful mourners. All black like a moving pillar box the woman was—& the man in a black cardboard casing. A story dont come—no but I may unfurl a metaphor—No. The windows very dove grey and & dim blue islanded—a rust red on L. & V. & the marsh green & dark like the floor of the sea. At the back of my head the string is still wound tight. I will unwind it playing bowls. To carry the virtues of the sketch—its random reaches its happy finds—into the finished work is probably beyond me.[98]

Woolf stills the originary modulation of Anon's "me rejoice" into "me thought," resisting the translation of the "cheep cheep" of birdsong and of Botten's metabolism with earth into temporality, into the strong narrative forms of first "story" and then "metaphor." A bird life and a local farmer's death instead are unfinished gatherings of sensate particulars, too fragile to totalize into *symbolon* yet still co-vibrational in their proximity to one another in the space of the diary entry. In the decreative movement from story to metaphor to sketch, Woolf imagines bird and earth life through a series of abandoned enfleshments, eventually abandoning thought itself to the piling up of sound and color: *epiphorae* that insist beyond both the "finished work" and the powers of the thinker.

In the wake of a story that does not come and a metaphor that will not unfurl, all that remains is the intensity and vividness of gathered colors—all black, dove grey, dim blue, rust red, green & dark—in inconceptual communion with one another. Their gathering is also a means of holding together sky, forest (marsh and trees), and sea within the weak (non)form of the sketch.[99] Another metaphorics becomes possible in the refusal of metaphor as enclosure, as the giving of names and the provision of place to nameless things. Woolf's is a metaphorics made of the syllablings of color, a means of imagining the earth communicating to itself, in which the sensations of a bird and the muddy metabolism of a corpse (sky and earth) are joined in an ongoing yet inarticulable inhuman comedy, and the still green scar of earth and marsh rhymes with a greened seafloor. The unmodulated vividness of color—color as noise that "unsettles the literary text, threatens to collapse it [*s'effondrer*] into an impossibility, an incommensurability," in Louvel's words—gives another articulation of metaphor as a gathering of groundless noise.[100] Beyond concept and beyond what can be carried into the finished work is this noise, these vividnesses, as a means through which to imagine the earth's continual self-expression.

Does the earth hear itself, its own expressivities, the dissonance of its own colors? In Karen Barad's formulation, earth acts (or "acts of nature") are unpredictable, "nonlocal" communications distributed across the vast continua of space and time.[101] Earth agency is thinkable, then, as the intra-active dynamism of "part of the world making itself intelligible to another part."[102] The earth communes with itself, as young Virginia Stephen's attempted co-vibration with the earth's unmodulable breath-sounds attests to. However, in the later diary entry, Woolf gives these closing earth acts as the comedic unimaginability of the earth's future generation, and as joyful surrender to the anarranging vibrancy of color.

We might, finally, think the earth's self-communion through the terms of Derridean autoaffection. As discussed in chapter 3, autoaffection refers to the mythical condition of "hearing oneself speak [*s'entendre parler*]" by which the speaker's speech is the pure expression—or "the mastered ideality"—of one's sensical interiority, "without any detour through the agency of exteriority, of the world, of the non-proper in general."[103] This is impossible for humans because there is no phonic expression (and no act of seeing or act of touching, even) that is unrouted through the temporality and material-affectivity of the external world—the nonself or nonproper—through which all expressivity circulates. However, we might intuit with Woolf a gesture toward thinking the auto-(material-)affectivity of the earth itself, in which the earth ongoingly hears the sound and color that it gives to itself without the mediation of humans and humanistic conceptualities, passing unmodulated from breathing earth mouth to gigantic earth ear. Insisting beyond human extinction, perhaps, is an impersonalized metaphorics or *symbolon* that the earth makes by relating (or throwing) itself (in)to itself. A (self-)relational earth fills the dark and the disastrous twilight of oblivion with an epiphoric gathering of noises and colors planted or breathed into its gigantic ear.

These personifications of earth—an earth troped variously as sounding, hearing, coloring, and breathing by Woolf/Stephen, Swithin, La Trobe, and

Anon—are not earth. And yet in their obliqueness to earth, in the post-extinction imaginaries toward which they strive, such articulations of earth may give us a means of comporting ourselves toward what can be neither heard nor thought but only surrendered to: inhuman comedy, the unmodulable tonality of the earth, forest song before the ground was opened to sky and scarred green, the syllabling *ainigmata* of oblivion.

Chapter 6

Forest, Ocean, Planet (Green & Blue)

Metamodernism's Post-extinction Archives

what the coral said.
once. we were all singing. somewhere. we are still. moving. as something huge, vibrational, wet. we dance and keep the world in place. we shiver and know the orbit. if you let the body undulate you will remember.

—Alexis Pauline Gumbs, *Dub: Finding Ceremony*

precambrian multiplicity in the densing
postapocalyptic forest wolf me down.

—Cody-Rose Clevidence, *~~Flung~~ Throne*

~~Flung~~ Throne's postapocalyptic forest grows from the "Universe of death" found in the underworlds of the earth in Book II of *Paradise Lost*:

> Where all life dies, death lives, and Nature breeds,
> Perverse, all monstrous, all prodigious things,
> Abominable, unutterable, and worse.[1]

Death is an unsayable relationality, growing ground for a hyperproductive and perverse earth, along with the "Gorgons, and Hydras, and Chimeras dire" to emerge in Milton's subsequent lines. Alongside its moral connotation as "to turn aside" or "to subvert," "perverse" combines the Latin *per* ("thoroughly") with *vertere* ("to turn," as well as the root of "verse" in *vert*) to give "thoroughly turned": a way of describing the action of poetry and the creative ongoingness of the earth itself. Perverse turning is an intensification of turning—a "thorough turning" which nonetheless implies the possibility of further turning—and names something beyond traditional ideas of what it means (for the earth, or the poem, or anything else) to turn. We might even imagine into being an alternative etymology that associates *vertere* with *vert*, to give us a "green turning." Imaginatively greened or not, turning takes on an even more monstrous dimension when coupled with the indeterminate preposition from which it can seem inseparable: "into." Writes Clevidence in an untitled poem, "& turning, turn into, mutant-child of the first fern, first fish, lisping / thru mud u ugly ugly pulse, eroded, honed, plated & horned."[2] Following Milton, the abstract "turning" simultaneously stands alone, pure of determinants, and becomes impurified (or perverted) by aggregating with its neighboring phrases. In the earth's turning arises a "turn into" that lays ground for the indeterminate generation of life forms: vegetal, aquatic, reptilic for Clevidence; Gorgons, Hydras, and Chimeras for Milton. Less than one and multiplicate, the human appears almost as an afterthought in this poem, an indeterminate lyric I that is immediately a "green w living" manifold: "I am dendritic & lonely, a swarm."[3]

Fractured between loneliness and dendritic sociality (dendritic: grown from forest, a fractal interpretation of a tree), the human also comes to be split between lyric I/swarm and the apostrophic "u" that it turns into, both in this poem and in the poem "<>":

& then u creatures
of the ~~trees~~
valley u

peel the throat
from the throat
learn skin from kin[4]

As with the origins of Anon, a self must be crafted from felled ~~trees~~, so that bodies may be skin-enclosed and protected from the methexic contagion of the forest. And further, in "THY VIRAL ORCHARDING || ESCAPE MY": "U INCARNATION OF MOLECULAR STRUCTURE U WARM BLOODED MYTHOLOGY OF LINEAGE. . . . U EVOLUTIONARY MONSTER ERUPTING FROM THE IMAGINARY FOREST."[5] To "learn skin from kin" is to supply the requisite imaginary and linguistic structures ("MYTHOLOGY OF LINEAGE") that enflesh swarm or evolutionary flux into bounded form; it is to make dendritic being into loneliness. Form itself becomes a kind of original sin in *~~Flung~~ Throne's* anti-mythology, in which the present order is structured by the "curse of the formless made form."[6] Form is a metaphorical death—fulfillment of an impoverished teleology—against which beings agitate, as in these lines: "a pattern climaxes at a form / a living form is restless within itself."[7] In another cryptic meditation:

a form is an answer
to paradise looming

a form is a loom
there is no paradise[8]

Form directs toward a (shrouded, unshaped) paradise to come. Yet form also is a persistent, Odradekian weaving directed toward nothing but its own expressive unfurling in the world, the continued generation of relational "patterns" before their calcification into skin-bound form: no paradise possible in the absence of kin.

We can also think of perversion, as both thorough turning and wrong turning, in terms of its deviations from the humanistic conceptions of biological and evolutionary form. As Sara Ahmed writes in *Queer Phenomenology: Objects, Orientations, Others*,

> Perversion is also a spatial term, which can refer to the willful determination to counter or go against orthodoxy, but also to what is wayward and thus "*turned away* from what is right, good, and proper." For some queer theorists, this is what makes "the perverse" a useful starting point for thinking about the "disorientations" of queer, and how it can contest not only heteronormative assumptions, but also social conventions and orthodoxies in general. . . . It is worth, then, rereading the "perverted" as that which "turns astray" or moves off the straight line.[9]

Queerness, following Ahmed, is an anarrangement of linearity itself, and one that speaks as well to the disorientation and perversion of stable, linear descent in Darwin's Tree of Life diagram, a theme that Clevidence draws on profusely in both *~~Flung~~ Throne* and in their earlier *Beast Feast* (2014). What has recently been called "the natural attitude toward gender" encapsulates some of the most damaging misreadings of both gender and nature ever propagated. According to contemporary philosopher Robin Dembroff, the "natural attitude toward gender" holds that "there are two universal genders, every person immutably has one, and all this is vindicated by science. Other ways of seeing the world are, to someone with the natural attitude, 'ideological,' 'primitive,' or 'unscientific.'"[10] Dembroff's critique of the transphobic universalization of binary gender through an appeal to biology or nature is written largely in response to Alex Byrne's essay, "Are Women Adult Human Females?" in which Byrne claims that "woman" is definitively a "biological (and not social) category."[11] The imposition of dimorphic sexuality on humans, as the biological substrate on which the "social" category of gender operates, works through the assumption that nature/biology/life itself is structured according to strict divisions between male and female members of a species. As evidence, Byrne notes that because most languages have terms to differentiate "male and female animals"—such as the English words "doe, sow, hen, goose, mare, peahen, [and] queen" (i.e., mostly "hunting and farming" animals)—this must mean that binary constructions of gender are a universal biological norm from which humans are not excepted. There is an implicit, even agrilogistic, animal metaphorics at work here: a kind of feedback loop wherein gender categories that have been imposed on domesticated farm animals are then used as analogous proof of the biological universality of "woman," or "adult human female." Humanism metaphysically guarantees both itself and the universality of its concepts through the transit of the nonhuman.

While the deeply misogynistic terms of an analogy between domesticated farm animals and women (as critiqued in Joyce's "Circe") deployed as a form of philosophical proof should be grounds enough to dismiss this line of argument, Byrne's essay also functions as a spectacular, though perhaps instructive misreading of the "biological" to which it consistently appeals. Dimorphic sexuality, according to science studies scholars such as Myra J. Hird, Katrina Karzakis, and Catriona Mortimer-Sandilands forms only a minuscule portion of biological existence.[12] Even within the class *mammalia* that Byrne seemingly uses to encapsulate the entirety of the "biological," several studies and observations have demonstrated intersexuality and hermaphroditism in pigs and other domestic mammals.[13] And if we dare to look outside of *mammalia* for our definition of the biological, we find a vast, perverse field of behaviors, practices, and relationalities unfolding in complete indifference to the dimorphisms and linearities imposed upon them. In "Naturally Queer," for example, Hird writes,

> Human bodies, like those of other living organisms, are only "sexed" from a particularly narrow perspective. The vast majority of cells in human bodies are intersex (and this category itself is only possible by maintaining a division between "female" and "male" chromosomes), with only egg and sperm cells counting as sexually dimorphic. Most of the reproduction that we undertake in our lifetimes has nothing to do with "sex." The cells in our bodies engage in constant, energetic reproduction in the form of recombination (cutting and patching of DNA strands), merging (fertilization of cells), meiosis (cell division by halving chromosome number, for instance in making sperm and eggs), and mitosis (cell division with maintenance of cell number).[14]

Hird continues,

> Nor does reproduction take place between discrete "selves," as many cultural analyses would have it. Indeed, only by taking our skin as a

> definitive impenetrable boundary are we able to see our bodies as discrete selves. Our human bodies are more accurately "built from a mass of interacting selves . . . the self is not only corporeal but corporate" [Carl Sagan]. Our cells also provide asylum for a variety of bacteria, viruses and countless genetic fragments. And none of this reproduction requires any bodily contact with another human being. Moreover, there is no linear relationship between sexual dimorphism and sexual reproduction. Male sea horses, pipe fish and hares get pregnant. Many species are male and female simultaneously, or sequentially. Many types of fish change sex back and forth depending on environmental conditions.[15]

In this sense, biological processes are inhuman in their nonlinearity and in their irreducibility to the normative conceptions of the natural and nature projected into them. We might then think biology itself as an enigmatic field of (queer) relations and generativities unsolvable with either the name of Man or the name of adult human female.

I.

I want to return to Clevidence's twinning of nonbinary and multimorphic sexualities with the inconceptual monstrosity of biological and ecological being, but only after tarrying further with some of the larger implications of an inhuman modernist reverberation in the contemporary moment. How might, in other words, the relations between Anon's forest and ~~*Flung*~~ *Throne*'s forest metonymize an ongoing dialogue between inhuman modernist and contemporary metamodernist ecologies? In this vein, Clevidence's queer dendritic poetics work as part of a broader, metamodernistic movement toward reanimating the creative and experimental energies of modernisms past to imagine another future. As discussed in the introduction, David James and Urmila Seshagiri define literary metamodernism as the contemporary striving—primarily among novelists like Ian McEwan, J. M. Coetzee, Zadie Smith, and Tom McCarthy—toward "alternative futures for fiction through engagements with their modernist past."[16] "In turn," they ask, "what fresh light does today's writing cast on modernism's own generative moment, its always vexed beginnings, middles, and ends?"[17] For inaugural theorists of the metamodern Timotheus Vermeulen and Robin van den Akker,

> metamodernism oscillates between the modern and the postmodern. It oscillates between *naïveté* and knowingness, empathy and apathy, unity and plurality, totality and fragmentation, purity and ambiguity. Indeed,

> by oscillating to and fro or back and forth, the metamodern negotiates between the modern and the postmodern. One should be careful not to think of this oscillation as a balance however; rather, it is a pendulum swinging between 2, 3, 5, 10, innumerable poles.[18]

Building from both of these accounts, James Brunton has recently argued for the need to attend to the political as well as the formal dimensions of metamodernism's reactivation of modernist aesthetics. "Metamodernism," Brunton writes, "ought to be understood as political/politicized nostalgia" that "establish[es] a space of dissent in the present precisely by reaching back into the aesthetic toolbox of the past."[19]

In broad terms, then, metamodernism sets terms, periods, affects, grounds into an imbalanced oscillation, whether in service of imagining new futures—along with the very possibility of a future—or, in Brunton's analysis, of carving a "livable now" from the reactivation of modernism's "incomplete [political and aesthetic] revolutions."[20] I am interested in how a metamodernistic analytic might help us think with contemporary poetic imaginings of extinction—from Clevidence as well as Alexis Pauline Gumbs, Brenda Shaughnessey, Anne Waldman, Jody Gladding, and Sawako Nakayasu—that are routed through (inhuman) modernist aesthetics. How do contemporary poets re- or interanimate the inhuman modernisms of the nineteenth and twentieth centuries—including their epistemic models of unpossessability and inconceptual relationality through which to think oblivion—for an Anthropocenic present, or a world already ended? What does inhuman metamodernism hear from the depths (and heights) of a scarred earth, its still-expressive sediments, dust, soils, skies? And what metaphorics are left to be made in and from humanism's ruins?

In *M Archive: After the End of the World*, Alexis Pauline Gumbs writes, "When she said the mud mothers she meant that energy close to the core of the earth where the planet felt more alive, soft, hot, and in production."[21] Mud mothers, later called "the rooted ones" and the "tree people," are both a mythology and a mode of being in Gumbs' postapocalyptic future archive: those whose "intimacy with the particular soil of their commitment allowed them to communicate underground" after ecological catastrophe has driven all humans—including the gentrifying, neoimperialist class of "mobile people"—into the earth's undergrounds.[22] *M* is both mud and mother, mud-made Eve or Sycorax "planted" in an underground "abyss" and growing/"branching" beyond what is conceptually legible to the surface people.[23]

As discussed in the introduction, we might also read the "after" of *M Archive: After the End of the World* as an after *to* the end of the world, as striving toward a language and a form through which to think the futurities grown from an ended world. Writing after is also writing with and alongside the end of the world, and in the multiplicate diffractions of this ending across the impossibly vast temporalities and spatialities of an inhuman (meta)modernism. As Kathryn Yusoff reminds us in *A Billion Black Anthropocenes or None*, "The Anthropocene might seem to offer a dystopic future that laments the end of the world, but imperialism and ongoing (settler) colonialisms have been ending worlds for as long as they have been in existence."[24] Gumbs calls *M Archive* a work of "speculative documentary," an aesthetic form as well as a strange temporality that reaches (or hears) back toward the histories effaced, mutilated, and drowned in the Atlantic slave trade, and forward to the oblivions beyond humanistic imaginability. For *M Archive* is both "ancestrally cowritten" (in the wake of M. Jacqui Alexander's *Pedagogies of Crossing*, one of its central inspirations) *and* "written in collaboration with the survivors, the far-into-the-future witnesses to the realities we are making possible or impossible with our present apocalypse."[25] The mud mothers gather such unimaginable histories and futures into themselves, into the material-affective recesses of the body, and ultimately continue the possibility of being beyond the ending of this world. In the experience of an unnamed survivor of apocalypse—the "she" around which *M Archive* is focused—the mud mothers inspire an alternative mode of somatic thought and memory: "she was aware of her function as a technology for remembering. specifically, she activated her body as a connection site for all intergenerational knowing and reveled in the edges of herself."[26] The mud mothers reconceptualize (or even inconceptualize) the body in terms of its relationality, its gathered generations, its capacity to hold "the dark feminine"—or "the black simultaneity of the universe"—without dispelling its everything into the terms of the individualized self.[27] The title of this section, "Archive of Dirt: *What We Did*," suggests a dirt that is both the expressive underground sediments heard by the critical geologists and the creative substance (*adamah*) that shaped the first earthlings, and that might enflesh beyond and otherwise than Man.

In this sense, there is a counter-lineage and an ante-knowledge, another relation to earth, rhythmed into being by the mud mothers: "the pressure of their pacing in the same spaces over again had a resounding impact" such that "depending on the direction and rhythm of the steps, the speed of the turn, the weight, the rooted people elsewhere could know the message."[28] This is what colonial (geo-)archives do not and have never been able to hear: the intergenerational knowledges and modes of communion of the rooted people; the possibility of something other than the image of Man persisting beyond apocalypse; alternative arrangements of matter that are no longer oriented around what Zakiyyah Iman Jackson (following Sylvia Wynter) calls the anti-Blackness around which "Eurocentric humanism" has defined itself.[29] For the counter-archiving of *M Archive* is also a metamodernistic "politicized nostalgia."[30] *M Archive* draws on modernism's ungrounding of linear models of time—along with modernism's striving toward a means of expressing inhumanist modes of being and gathering—to avoid capture and to craft a livable present. *M Archive*, like *Between the Acts*, as well as the modernist science fiction works discussed in the introduction (from Stapledon, Lovecraft, and Hodgson), modulates extinction through comedy, opening beings to indeterminate futures and supplying a new, decolonial archive for the speculative remaking of the present. In Gumbs's subsequent volume, *Dub: Finding Ceremony* (2020), the ancestral histories and ongoing kinship relations gathered into the postapocalyptic body come to include nonhuman as well as human beings, all ceremonially entangled: "if you gathered them they would be everyone," all "seeking what but this multitude."[31] Whale and coral choruses sing the persistence of breath, body, and dance "across generations" in remembrance and in ongoing celebration of the rhythms the ocean has given:

> *what the coral said.*
>
> once. we were all singing. somewhere. we are still. moving. as something huge, vibrational, wet. we dance and keep the world in place. we shiver and know the orbit. if you let the body undulate you will remember.[32]

Corals narrate geo- and aquahistories unholdable by concepts, uninscribable within colonialist archives and forms of memory. We can only experience such histories vibrationally, through the undulating (from the Latin *unda*: "wave") body as it dances remembrance of the entangled ancestral, coral, cetacean past from which it emerged.

Brenda Shaughnessy's *The Octopus Museum* (2019), another hybrid speculative-documentarian and apocalyptic-comic work, imagines the recomposition of oceanic and terran futures by an intelligent species of octopi (or "Cephalopod Electro-Overlords" [CEO], as they have termed themselves) whose habitats have been destroyed by anthropogenic climate change and ocean acidification. In Shaughnessy's comic vision, climate change drives octopi first to migrate and adapt to land, and then to take control of human civilization to prevent the fulfillment of the Anthropocene. This protective instinct applies to both planetary and intrahuman violence: humans have to be kept in pods to protect them from killing and oppressing one another. Podded humans are fed rations to keep them alive as laborers for the ruling cephalopods, who now gather knowledge of the human species through the archival apparatus of the museum. Cephalopod scholars and archivists study humans and their humanisms—their findings gathered in five main "galleries" of the museum—as a means through which to understand their own evolutionary and sociogenic histories. Many of the poems archived in each of the museum's galleries mourn the ecological and social devastation and near-extinction of *anthropos* brought on by agrilogistic, late-capitalist humans. Others imagine a cephalopodism that might sustainably continue interspecies and planetary being beyond the failures of humanism: Haraway's tentacular Chthulucene both literalized and comedized. As one human observer notes in a series of "Letters from an Elder," after the octopi have dissembled all of the humans' guns and other instruments of warfare, "The lovely planet may be salvaged with our extinction—I won't live to know, but it would be some last light."[33] For with the return of ocean to land comes a new geography or form of earth-writing: waves, octopi, and other ocean-brought beings now furrow their own determinations into land, reworlding the colonialist structures and humanist conceptualities that have overdetermined themselves as the only mode of having world. In the words of the letter-writing Elder, "The Octopodes" are "an ink species. They overwrote us. . . . The writing was never on the wall, it was in the water."[34]

Although the vast majority of the archived writing consists of the preserved remains and fragments of post-Anthropocenic humans, the octopi do include in one of the galleries a document written by the cephalopod scholars of the "Department of Human Studies" titled "Are Women People?" While the title reads as a prescient reversal of Alex Byrne's glib rhetorical questioning of transgender existence, the report as a whole explores how "personhood" has historically operated among human civilizations. Never a "natural" biological or a species distinction, humanist personhood, according to the scholars, has been constructed through the exclusionary legal and political machinery that distributes beings into the categories of Man and non-Man, the latter comprised of people of color, undocumented people, disabled people, children, and finally, women. According to the report, "*Man* claims to include women, but doesn't. . . . *Man* is plural, encompassing humanity (which, clearly, serves man). *Woman* is singular, individual. To each her own."[35] Yet *Woman* is also the transit through which humanist universality secures itself, the vanishing—but never disembodied—mediator for a humanity from which woman is constitutively barred. Insofar as the category woman has been defined (by humans) in terms of its members' possession of particular reproductive organs, women are deemed, finally, as "too essential in their reproductive capacity to be allowed full personhood—their bodies must be held in trust by the state (which confers personhood) in pre-status from ages thirteen to fifty in order to preserve the future of potential people."[36] Extrapolating from the terms of this report, we can read cephalopodism, in part, as an attempted unfuturing of the institutional machinery—and the forms of gender, racial, xenophobic, and ableist oppression it produces—by which Man has been written into the being of both earth and ocean. A Cephalocene then might then be imaginable as a differently futured ocean and an alternative joining of sea to land, the writing or inking of new intra- and interspecies relationalities, a tentacular planetarity.

Like *The Octopus Museum* in its wake, Anne Waldman's *Manatee/Humanity* (2009) takes as its subject the relations between humans and intelligent sea animals amidst a damaged, acidified planet. Yet *Manatee/Humanity* strives not toward something like a sirenianism or manateeism to be built upon the wreckage of humanisms past, but rather toward a way to rethink the linguistic and conceptual structures by which worlds, relations, and futures have been made determinable in the first place. For both Waldman and Clevidence, such rethinking begins at the levels of the contingent evolutionary path of the speech organs and the rudimentary grammars imposed upon the sounds of speech to resolve (or to re-mark) them into significance.[37] "& important changes / in the throat structure allowed the sounds / uttered by *homo erectus* to approximate human speech," writes Waldman.[38] In both Clevidence's strained, restless biogrammar of the forest and Waldman's sirenianist "grammar of sound" caught in and emitted from "manatee-larynx," we hear toward another, human-indifferent possibility of tonal and grammatical organization.[39] In *Between the Acts*, humanism self-expressed in part as a collective silencing tone (of nonhumans and their noise worlds). And in *Forschungen eines Hundes*, humanist conceptuality was theorized as privation from a fundamental, joy-giving philosociality with other dogs and with the earth itself. Forest- and sea-grammars are perhaps the expressive outgrowths of an inhuman metamodernism, one continually longing—with Woolf, with Kafka, and with the (perverse) inhuman modernist counter-lineage as a whole—for ways to think a future beyond both humanistic determination and the possessionism of our languages.

Worlds, for Waldman, are perceptive enclosures (or Uexküllian bubbles), each with a particular temporality and tonal system by which environmental phenomena are made meaningful: "seeing is always 'seeing as,'" writes Waldman.[40] And "seeing" is perhaps also "sea-ing": a subset of worlding, a making into sea, a means of conceptualizing the particulars of one's environment through the tonalities and temporalities of the sea. To sea, and to sea as a manatee, is to unmake the time of landedness—specifically, the temporality of individualistic, capitalist, bubbled and landed humanity—or to flood it with the multiplicate universes, the "fractals in chaos," of a conceptless ocean time. From ocean time emerges ocean thought: "the manatee is perhaps thinking archivally deeper than man," Waldman writes; for Shaughnessy, "as a species we [octopodes] have extensive, meshed, intergenerational memories."[41] For both sirenian and cephalopod, species memory is simultaneously cyclical and ancestral, intergenerationally and oceanically meshed beyond the narrow time of the (bubbled or podded) individual: "& what kind of anchor, then, is time? / if in the grammar of sound it's always revolving around / 'what is the time of manatee?' or / 'what sound caught in waves in manatee-larynx?'"[42] Manatee sea-ing is the composition of another grammar, an organization of sounds into something other than (landed-)humanistic meaning, unholdable by a noncyclical time.[43]

"it=was=a=time=of=fossil=fuel=priorities=of=precious=business=time": ecocide's syntax is, for Waldman, the arrangement of words, tones, and beings into the abstract equivalence of capitalist-humanist universality.[44] Extinction would then be the completed subsumption of every nonhuman and human temporality and tonal system to the grammar of agrilogistic, linear progress, the overwriting of every future—all time and times—with the name of Man. As Waldman continues, "that's what they'll say about us: what were they thinking? stupid fuckers it was commodification fun-hog time, time-modification time, got on with time we killed time they fucked us over in our future time"; "*& what of animals on desperate clocks of survival & flood, famine, ice-caps / melting.*"[45] The time of cephalopods, as given in Shaughnessy's octopode-narrated poem "New Time Change," is the imagined arrest of ecocide/progress, both a new time and time change, and a further overwriting of this humanistically grammared future with new ink: "You had your time you took your time after time you had your cake by the ocean and ate it too but now the tide has turned the times tables too when it's time to change you've got to rearrange #timesup."[46] But sirenianism sees/seas through the terms of an even more expansive cyclicality in which the intergenerational entanglement of beings persists beyond the time and tense of any single species. Manatee time vibrates with the coral song of Gumbs's *Dub: Finding Ceremony*, in which ancestral memory is also interspecied memory, felt and carried in the undulative rhythms of the body: "we dance and keep the world in place. we shiver and know the orbit. if you let the body undulate you will remember."[47]

Similarly, in striving toward kinship with both the "time" and "*the mind of manatee,*" Waldman writes, "Manatee reminded me ~" (to remind also being to *re-mind*):

that multiple
hydra-headed
        universes,
all fractals in
        chaos
including more cycles
        will emerge[48]

Sirenian futures are grammared cyclically, through an unshapable but also an endlessly generative chaos. (In *Paradise Lost*, as discussed in the previous chapter, chaos is described as both a "dark / Illimitable ocean without bound" and as the "womb of Nature," preceding creation yet also holding the indeterminate "dark materials" or "first matter" out of which God creates by ordaining their multitude into form.[49]) Fractals, cycles, universes form and disintegrate ongoingly, giving rise to futures without any (human or sirenian) selves there to perceive them. In this sense, there may be no such thing as a sirenianism, since manatee thought and time (as imagined here) unfolds indifferently to possession—of any self, species, world, or future—and in c(h)oral community with all beings co-undulatively gathered in chaos.

An entry from the *Voyages of Christopher Columbus* reads, "On the previous day [January 8, 1493], when the Admiral [Columbus] went to the *Rio del Oro* [Haiti], he saw three mermaids [*serenas*], which rose well out of the sea; but they are not so beautiful as they are painted, though to some extent they have the form of a human face."[50] Through the conquistador humanist gaze of Columbus,[51] the manatee is only thinkable either in terms of its proximity to the European human or as mythical *álogon*: a riddle rising out of sea, a troubling likeness, solved with the naming of Man. The colonialist metaphoricity imposed upon the manatee—along with its close relation the dugong, also of the order *Sirenia*—would persist well into the nineteenth century, as evidenced by an account on "Mermaids" from the *Arcana of Science and Art: Or an Annual Register of Useful Inventions and Improvements* published in London in 1831. An unnamed "Correspondent of the Magazine of Natural History" writes,

> A few years back a mermaid was shown in London. . . . The impression on my mind was, that it was an artificial compound of the upper part of a small ape with the lower half of a fish; and being allowed to examine it as closely as I pleased externally, my attention was directed, by the aid of a powerful glass, to ascertain the point of union between the two parts. I confess I was somewhat staggered to find that this was so neatly effected, that the precise line of junction was not satisfactorily apparent.[52]

And further,

> A short time back the skeleton of a mermaid, as it was called, was brought to Portsmouth. . . . To those who came to the examination with preconceived notions of the fabulous mermaid, [the dugong] certainly presented, as it lay on the lecture-table, a singular appearance. It was . . . about 6 ft long: the lower dorsal vertebrae, with the broad caudal extremity, suggested the idea of a powerful fish-like termination; whilst the fore legs, from the scapula to the extremities of the phalanges, presented to the unskilful eye an exact resemblance to the bones of a small female arm. The cranium, however, had such an *outré* brutal form, that even the most sportive imagination could never have supposed it to have borne the lineaments of the "human face divine."[53]

The dissected sirenian is a monstrous amalgam of ape, fish, human female, mythical and naturistically "brutal" mermaid: a hybrid, many-concepted being without proper linguistic, taxonomic, or mythological place. (A being as or more enigmatic even than Darwin's monstrous bear-whale.) Both the conquistador-humanist gaze and the vivisectional blade seek after, and fail to find, a "point of union" or "line of junction," a linearity—or what Clevidence calls a "MYTHOLOGY OF LINEAGE"—that will solve the *ainigma* of sirenian being and make legible its unpossessable archival depths and unimaginable futures.

"The descriptive apparatuses of natural history," as Mary Louise Pratt claims in *Imperial Eyes*, went hand-in-hand with mid- to late eighteenth-century European imperial conquest and expansion: a European "planetary consciousness" and its networks of "global-scale meaning" were made possible by the coupling of the natural historical and colonialist gazes.[54] The "seeing-man" is Pratt's term for "the white male subject of European landscape discourses—he whose imperial eyes passively look out and possess."[55] Yet not all is reducible to order, and the possessivity of the gaze is not absolute. "When it comes to entering the contact zone and confronting the object of extermination," Pratt writes in relation to Argentinian poet Esteban Echeverría's representations of the Pampas Indians in *The Captives* (1830), "the visual code and seeing-man's imperial authority break down into sound, blindness, confusion."[56] In this sense, there is perhaps something to be retained in both the anonymous natural historian's and Columbus's perplexions, in which the *serena*/sirenian remains opaque to the sight—along with the ordering mechanisms and technologies—of the seeing-man in each of its various iterations: conquistador, imperialist, natural historian, vivisectionist, and as we will discuss shortly, eugenicist. The grammars of manatee and dugong, then, are those which unsettle an (imperialist, landed) conceptual regime grounded in rigid separations between land and sea, colonialist myth and intergenerational archive, divine and outré, determinate creation and the confused materiality of chaos. Sirenian sea-ing is an alternative intimacy of land and sea; sirenians heralds of another, inhumanist and decolonial planetarity to be made from the rising seas and shrinking terra firma of an increasingly oceanic future.

Imperial planetarity is also rooted in a landed and territorialist imagination that, in the words of literary scholar Jonathan Howard, has lost sight of the sea. In "Swim Your Ground: Towards a Black and Blue Humanities," Howard compellingly critiques the possession-based ideology of landedness, arguing that

> humanity's prevailing plan for life on this watery planet of ours is decidedly terracentric. But if on terra firma, it is possible to interpellate the surface of the planet as the guaranteed and given instrument of our walking, to sterilize it of all alterity, to domesticate and delineate it, to reduce it to private property, to enclose it in walls or national borders, the ocean, by contrast, elicits our restless and conscious negotiation of not only space but our relations with one another in space. Unlike the modern plan for living presently exhibiting all of the signs of pending collapse, it indexes a contingent and radically inclusive practice of life together, whose first principle is neither property nor the individual but rather the irreducible fact of relation both human and nonhuman in its scope.[57]

The "oceanic turn" in the humanities, according to Howard, constellates increasing interest in the "nature of the ocean and the prospect of human life on a blue planet," along with oceanic ways of being and knowing—or what Karin Amimoto Ingersoll calls a "seascape epistemology"—with the geohistorical framework of climate catastrophe and rising sea levels.[58] Global sea-level rise, as Elizabeth DeLoughrey argues in "Submarine Futures of the Anthropocene," "produc[es] a new sense of planetary scale and interconnectedness through the rising of a world ocean."[59] DeLoughrey continues, "The Anthropocene has catalyzed a new oceanic imaginary in which, due to the visibility of sea level rise, the largest space on earth is suddenly not so external and alien to human experience."[60] Such a monstrous imbrication of oceanic with terran futures is what is indicated, and perhaps even prophesied, by manatee. Manatee's interspecied littorality—imagined as belonging to both water and land, as given in its colloquial name, "sea cow"—blurs the line demarcating green from blue planet. In this vein, Ingersoll's seascape epistemology, theorized in part through Indigenous Hawaiian (Kānaka Maoli) cosmologies, does not simply shift the terms of inquiry from land to sea, but rather "incorporates knowledge of both land and ocean. Kānaka Maoli perceive the ocean as an extension of the land, a perception reflected in the fact that activities that take place on the land affect the sea, just as oceanic activities have effects on the health of the land."[61] Seascape, oceanist, cephalopodist, and sirenianist epistemologies, then, are perhaps formulated not in and as opposition to land itself, but rather to the landednesses or terracentrisms that, because they cannot imagine a multispecied kinship of the submerged, continue to mythologize the line, the gaze, the furrow, and the forms of possession enabled by them.

II.

If oceanist thought is in part defined as a perversion of linearity—or of the teleological line of capitalist-imperialist progress and the littoral line demarcating *terra* from *oceania*—then are land-dwelling beings also oceanic? And is the ocean also a forest? Clevidence theorizes forest as an inhumanly grammared evolutionary flux entangling "fern" and "fish," beings whose primordial relation preexists their conceptual differentiation: "& turning, turn into, mutant-child of the first fern, first fish, lisping."[62] We can thus take planetarity both literally and littorally, as the thought of a planet made of blurred forestial and oceanic space, where the branchings of trees and of coral reefs are indistinguishable from one another. Flaubert literalizes this communion in *La tentation* during the scene where Anthony accompanies the "Beasts of the Sea" to the lost "countries of the Ocean": "Polyparies that seem like trees have arms upon their branches" (189).[63] (Meta)modernist planetarity, from this perspective, is immersion in the epistemologies of forest and ocean, in which linearity is turned astray, loses ground, enters the unpossessable cyclicality of another, blue-green grammar.

Jody Gladding's *Translations from Bark Beetle* imagines forestial thought through the material geographies and grammars of the bark beetle. The book's conceit is that the poet is "translating" the language of bark beetles by examining the traces that they leave from first digging tunnels through trees, then making graphite etchings of these traces, and finally converting these patterns into an approximation of language.[64] A "translator's note" explains, "Certain elements of the grammar make translating Bark Beetle problematic. There are only two verb tenses: the cyclical and the radiant. Prepositional phrases figure prominently and seem necessary for a complete syntactical unit. The same pronoun form (indicated as •) is used for first and second person in singular, plural, and all cases."[65] (Similarly, the cephalopods in *The Octopus Museum* speak only as a collective, through a communal grammar that has to be retroactively converted into the terms of subject- and individual-oriented human languages for us to read it.) Bark Beetlian, as imagined by the poet, is an epiphoric, swarm-made grammar. Each expressive trace gathers and co-implicates self and species, I and you (or should we call these perspectives "first and second beetle"?), and past and present time. Beetle traces are a form of geo- or sylvan writing, an intergenerational archive made of the strivings of beetles with and through the inscribable and re-markable surface of bark. The cyclical and the radiant tenses, while ultimately opaque to humanist conceptualities, are time-indifferent, and instead name something of the particular qualities of each environing trace or tunnel: the difference, perhaps, lies between burrowing inward (cyclical "turning") and reverberating outward (radiant "growing"). Waldman's sirenianism, too, is cyclically tensed: "& what kind of anchor, then, is time? / if in the grammar of sound it's always revolving around."[66] Read relationally, these poetic imaginings of sirenian, cephalopodist, and coleopteric thought suggest a collectivist, planetary grammar made of the inter-implication of ocean and forest.

Bark Beetle language, as Gladding imagines it, also approximates what Eduardo Kohn has called "sylvan thinking," which is both a mode of thinking particular to forests and a spirited wager against the Western metaphysical emphasis on thought as an exclusively human property. Kohn's *How Forests Think: Towards an Anthropology Beyond the Human* is an ethnographic and philosophical exploration of the sylvan epistemologies of the Quichua-speaking people of the Runa village of Ávila (in Ecuador's Upper Amazon). Kohn argues that thought itself, or at least many of the properties that constitute what is called thought, may be forestial:

> If thoughts are alive and if that which lives thinks, then perhaps the living world is enchanted. What I mean is that the world beyond the human is not a meaningless one made meaningful by humans. Rather, mean-ings—means-ends relations, strivings, purposes, telos, intentions, functions and significance—emerge in a world of living thoughts beyond the human in ways that are not fully exhausted by our all-too-human attempts to define and control these. More precisely, the forests around Ávila are animate. That is, these forests house other emergent loci of mean-ings, ones that do not necessarily revolve around, or originate from, humans. This is what I'm getting at when I say that forests think.[67]

Mean(-)ing emerges from the animate relations of a forest that is also many forests, and we can also read forest as any place or any imaginary that persists in the making of meanings (or the crafting of signs, in Kohn's Peircian semiotic frame) in indifference to the humanist determinations imposed on it. "Nonhuman life-forms also represent the world," Kohn continues, and "signs exist well beyond the human. . . . What we share with nonhuman living creatures, then, is not our embodiment . . . but the fact that we all live with and through signs."[68] Similarly, "biosemiosis" is the term used by Jesper Hoffmeyer and Wendy Wheeler to encapsulate the myriad of nonhuman forms of representation and sign-making found amidst "the richly communicative global web teeming with meanings."[69] And in John Durham Peters's related assertion at the conclusion of *The Marvelous Clouds*: "We make meanings, but do not do so in media of our own making; our bodies are embedded in climate history, fire regimes, the spin of the earth, north and south, and relations with plants, artifacts, and organisms of all kinds, especially each other."[70] In this sense, forests both think and are thought through, and the inseparability of these processes is perhaps constitutive of what we might call the thought of forest (as subjective and objective genitive: forest thinking and forest being thought). Bark Beetlian glyphs (from the Greek γλυφή [*glyphē*], "to carve") are one manifestation of such sylvan thinking, a potential sign system unfolding through the dynamic, intra-active, simultaneously thinking and thought medium of the forest.

In an early poem, "Spending Most of Their Time in Galleries, Adults Come into the Open on Warm Sunny Days: Translations from Bark Beetle," Gladding writes,

> • 've learned through wood
> yo• can only travel in one direction
> but turn again with m• there love

Further,

> there are other ways in the wood's
> growing
> if not for m•—
>   find hollow
>   find spell[71]

Turning (a cyclical verb), and turning again, is an act of faith, a devotional tunneling in and through what "very early" Bark Beetle language calls the life-giving "matrix" of "wood" to achieve self- and species expression.[72] It is only possible to travel in one direction, but one can turn infinitely and indestructibly along the path taken in this direction, which leads simultaneously toward the tree's nourishing sap, toward one's inner essence, and toward the being of the beloved. For love is conjugated in Bark Beetle grammar as cyclical co-striving, a turning together that modulates being into a communal interpretation or glyphing of wood. ("My feeling," Gladding remarks in an interview about *Translations*, "is that they are love poems. Like many of our poems, they speak of longing."[73])

And yet m• is not the only way, path, or direction of growing in and with wood. The beetle poet(s) perhaps intuit something like what Kata Gellen calls the "species transcendence" of Kafka's nonhuman narrators, in which "an animal or group of animals begins to grasp how circumscribed its perspective on the world really is" when perceiving something from "beyond" the epistemological limits of their species.[74] But Gellen's species transcendence focuses almost exclusively on the misunderstandings and interpretive errors that these animals make when confronted with extra-species forms of experience or knowledge. What bark beetles intuit instead is more akin to an interspecies sociality based on the awareness of many beyonds—of other, non-coleopteric grammars of selving, communing, glyphing, loving, through wood and forest—coupled with an epistemology weak enough to cede possession of every beyond.

A spell must be found, as this poem's last line enjoins, spell being both the material origin of Bark Beetlian language—the gathering of letters into words—and the primordial incantation by which enigmatic scribbles or marks on a surface are re-marked into a dynamic system of meaning. In "The Origins of Bark Beetle as a Written Language," Gladding translates—or spells/glyphs into being—an alternative origin of poetic form. Poetry begins in the forest, not with Woolf's Anon hearing and modulating birdsong into the first rhapsodic words, but rather with the first encounter of beetle with wood:

> They came down out of the hills, stepping over the cracks where the light broke through the wood.
> Wood said one and they remembered how to name the matrix
> was also to address it.[75]

As with Anon's song, bark beetle *poieses*—as the material-affective traces or biosemiotic signs made through the matrix of wood—can only come into being after light has broken into wood, after heaven's gaze has scarred both forest floor and the insides of trees. The languaging of beetles begins with an address to wood that is also a means of violence: the condition and the cost of bark beetlian (self-)expression seems to be the destruction of the wood, trees, and forest itself. Address to wood is prefigured in an earlier poem as address to "wound": "**how do you (ad)dress the wound?**)"[76] From the Latin *ad* (toward) + *directus* (direction) and inflected by the Old French *dresser* (to arrange or put straight), to "(ad)dress" is both to turn toward or in the direction of the object and to attempt to heal wound/wood after violation. (The "wound" of the earlier poem, "(lucidity is the wound," is the rape of a young woman addressed by the poetic speaker.) As Anon travels along a not-yet-healed green scar in one origin story, so bark beetles emerge from the wounded matrix of wood in another, glyphing an existence through strivings in and with forest.

At the time of this writing, bark beetles are devastating forests across the globe. As ecologist Cheryl Katz writes in "Small Pests, Big Problems: The Global Spread of Bark Beetles" (2017),

> With their advance fueled by climate change, bark beetles have ravaged 85,000 square miles of forest in the western United States—an area the size of Utah—since 2000. . . . The beetles are now advancing up the Atlantic coast, reaching New York's Long Island in 2014 and Connecticut the following year. . . . And from Europe to Siberia, bark beetle outbreaks are erupting with increasing frequency in woodlands weakened by rising heat and drought.[77]

Yet there is a cyclicality to bark beetle consumption of wood, as Katz goes on to suggest, that is only considerable as violence or destruction from a linear, nonforestial perspective that is grammatically and epistemologically alien to Bark Beetle. "Bark beetles," Katz continues, form

> a natural part of the conifer forest life cycle, regularly flaring and fading like fireworks. But the scope and intensity in the past two decades is anything but normal, scientists say, in large part because rising temperatures are preventing the widespread winter die-off of beetle larvae, while also enhancing the beetles' killing power. Not only are the insects expanding into new territory, they're also hatching earlier and reproducing more frequently.[78]

Anthopogenic climate change makes this cycle into a line, turns bark beetles and their forestial glyphs into heralds of extinction. *Translations from Bark Beetle* is an attempt, and a provocation, to inhabit the grammars and temporalities of cyclicality and radiance, in which there is no time and no self apart from what can be made in wood. Dendritic creation and decreation feed into one another. This is a dynamic that also informs Clevidence's reframing of nature as the queer conjugative assemblage prophesied in Milton's depiction of the underworld in *Paradise Lost*: "Where all life dies, death lives, and Nature breeds, / Perverse, all monstrous, all prodigious things."[79]

A similar perverse, cyclical forestialism inheres in Darwin's Tree of Life, as discussed in chapter 2:

> As buds give rise by growth to fresh buds, and these, if vigorous, branch out and overtop on all sides many a feebler branch, so by generation I believe it has been with the great Tree of Life, which fills with its dead and broken branches the crust of the earth, and covers the surface with its ever branching and beautiful ramifications. (177)

Forestial *epiphora*, or nature's hyperproductivity, is multivalent: the same, ever-branching matrix of wood lets some branches grow and sediment underground and others radiate upon the surface. It is only through the imposition of time, however, that the cyclical flux of the forest—a place and an imaginary where "all life dies, death lives"—can be given distinct directionality: in Bark Beetle, there is only one direction to travel and to turn in. Anthropocene, in this sense, is also the overwriting of sylvan glyphs and selvings with the name and the time of Man, reducing each expressive trace to a unidirectional and univocal sign of the extinctions to come.

Yet a grammar of radiance is also one that insists beyond extinction, indifferent to furrowed lines: whether the lines furrowing the forest, the line dividing forest from ocean, or the line leading from being to oblivion. Sawako Nakayasu's *The Ants* (2014) is an attempt to inhabit the thought—as well as what Julian Huxley has called the "philosophy"—of another radiantly grammared species. Ants saturate the air and the pages that they materially and metaphorically stand on (or radiate across): one ant in particular is able to "write . . . with no implement but its rear end" on a folded paper while traveling in the hold of a cargo ship; later a colony of underground-dwelling "Korean ants" manages to "draw—with the thickness, turns, curves of the passageways—a particular Chinese character."[80] *The Ants* is not an (attempted) translation of these and other ant traces, however, but rather an archive of a language come under the contagious spell of myrmecine expressivities—or, in Huxley's words, the "myrmecine sacred formulae."[81] Ants exist simultaneously as individuals and colonies, metaphors and nonmetaphors, lines and spirals. Each poem is a trace of language's infection by monstrous, ant-wrought irrealities; each revealing the inability of humanistic language to contain what radiates in excess of its own concepts.

In "Ants in the Wind," for example, ants are both the content and the medium of transfer through the air that they both fill and are: "It's that time of year again and ants are filling the air, both the physical air as well as the radio and TV airwaves, not to mention the internet." The "ant-density levels" are up to "72%," meaning that they are "fogging up vision throughout the world over."[82] Ants are metaphors for, among other things, invasion, strange weather, chaos, and perplexing eclipse. Ants are also the saturated air(waves) through which they travel, material disruptions of metaphoricity itself, if metaphor is considered in its traditional sense as a medium for the transfer of meanings across contexts and the giving of names to that which is unfamiliar. The sky is a locus for the making of meaning, or perhaps a metaphor-like medium that enables the spiritualization of the particulars of being into concept. Yet the sky is also defined by what fills it—and the medium by what it mediates—as John Durham Peters suggests in his discussion of clouds in and as "sky media": "What would it mean to ask if clouds have meaning? Are clouds media? They are the ultimate test of the idea that there could be natural media. But first, clearly clouds are full of meaning—ask any farmer, pilot, or sailor."[83] To fill the field of meaning and to be full of meaning: a condition shared by clouds and floating ant colonies, each an enigmatic gathering that concepts cannot divine.

Accordingly, ants descend from what seems to be an invisible ant cloud in the opening of a later poem, "Rain": "It begins to rain ants, out of nowhere but the clear cliché of a bright blue sky. We are instantly distraught, confused, and frightened at this new meteorological development."[84] Like the "horrible phaenomena" of global weather events caused by Lakagígar's eruption in 1783, both rain-ants and sky-ants disrupt the comprehensibility of the air, materially saturating sky with their own bodies—"fogging up vision throughout the world over"—and preventing the passage from being to concept. The soaring ants, we are told, have "studied countless tutorials, textbooks, videotapes, DVDs, podcasts, everything but the real thing, all invested in the sole purpose of helping to train young ants to prepare for the ride up . . . as well as the inescapably treacherous way down."[85] Ants travel as ants and as metaphors for immigration and displacement—along with confounding weather—without a clear line of separation dividing their materiality from their metaphoricity. In this way, material-metaphoric ants confound the agrilogism and humanism of traditional metaphor, which depends on the metaphysical separability between what Hegel calls the "sensuous content" and "spiritual meaning" of a name made meaningful. And yet, as we have explored, this disruption of metaphorical meaning through the provision of an epiphoric excess of meanings is also internal to metaphor itself. In other words, metaphor itself *is* this split between the transfer of names across contexts (*metaphora*) and the piling up of names and bodies beyond legibility (*epiphora*).

Nakayasu's generation of material-metaphoric ants has a set of modernist predecessors in British naturalist and evolutionary biologist Julian Huxley's extensive writings on ants, alluded to above. These include, most notably, the essay "Philosophical Ants: A Biologic Fantasy" (1923) and the popular science work *Ants* (1930). Both texts are, as the former title suggests, fantasies of inhabiting myrmecine thought while also resisting what Caroline Hovanec calls firm "analogies to human society." Most particularly, Hovanec writes, Huxley wants to avoid "the 'fallacy' that with increasing mechanization and specialization, we are becoming more like the highly specialized and rigidly hierarchical ant."[86] For Huxley, ant thought or "philosophy" begins when a single precocious ant performs a set of scientific experiments to calculate the length of the days in summer and winter to explain why an ant will, "on a warm day . . . get through a task four or five times as heavy as she will on a cold one."[87] These findings, however, contradict the sacred myrmecine dogma that "Ants . . . were made in the image of God" and that "the Ant is the norm of all"; following from these, it must mean that the "cause" of the "strange alteration of diurnal length . . . lay in the Ants themselves."[88] The philosopher-ant is banished from society for heresy, his discoveries only accepted among "the educated classes" (and not the "common people") once they are confirmed by later, more secularized generations of philosophers.[89]

Huxley's biologic fantasy thus operates upon an implicit metaphorical structure in which pre-Enlightenment humans and the present day's "common people" are myrmecinic in their slavish devotion to dogmatic belief systems and rejection of scientific knowledge.[90] Once they have fulfilled this trajectory toward metaphysical concept, however, Huxley's ants vanish. When they reemerge, in Huxley's review of Auguste Forel's *The Social World of Ants Compared with Man* (1928), ants are no longer capable of philosophy and now sundered from the metaphorics through which they were briefly and fantastically joined to humans. Much of the review is crafted as a disagreement with British linguist and translator C. K. Ogden's interpretation of Forel, and especially with the gnomic opening words of Ogden's foreword to the volume: "There are scientists who hold that in due course Man will yield to the Ant the mastery of a planet grown less hospitable to the relatively idle and unorganised."[91] According to Huxley's counter-reading of Forel, ants have "no tradition in the broad biological sense, no transmission of experience from generation to generation, no real education" but only an existence that is "based primarily on instinct, while ours owes its distinctive qualities and its biological success to the capacity for conceptual thought or reason."[92] Yet in his attempt to think ants nonmetaphorically and nonfantastically, Huxley ends up mediating ants through another metaphoricity, one inflected by the strident eugenicism informing his approach to evolutionary biology.[93] No "conceptual thought," no culture, no reason, no "rapid learning" or educational and generational improvement: these are colonialist as well as speciesist and eugenicist tropes that have historically been used to secure a normative conception of the human as—white,

Western, propertied, straight, cis-gendered, able-bodied—Man through the transit of the non-Man and the non- or less-than human. As Huxley continues, "Man can consciously envisage improvement in his conditions, and deliberately set about the control of his racial destiny; ants can do nothing of the sort."[94]

It is this particular animacy hierarchy, and the colonialist-eugenicist metaphorics that stem from it, that Nakayasu's metamodernist ants complicate. Ungraspable and oscillating between the categories of individual and collective, material and metaphor, mediated and medium, Nakayasu's ants anarrange the modes of knowledge through which they have been made visible to humanist and imperial eyes. In so doing, they reclaim the lost promise of an ant philosophy, though theirs is a philosophy made through the labors of an impersonal "collective" rather than in the Huxleian image of individualist dissent and solitary (male) genius. We see this dynamic at work in the poem "No Collective":

> Believing themselves to be quite progressive for their species, a group of ants gets together and decides to form a collective. They gather the necessary documentation, fill out all the proper information in the correct little boxes, get photos taken in the right size and dimensions and angle, and step precisely through every single hoop required of them to become an officially recognized collective.
>
> Their application is denied, however, on the grounds that ants are an inherently collective species, and this designation would be redundant and downright unnecessary.
>
> One ant is so upset by this verdict that it begins to cry, thereby forging a breach in the collective emotional unity of the group. This very breach, however, makes the officer falter, reconsider for a brief moment, entertain the possibility of a radical change of heart, but this very possibility of a change in the officer's heart makes the ant's tears dry up, which lands them all back at their original, inherently collective state, and that's the end of that story.[95]

In Nakayasu's Kafka-inflected parable ("Before the Collective," perhaps?), a being or species cannot be designated as what they inherently are, or cannot become their own concept. Philosopher Massimo Cacciari's reading of the "already open" door to the law in Kafka's "Before the Law" might also speak to the possibility or impossibility of an ant collective: "How can we hope to 'open' if the door is already open? . . . We can enter only there where we can open. The already-open [*il già-aperto*] immobilizes. The man from the country cannot enter [the law], because entering into the already open is ontologically impossible."[96]

Yet the collective, like the law, is only thinkable in its abstraction from the material relations out of which the (ur-)concepts of law and collective are composed. In Derrida's interpretation of Kafka's parable,

> To be invested with its categorical authority, the law must be without history, genesis, or any possible derivation. That would be *the law of the law*. . . . Narrative accounts would try to approach the law and make it present, to enter into a relation with it, indeed, to enter it and become *intrinsic* to it, but none of these things can be accomplished.[97]

Both Kafka's man from the country and Nakayasu's weeping ant attempt to touch (and to be touched by) what has removed itself from history, genesis, derivation, from its own absented or annulled foundations, and from the material-affectivity of the world itself. And yet the law, and the collective, must continually enact this sublation from its material foundations to function as a universal medium for the giving of law or the giving of collectivity. In this way, the ant's tears are a "breach" of both the collectivity of the species and the law of collectivity itself. Following the metaphor that the poem both is and is not, a similar set of institutional practices might deny collective status, statehood, sovereignty, or other forms of recognition to a people or community that already "inherently" constitute a collective/state/nation without being designated as such. (Interestingly, there is a moment in Kafka's parable when the man from the country, who has stood by the door for so many years that "he has come to know even the fleas in his fur collar," "begs the fleas as well to help him and to change the gatekeeper's mind": the man from the country is himself an interspecied collective, appealing to the law simultaneously as a self and as a multitude.[98]) With no individuality, no tradition, no conceptual thought, no intergenerational knowledge, no "control of [their] racial destiny," ants and other nonhuman and human collectives are systematically excluded from the province of Man. The crying of the ant, however, disrupts the mechanisms of inclusion and exclusion upon which law, collectivity, and meaning are based. The ant's tears open the possibility of an enigmatic, knotted being or collective of beings that means both itself and what is in excess of its own concept, and what is in excess of every law of determination. In the words of an earlier poem, "the ant is, and isn't, and is."[99]

The ants of "No Collective" are a gathering that (momentarily) arrests conceptuality. As we have discussed, there is another commons to be thought when the throwing together or thrown-togetherness of bodies is not determined by the metaphysical bonds of *symbolon*. And we have theorized this commons through the logics of *epiphora*, or the profusion of names and meanings beyond determinability, in relation to and as proper to the work of metaphor itself. Fittingly, then, another definition of the Greek *epiphora* (επιφορά) is a "persistent flow of tears," a meaning that is retained in contemporary medical terminology. *Epiphora*, as material-affective flux, allows us to think the arresting ant tear as a sign or glyph of what in ants and other beings is beyond humanistic thought: an indeterminate grammar; another metaphoricity; another law or logic of joining being to being, being to earth, and earth to earth.

Huxley's review of Forel's *The Social World of Ants* is uncannily intuitive of such a beyond. Writing in 1928: "It is difficult for anyone, scientist or no, to believe that man will yield his supremacy to ants unless there were to be some very radical and peculiar changes in the conditions of this planet."[100] (LOL, the present generation comprised of over 20 quadrillion ants and growing might respond![101]) Huxley unwittingly prophesies the weird planetarities to come over the next century, though his stubbornly humanist imaginary clings not only to the supremacy of man but also to the persistence of the structural position of planetary mastery into any possible future, whether yielded to or supplanted by another species. However, the metamodern futures we have been exploring—sirenian, oceanic, cnidarian, glyphic, forestial, myrmecinic—all follow their modernist precursors in abandoning the strong forms of subjective sovereignty and humanist determination. The inhuman metamodernism collectively articulated here is, rather, a radiant cycling through chaos in which mastery remains unthought and other, weaker planetarities become possible in and around its voiding.

III.

What, then, grows and gathers in the exhaustion of human(ist) meanings and futures? The comedic ongoingness of the earth beyond oblivion, as it breathes into itself its own planted noises, colors, and radiances, for Woolf; inhuman philosocialities made in and of earth, for Kafka; material-metaphoric contagions among expressive bodies, parts, histories, and lineages for Joyce; alternative shapings of *adamah* and the confounding of stone, plant, animal, dust, and light for Flaubert; the fertile ungrounds grown from the nonsense and the noise of the earth's jabbering sediments, for Carroll and Darwin. Inhuman metamodernisms, as we have seen, spring from modernism's anarranged temporalities, spatialities, grounds, and worlds: an indeterminate generation grown from an indeterminate ground. Metamodernistic futures, however, approach with even more exigency the prospect of human extinction amidst an increasingly and almost irreparably damaged, warming, and acidifying planet: planetary conditions that, for Huxley, were only conceivable as a thought experiment conducted to emphasize the present and future supremacy of man. In James Brunton's terms, metamodernism reaches back into the "aesthetic toolbox of the past," either in the service of imagining new futures (James and Seshagiri) or crafting livable worlds in the present (Brunton). An inhuman metamodernism might also include among these an attempt to imagine extinction—along with the question what comes after Man and the atrophying of his humanisms—through the aesthetic and ecological modalities that it has inherited from inhuman modernism.

Yet the question of inheritance is a vexed one, as both Vermuelen and van den Akker and David James emphasize in their accounts. In *Modernist Futures* (2012), James argues convincingly that contemporary fiction "partakes of an interaction between innovation and inheritance that is entirely consonant with what modernists themselves were doing more than a century ago, an interaction that enables writers to work *with* their lineage in the process of attempting new experiments with form."[102] In this way, the contemporary—or what James, with Urmila Seshagiri, would later include under the banner of metamodernism—is an ongoing collaboration with lineage, and one that resists the idioms of both agonistic rupture and uncritical assimilation of the past. For contemporary writers, James adds, "the *prospect* of the unrealised potentialities of modernist fiction exists not simply to be used but to be continually tested, redesigned and remade."[103]

As discussed earlier, Vermuelen and van den Akker's word for this collaborative relation between present and past is "oscillation": metamodernism, for them, "oscillates between the modern and the postmodern. It oscillates between *naïveté* and knowingness, empathy and apathy, unity and plurality, totality and fragmentation, purity and ambiguity."[104] Ultimately, metamodernism grasps toward what they call an *"atopos,"* a place that is no place and a present that is permeated by futurelessness:

> Metamodernism displaces the parameters of the present with those of a future presence that is futureless; and it displaces the boundaries of our place with those of a surreal place that is placeless. For indeed, that is the "destiny" of the metamodern wo/man: to pursue a horizon that is forever receding.[105]

We might then think an inhuman metamodernism, in part, as an attempted rhyme between *atopos* and Anthropocene. In other words, the displacement and futurelessness of Vermuelen and van den Akker's *atopos* also name something essential to an Anthropocenic present structured by the unsettling of place—which includes the material and epistemological collapse of sea/land distinctions—and the inhumanization of temporal scale as it confronts its limit point in the prospect of human extinction.[106]

Atopos, too, names the impersonal nonplace of La Trobe's imagining—the "high ground at midnight" with its "two scarcely perceptible figures" in her starling- and oblivion-inspired future play—as well as the underground "abyss" of *M Archive*. Here, "close to the core of the earth," the mud mothers craft new subterranean intimacies beyond what colonial geo-archives and geo-knowledges can capture. Through *M Archive*'s post-extinction imaginary, Gumbs modulates modernist impersonality—along with the late Woolfian passage between the impersonal and the interspecies—into forms of livingness or survivance for a world already ended. Gumbs's metamodernist impersonality, like Woolf's comedy of an inhumanly generative earth, is also an opening to other (human and nonhuman) relations, which only become possible with the cession of landed place and time to the unimaginable futures on the other side of what Gumbs calls "our present apocalypse."[107] Similarly, the modernist and metamodernist *atopoi* of oblivion, underground, abyss, night joined to night, syllabling forest, ocean archive, are the impersonalized nonplaces from which other, postapocalyptic planetarities might emerge.

In a subsequent essay, "Periodising the 2000s, or, the Emergence of Metamodernism" (2017), Vermuelen and van den Akker theorize metamodernism's relation to the past as the "bending of history," which "may simultaneously imply forcing History into a different direction or shape as well as causing History to deflect from the more or less straight line of teleological narrative."[108] We might understand this bend as the modulation of both history *and* futurity into new forms of alinear earth-writing, or the perverse, monstrous, dendritic, oceanist poetics by which inhuman metamodernists imagine *with* the ending of a world. Mud mothers, corals, octopodes, sirenians, oceans, forests, bark beetles, ants—each a metaphoricity unto itself, each a grammar by which extinction might be modulated into new forms of continuity and creation.

The seemingly paradoxical generativity of (meta)modernist *atopos*, in which a place outside of place and a present without future is also a site of unimaginable creation, has its precedent in the Miltonic realms of Chaos and Night, as discussed in chapter 5. "eldest Night / and Chaos, Ancestors of Nature / hold eternal anarchy" in the

hoary deep, a dark
Illimitable ocean without bound,
Without dimension, where length, breadth, and highth
And time and place are lost.[109]

Night and Chaos, *atopoi* outside of God's creation, are also the matrices—"the wide womb of uncreated Night / Devoid of sense and motion"—through which creation occurs.[110] Creation is the giving of form to the confounded matter of Chaos and Night, "the womb of Nature, and perhaps her grave": "Of neither sea, nor shore, nor air, nor fire, / But all these in their pregnant causes mixed / Confus'dly."[111] Milton's account of the confused generativity of what precedes creation—mingled sea, shore, air, fire—might also suggest something of the modernist and metamodernist grasping toward the inhumanisms emerging from creation's ruins. In *M Archive*'s Miltonic (post-)creation narrative, the pre-apocalyptic humanist worlds we inhabited were formed from "the constitutive element of individualism," a tragic reduction of being and "the scale of breathing" to "one body. the unitary body." "Everything could have been different," writes Gumbs, if there were another way of giving form to matter that did not individualize, or if there had been a recognition of "the primary incompatibility" between "the constitutive element of individualism" and "the dark feminine, which is to say, everything."[112]

From the element of individualism emerges a breathing scaled to the level of the individual body, as given in Genesis 2:7 (*The Contemporary Torah*): "Then the Lord God fashioned the earthling [*ha-adam*] from the soil of the earth [*ha-adama*] and breathed into his nostrils the breath of life, so the earthling became a living soul." Adam/earthling is breathed into earth, individuated soil, mud-fathered mud father. The "dark feminine," also called "the black simultaneity of the universe," is another origin of—and another future for—both breathing and creating, rooted in the ontological sociality of "everything" and "the radical black porousness of love" joining "everything to everything."[113] Predating and postdating individualism, the dark feminine, like the wombs of Night and Chaos, generates outside of the dominant logics of creation, while also being the primordial material without which creation itself would be impossible.[114] In the postapocalyptic revelry of *Dub: Finding Ceremony*, breathing is rescaled (or scaled anew) to the illimitable oceanic of the dark feminine, its inhumanist joining of everything to everything:

the trees knew. the trees and the ferns and the moss and the lichen knew. the rocks knew and seemed to do nothing about it. . . . the bats knew with their sensitivity of shape, the mosquitos knew. . . . the whales sang

> of what it would mean. and the smallest plankton had to get ready after centuries of making life out of sun. the coral had to get ready after perfecting collaboration bright solid and grown. . . . once a species could do that to itself. do this to itself. what we do. get ready for the influx of carbon dioxide, the mountainous islands of trash. the unearned permanence of plastic. all life would have to change the meaning of breathing. they all knew.[115]

The ecological apocalypse intuited by "all life" (terrestrial, aquatic, forestial, and more) calls for a collective modulation of breathing into a new form, one capable of surviving the planetary damage wrought by the scaling of all life and all breath to the being of Man. Perhaps this modulation is a return to the grounding tonic of the earth's breath, as given in Woolf's/Stephen's early diary entry. Or perhaps it is a transformation of the "meaning of breathing" beyond humanistic imaginability, of the earth or anything else.

For Woolf/Stephen, such planetary breathing is thinkable in terms of the self-inspiration of earth: "If you lie on the earth somewhere you hear a sound like a vast breath, as though it were the very inspiration of earth itself, & all the living things on her."[116] This is also an inspiration oblique to thought, for Stephen allows that the "sound like a vast breath" may be both another sound and another breathing. And as suggested at the close of the previous chapter, the earth's inspiration may, finally, be imaginable as the autoaffectivity of the earth, the earth hearing itself breathing, gifting back to itself its own sensate givens in the form of inhumanly gathered noise, color, syllabling. Inhuman metamodernism, I offer in closing, may be the abandonment of thought and futurity to the earth's ongoing revelry in itself.

Is there such thing as an impersonal and inhumanist joy to be thought, a joy that inheres in the earth that is both an (inevitable) personification as well as something beyond the measure, scale, and breath of the person? The penultimate section from Clevidence's "[AGATE/ALGAE]" seems to suggest something of this possibility:

> *citrine & milky & rose, jasper, feldspar, lapis & flint, tigers-eye, olivine & smoky, calcite & tourmaline, onyx, beryllium, emerald, chalcedony & chert. slate slate slate salt ruby gold lead dead sea.*
>
> [what gaze were the colors for, then, what plumage, / display, nemesis of sight—or was it a joy][117]

An *epiphora* of color, ungraspable vividnesses, turns the gaze back on itself: the earth's lithic colors have become the "nemesis of sight." Yet what is color without a gaze? Or "a joy" that is not the possession of a (human or nonhuman) subject? Colors are "nemesis of sight" in at least two ways. First, as the antagonist of sight, or that which overwhelms the deterministic and possessive power of the gaze. And taking the origin of "nemesis" in the ancient Greek *némein* ("retribution"), the colors of minerals would also be the giving back what is due to earth—the expressive noise of the earth's own colors—once both color and mineral are no longer enclosed within the humanistic conceptuality of the gaze, the imperialism of eyes.[118]

The earth's reception of its own sensate givens, of the colors and colorful minerals it has birthed, is one way to imagine (with) the other side of oblivion. Clevidence's lines themselves are a modulation, from the celestial to the terran, of Eve's cosmic riddle in Book IV of *Paradise Lost*, as discussed in chapter 4: "But wherefore all night long shine these, for whom / This glorious sight, when sleep hath shut all eyes?"[119] In the cession of gaze, Eve seems to intuit something of the inhumanist persistence of night, before her wayward thoughts are once more captured by the visual and conceptual regime of Man. In Adam's explanation,

> These have their course to finish, round the Earth,
> By morrow evening, and from land to land
> In order, though to nations yet unborn,
> Minist'ring light prepared, they set and rise;
> Lest total darkness should by night regain
> Her old possession, and extinguish life
> In nature and all things [120]

Yet Eve's *ainigma* insists beyond the determinations by which it is solved in the name of a particular humanism: starlight, for Adam, is a claim to possession of the sky that counters that of Night. Eve, however, asks after a beyond to possession itself, a beyond made of something other than warring sky-conceptualities, in which what is given is given freely and indifferently to orders of meaning. (A beyond and a seeing by which to set free the stars, in Kafka's terms.) The joyous impersonality of Eve's "glittering starlight" is overwritten by the "Minist'ring light" of a fixed and ordered cosmos refracting Man's image into the dark places of the earth.[121]

If we think *ainigma* according to its Aristotelian definition as a "an impossible combination of names," then Clevidence's lithic *epiphora* might be another irresolute riddle, the glittering of earth colors beyond the capture of the gaze: "*citrine & milky & rose, jasper, feldspar, lapis & flint, tigers-eye, olivine & smoky, calcite & tourmaline, onyx, beryllium, emerald, chalcedony & chert. slate slate slate salt ruby gold lead dead sea.*" In their fugitivity from both concept and determination, colors return once more to the earth, a relation that, for

Clevidence, achieves expression through the indefinite impersonality of "*a* joy." On the other side of oblivion, there is a terrifying joy that is not and has never been humanistically imaginable: a joy that, to paraphrase the late Deleuzian definition of immanence, "does not belong to a subject or depend on an object."[122] In Part III of the *Ethics*, to which we will turn for a final time, Spinoza suggests something of the impersonal dimensions of a joy that, like Substance itself, inheres in beings (or modes) while also preexisting and persisting beyond the beings that express it. In this way, joy, along with all of the affects (*affectus*), emanates from the world itself. In Spinoza's definition, "Joy is pleasure accompanied by the idea of something past, which has had an issue beyond our expectation [*Gaudium est laetitia concomitante idea rei praeteritae, quae praeter spem evenit*]" (*E III* DXVI; translation modified). Joy's is a multifaceted temporality: joy is grounded in the body's memory of its past affections, of what of the world has imprinted itself on both body and mind—which Spinoza defines as the common notion, and which we might think as intergenerational and interspecied memory—but joy also exceeds what both bodies and minds can contain. Joy, in other words, is also a creation—also a futurity—that issues forth beyond the constraints of (mental and bodily) imagination, beyond the small terms of our "expectation" or "hope" (*spem*).

An excessively historied and futured joy, then, may be a modality through which to imagine after—in the sense of taking inspiration from—the ending of a world. After thought is another breathing, scaled to the level of the planetary. Entangled forest and ocean, noise worlds and color worlds made in and through earth, the glittering of mineral and starlight: gathered, these amount to something like a planetary breathing, the vast sound obliquely intuited by Virginia Stephen and reverberating beyond extinction. And the joy of the earth, in closing, is perhaps the ongoingness of breathing, a multispecied breathing shared in common by "all the beings upon her" (Woolf) or "all life" (Gumbs), and a breathing that might inspire *adamah*—the breathed-into sediments, minerals, and colors of earth—into forms and futures otherwise than Man.

Gumbs's poem, "cycle," gives the post-(humanist-)apocalyptic ongoingness of earth in terms of an Eve-like attunement to the inhumanism of the cosmos:

> *what the star said to the earth:*
>
> sometimes i wish i was like you. solid and findable. think and round. cool and stable. sometimes i look at you when you don't think i'm looking at you. you are there in the dark of me and i see mysteries that make me want to sing. but would you hear?

> *what the earth said to the star:*
>
> sometimes i wish i was like you. bright and hot. fast as light and always lit. seen even after you are gone. . . . sometimes i listen to you when you think i am not listening to you and i don't know how to feel about your song. so i tilt. so i keep spinning. so i stay around.[123]

Gumbs locates metamodernist *atopos* somewhere between singing star and sung-into earth, or in the co-vibration between seen earth and heard star. And this conversation (from the Latin *con* + *vertere*: "to turn together") itself is a turning toward modernist impersonality, a resonance with the enigmatic syllabling of what the thunder once said. For a similar song is given in the eponymous section of Eliot's *The Waste Land*:

> In this decayed hole among the mountains
> In the faint moonlight, the grass is singing
> Over the tumbled graves, about the chapel
> There is the empty chapel, only the wind's home.[124]

As in the ("empty") Noble Barn of *Between the Acts*, humanism's ruins allow inhumanist expressivities to flourish. Gumbs transposes grass song—sounding into the human oblivions of "tumbled graves" and "empty chapel" at night for Eliot—into its inverse: the song of a star sung into earth. In seeking an affective comportment toward the *ainigmata* of star speech and song, the earth becomes incongruous with itself, opened to an affectivity both beyond itself and beyond all understanding. We might then think of this relation as a kind of inhumanist apostrophe in which earth and star interanimate one another, and in so doing attempt to constitute themselves through the gaze or the audition of the other.[125]

Yet the earth perhaps wants not self- or subjective constitution from this relation but rather an ongoing striving toward what in itself is indestructible: the terran mysteries that inspire star song, yet remain opaque to both star and earth. "so i tilt. so i keep spinning. so i stay around": the earth's expressivities, its modes of turning and stilling, are what creatively and indifferently insist through humanist and other oblivions. The earth metaphorizes itself—and its own mysteries—to itself through the medium of cosmic song, the inconceptual generativity of night as it passes from star mouth to earth ear. What name to be given to such an inhumanist *symbolon*, made of the self-unfolding of the earth in relation to what of the cosmos has been unfolded into it? A metaphorics of joy, perhaps, joining the *metaphora* of transfer or passage from cosmos to earth with the *epiphora* of what grows and gathers in excess of all imaginable extension and thought, a joy immenser than what even the earth can hold.

Notes

Introduction

1. Richards, *The Philosophy of Rhetoric*, 94.

2. Aristotle, *Poetics* 1457b1: 7–9. Quintilian adopts a version of the same argument in the *Institutio Oratoria*, where he divides metaphor into four distinct categories: (1) metaphors that "substitute one living thing for another"; (2) metaphors that substitute "inanimate things . . . for inanimate [things]"; (3) metaphors that substitute "inanimate things . . . for animate [things]"; and (4) metaphors that substitute "animate . . . for inanimate [things]" (bk. 7, chap. 6, 305–7). Quintilian adds that these four could be divided into "more species," such as "transference from rational beings to rational and from irrational to irrational and the reverse, in which the method is the same, and finally from the whole to its parts and from the parts to the whole" (bk. 7, chap. 6, 307).

3. Kirby, "Aristotle on Metaphor," 532.

4. Kirby, "Aristotle on Metaphor," 532.

5. Kirby, "Aristotle on Metaphor," 532. Additionally, in medical terminology, *epiphora* is the persistent overflow of tears.

6. Cervenak, *Black Gathering*, 3.

7. This notoriously uninterpretable text has generated a multitude of analyses that would be impossible to list here. Canonical early interpretations include Walter Benjamin's "Franz Kafka" (in which Odradek is "the form which things assume in oblivion" [133]) and Heinz Hillmann, *Das Sorgenkind Odradek*. Ian Thomas Fleishman's "The Rustle of the Anthropocene" (2017) compellingly puts into dialogue eco-critical and new materialist analyses of Odradek (primarily from Timothy Morton, Jane Bennett, and J. Hillis Miller). See also the essays in the volume *Kafka's Creatures*, especially Helms, "The Difficult Task of Being Real"; Reynaga, "Agents of the Forgotten"; and Bauer, "The Power of the Look."

8. Kafka, "The Cares of a Family Man" (trans. Edwin and Willa Muir), 428.

9. Kafka, "The Cares of a Family Man," 428.

10. Kafka, "The Cares of a Family Man," 428, 429 (translation modified).

11. Kafka, *Zürau Aphorisms*, 58. In Kafka's original: "Die Sprache kann für alles außerhalb der sinnlichen Welt nur andeutungsweise, aber niemals auch nur annähernd vergleichsweise gebraucht warden, da sie, entsprechend der sinnlichen Welt, nur vom Besitz, und seinen Beziehungen handelt." In Michael Hoffman's translation, cited here, "annähernd vergleichsweise" is translated as "metaphorically," which I have modified in line with Ernst Kaiser and Eithne Wilhelm's translation in their edition of the *Blue Octavo Notebooks*.

12. North, *The Yield*, 52.

13. Kafka, "Cares of a Family Man," 428. In Kafka's original: "Die Unsicherheit beider Deutungen aber läßt wohl mit Recht darauf schließen, daß keine zutrifft, zumal man auch mit keiner von ihnen einen Sinn des Wortes finden kann."

14. Aristotle, *Rhetoric* III.2.12 (trans. John Kirby).

15. Aristotle, *Poetics*, 1458a1: 26–27 (translation modified in accordance with Guy Stroumsa's translation in "Myth as Enigma" (272).

16. Deleuze, *Difference and Repetition*, 107.

17. Choi, "Early Whales Had Legs," n.p. Amidst a planetary future of glacier melt, uncontrollable sea-level rise, and ever-closer intimacies between the oceanic and the terran, as we will discuss further in chapter 6, humans might gradually morph once more into aquatic or hybrid aquatic-terran beings.

18. Specifically, from the Aboriginal people of the Victoria River region of Australia's Northern Territory.

19. Rose, "Multispecies Knots of Ethical Time," 129, 136.

20. Rose, "Multispecies Knots of Ethical Time," 136.

21. Friedman, *Planetary Modernisms*, 8.

22. The *Oxford English Dictionary* defines "inhuman" as "not pertaining to or in accordance with what is human, in form, nature, intelligence, etc.; not of the ordinary human type" (def. 2).

23. Colebrook and Weinstein, "Introduction," 5.

24. Colebrook and Weinstein, "Introduction," 5.

25. Colebrook and Weinstein, "Preface," xi. Jean-François Lyotard's canonical account in *The Inhuman* (1988) defines the liberatory potential of the inhuman through the figure of the child, which holds in itself a fundamental indeterminacy—an "unharmonizable" core, or a "senseless difference"—that also marks a radical openness to the future and a destabilization of thought as the making of (Enlightenment-humanist) sense. Adulthood is not the end of the child and its futures, but only a form in which the "traces" of the indeterminate and the nonsensical are held in provisional suspension (3). Thinking, then, becomes redefined as an openness to what cannot be thought, an act of "letting a givable come toward you" within a preexisting field or "world" of meaning, rather than the "selecting and tabulating of data" that is also identified with the predatory inhumanism of a late-stage, technocratic capitalist order (18).

Along the lines of Colebrook and Weinstein's critique, Lyotard still imagines the inhuman's relevance insofar as it more precisely limns the indeterminacy and radical openness/sublimity of human thinking, rather than that which exists *indifferently* to human being, thought, and determination. In contrast to Lyotard's liberatory and avant-gardist inhumanism, an inhumanism defined in terms of both indifference *and* indeterminacy asks us to consider what else happens in the universe—what other modes of thought, relation, and expression among nonhuman, inhuman, and nonliving entities—when what constitutes happening is not determined by its relation to human thought and human futures.

26. Yusoff, *A Billion Black Anthropocenes*, 5.

27. See J. Brown's *Black Utopias* as well for an interesting analysis of thinking "after the end of the world" as a means of articulating utopia. Brown defines utopia as "a

condition of being temporally estranged" that is "not accessible in standard linear time or in normative spaces" (8).

28. See, for example, Adkins, *The Modernist Anthropocene*; Hovanec, *Animal Subjects*; Rubenstein and Neuman, *Modernism and Its Environments*; Sultzbach, *Ecocriticism in the Modernist Imagination*.

29. "The warm comfort of being together," or perhaps more simply translated as "the warm togetherness" or "together-being."

30. V. Woolf, *A Passionate Apprentice*, 203.

31. On the intersections between multispecies studies and modernist literature, see (in addition to Adkins, Hovanec, Rubenstein and Neuman, and Sultzbach above), for example, De Bruyn, *The Novel and the Multispecies Soundscape*; Chow, *Nineteenth-Century American Literature and the Discourse of Natural History*; Danta, *Animal Fables after Darwin*; and McCarthy, *Green Modernism*.

32. As we will discuss shortly, these include such provisional epoch designations as Capitalocene, Plantationocene, Eurocene, Plasticene, and Chthulucene, among others.

33. Yeats, "The Song of Wandering Aengus," 48.

34. Van Dooren, Kirksey, and Münster, "Introduction," 3–4. See also the essays in Kirksey, *The Multispecies Salon*.

35. Dooren, Kirksey, and Münster, "Introduction," 2.

36. Tsing et al., "Introduction," M6.

37. *Oxford English Dictionary* 2nd ed. "Monster."

38. Kant, *Critique of Judgment*, 109. See Gigante, "Too Much Life," for a further analysis of the interaction between biological and aesthetic theories of the monstrous in the late eighteenth and early nineteenth centuries.

39. Tsing et al., "Introduction," M6.

40. Colebrook and Weinstein, "Introduction," 5. The term "critical life studies" is meant to serve as a gathering place for the theoretical "turns" and "studies" that are, in their words, "oriented around the continual modulations of the epistemology, ontology, and resituation of the status of the human" (2).

41. Barad, *Meeting the Universe Halfway*, 33.

42. Barad, *Meeting the Universe Halfway*, ix.

43. Barad, "Nature's Queer Performativity," 127, 129, 130. Barad also draws from Vicki Kirby's analysis of lightning's performativity, which Kirby theorizes as "an intriguing communication, a sort of stuttering chatter between the ground and the sky" in which "directional causality" is impossible to trace amidst the relational field opened between ground and sky in the space of the lightning flash (quoted in Barad, "Nature's Queer Performativity," 128).

44. Barad, *Meeting the Universe Halfway*, 185.

45. See, for example, Latour, *Politics of Nature*; Haraway, *Staying with the Trouble*; Alaimo, *Bodily Natures*; Kirksey, *The Multispecies Salon*.

46. Alaimo, *Bodily Natures*, 2.

47. Barad, *Meeting the Universe Halfway*, x.

48. Giraud, *What Comes after Entanglement?*, 2.

49. Giraud, *What Comes after Entanglement?*, 2; emphasis in the original.

50. Giraud, *What Comes after Entanglement?*, 10.

51. Giraud, *What Comes after Entanglement?*, 11.

52. In this sense, Tsing et al. define monstrosity also through its anthropocidal and ecocidal consequences in which modern industrial society's attempt to regulate monstrosity (by growing monocultural crops in plantations, using increasingly advanced forms of pesticides and preservatives, "raising supercharged breeds" in dairy and meat farms, burying nuclear and other forms of waste in the earth, etc.) paradoxically gives rise to "a new kind of monstrosity":

> Industrial campaigns exterminate impurities, undermining the co-ordinations that make life possible. . . . Our entanglements, blocked and concealed in these simplifications, return as virulent pathogens and spreading toxins. Industrial chemicals weave their way through our food webs; nuclear by-products sicken us not just through our human cells but also through our bacteria (Tsing et al., "Introduction," M4).

53. Giraud, *What Comes after Entanglement?*, 12.

54. Uexküll, *A Foray*, 54.

55. Uexküll, *A Foray*, 43.

56. Along similar lines, Ben de Bruyn in *The Novel and the Multispecies Soundscape* employs the term "acoustic contact zones" to theorize relations among (human and nonhuman) sound worlds (14).

57. Hovanec, *Animal Subjects*, 3. Hovanec expands on the connections between science and literature in the early twentieth century, which "coevolved" in "a common environment" (5) and "drew on the legacy of [late-Victorian] comparative psychology," a discipline that "aimed to understand human and animal minds from the inside" (160) and thus became associated with modernist strategies of defamiliarization and "a plurality of perspectives" (162). See also Herwig, "The Unwitting Muse," for an account of Uexküll's "subjective biology" on modernist literary and scientific figures, including Aldous Huxley, Rainer Maria Rilke, and Thomas Mann (571).

58. Hovanec adopts the definition of the subject at work in the discourses of nineteenth-century evolutionary biology and psychology as "any being capable of subjective experience." These fields additionally "emphasized continuity, not rupture, between humans and animals." Hovanec's historicist approach focuses on how British modernists engaged with this conception of subjectivity, even while pushing toward more contemporary theories (e.g., "intersubjectivity," "intercorporeality") that attempt to further complicate or to do away with prevailing notions of the subject (6).

59. Heidegger, *The Fundamental Concepts of Metaphysics*, 274; emphasis in the original.

60. Heidegger, *The Fundamental Concepts of Metaphysics*, 274, 198.

61. Heidegger, *The Fundamental Concepts of Metaphysics*, 198.

62. Heidegger, *The Fundamental Concepts of Metaphysics*, 311.

63. Blattner, "Ontology," 15. See also Calarco, *Zoographies*, for a further elaboration of the anthropocentrism of Heideggerian world. Calarco writes:

> Even if the animal has a relation with and access to the entities in its environment, this does not mean, Heidegger argues, that the animal and the human Dasein have the *same* relational and affective structure. In particular, no matter how rich and complex a given animal's world might be, that world never grants it access to another being *as such*, that is, to the Being of an entity. Only human Dasein is capable of relating to beings *as* beings—a tree *as* a tree, a dog *as* a dog. This "as" structure, which marks the uniquely human opening to world and Be-

ing, is something forever barred from animal life. And it is this "as" structure that the animal is deprived of, that the animal lacks, and that renders the animal poor in world. (25; emphases in the original.

64. Heidegger, *The Fundamental Concepts of Metaphysics*, 196.

65. Friedman, *Planetary Modernisms*, 52; emphases in the original.

66. Jameson, *A Singular Modernity*, 13.

67. Friedman, *Planetary Modernisms*, 52.

68. Doyle and Winkiel, *Geomodernisms*, 3.

69. Friedman, *Planetary Modernisms*, 186. Similarly, instead of reading postcolonial modernisms as either an outgrowth from or a belated response to Euro-American modernism, Friedman argues that the modernisms "produced in the matrix of empires in the long twentieth century" are "central and constitutive" rather than "adjunct to the modernities of the Western metropole" (216, 217).

70. Garcia, *Signs of the Americas*, 42.

71. Garcia, *Signs of the Americas*, 44–45.

72. I borrow this neologism from Moten, *In the Break*.

73. Friedman, it is important to note, defines the planetary through the terms of Deleuzian relationality: "A planetary epistemology of modernity . . . needs to be relational, delineating not a nominal set of characteristics (e.g., 'big' = x, y, and z), but rather defining one thing in relation to another (e.g., 'big' is not 'small'). It focuses on the dynamics of change, the self-consciousness of change, or even the illusion of change—a 'before' and 'after,' an 'old' and a 'new'" (*Planetary Modernisms* 56).

74. See Chakrabarty, "The Climate of History," as well as Baucom, "History 4°."

75. Crutzen and Stoermer, "The Anthropocene," 17–18.

76. See, for example, Davis and Todd, "On the Importance of a Date"; Hayman, "Future Rivers of the Anthropocene"; DeLoughrey, *Allegories of the Anthropocene*; Liboiron, *Pollution Is Colonialism*; King, *The Black Shoals*; Murphy, "Alterlife and Decolonial Chemical Relations"; Wynter, "Unsettling the Coloniality of Being"; Yusoff, *A Billion Black Anthropocenes*.

77. Tiffany Lethabo King coins the term "conquistador humanism" in *The Black Shoals* and defines it as "the crafting and sustaining of European human life and self-actualization through Black and Indigenous death" (84).

78. Chtulucene, Haraway continues, is "an ongoing temporality that resists figuration and dating and demands myriad names" and one in which "human beings are not the only important actors . . . with all other beings able simply to react" (51, 55). Rather, "the order is reknitted: human beings are with and of the earth, and the biotic and abiotic powers of this earth are the main story" (55).

79. Davis and Todd, "On the Importance of a Date," 763.

80. Yusoff, *A Billion Black Anthropocenes*. In the essay "Geologic Realism," Yusoff further defines "Black Anthropocenes" as "engagements with the Anthropocene that insist on a politics of recognition of the apocalypses of black and brown subjects within the ecocide of colonialism and Empire, and challenge the visibility of another inhuman history within the historicity of the ongoing coloniality of thought and relation in matter economies" (6). Yusoff's groundbreaking analyses of the imbrication of geology and European imperialism proceed in part from Sylvia Wynter's critique of "Man" as the universalization (and "overrepresentation") of one version of the human

(as white, Western, and male) that emerged primarily during the European encounter and ongoing conquest and imperialization of the New World. See Wynter, "Unsettling the Coloniality of Being." See also Karera, "Blackness and the Pitfalls."

81. See the essay collection *Anthropocene or Capitalocene? Nature, History, and the Crisis of Capitalism*, especially Crist, "On the Poverty of Our Nomenclature" and Moore, "The Rise of Cheap Nature."

82. In *Allegories of the Anthropocene*, Elizabeth DeLoughrey gives a provisional catalog of critical revisions of the term "Anthropocene": "Thus, a decade after the coining of the term 'Anthropocene' we begin to see the start of a robust dialogue about the origins of our environmental crisis—variously attributed to the dominance of capitalism (Capitalocene, Econocene, Necrocene), transatlantic empire (Plantationocene), patriarchy (Manthropocene), European/white settler colonialism (Eurocene), twentieth-century globalization and its regimes of disposability (Plasticene), or all of the above and their engagements with a frightening alterity (Chthulucene)" (22). Graham Redfearn cheekily coined the term "Trumpocene" in *The Guardian* in 2016 to give a name to a political era oriented around the convergence of populist rhetoric, conspiratorial thinking, anti-intellectualism, and climate change denial. See also Malcom Ferdinand's groundbreaking *Decolonial Ecology*.

83. Yusoff, *A Billion Black Anthropocenes*, xiv.

84. Yusoff, *A Billion Black Anthropocenes*, 2. In his discussion of the Atlantic slave trade in "Swim Your Ground," literary scholar Jonathan Howard provocatively queries regarding the 1,818,681 enslaved Africans thrown overboard during the transatlantic slave trade: "How much did the ocean have to rise to accommodate its nearly two million new residents? Statistically negligible I'm sure, but large masses always build momentum slowly, imperceptibly. Rise is rise" (311). We will return to the archival memory and alternative temporalities of the ocean in chapter 6.

85. Broswimmer, *Ecocide*, 9.

86. Broswimmer, *Ecocide*, 9. For Broswimmer, the development of language is the first of three critical stages of ecocide, the second being the development of agriculture, and the third the development of modernity, including the development of the nation-state and the capitalist mode of production.

87. Yusoff, *A Billion Black Anthropocenes*, xiii. As Tiffany Lethabo King notes in *The Black Shoals*, 1441 AD marks "the commencement of the Portuguese slave trade, as well as European voyages poised for the conquest of territory on the coast of West Africa" (1). In a similar vein, Elizabeth DeLoughrey writes of an Anthropocene already happened, one that has been "experienced through the violent processes of empire" and that only "continues because empire is process" (*Allegories of the Anthropocene*, 7). Her analysis builds from one of Davis and Todd's key arguments in "On the Importance of a Date": "The Anthropocene—or at least all of the anxiety produced around these realities for those in Euro-western contexts—is really the arrival of the reverberations of that seismic shockwave into the nations who introduced colonial, capitalist processes across the globe in the last half-millennium in the first place" (774).

88. The Fertile Crescent agriculture of Mesopotamia (along with Egypt, India, China, and Mesoamerica), according to Broswimmer, is the second major phase in the "etiology" of ecocide (the third being the growth and spread of modernity, including the rise of the nation-state and the capitalist mode of production).

89. Broswimmer, *Ecocide*, 9.

90. Morton, *Dark Ecology*, 43.

91. Morton, *Dark Ecology*, 44.

92. Morton, *Dark Ecology*, 63.

93. As Yusoff writes in *A Billion Black Anthropocenes or None*, "The division between the figures of the human and inhuman and its manifestations in subjective life exhibits one of the most terrible consequences of the division of materiality organized and practiced as a biopolitical tool of governance. The division of matter into nonlife and life pertains not only to matter but to the racial organization of life as foundational to New World geographies" (5).

94. Morton, *Dark Ecology*, 47.

95. Morton, *Dark Ecology*, 62.

96. Morton, *Dark Ecology*, 62.

97. See Rob Nixon's foundational *Slow Violence and the Environmentalism of the Poor* for an elaboration on these intersections.

98. Yusoff, *A Billion Black Anthropocenes*, 58–59, 85.

99. Morton, *Dark Ecology*, 81.

100. Quoted in Morton, *Dark Ecology*, 81.

101. Jacques Derrida, *Glas*. See also Nealon, *Plant Theory*, for further analysis of plant and flower being in Derridean theory.

102. Derrida writes, "The furrow of agriculture, we remind ourselves, opens nature to culture (cultivation). And one also knows that writing is born with agriculture which happens only with sedentarization" (quoted in Morton, *Dark Ecology*, 81).

103. Haraway, *Staying with the Trouble*, 54.

104. Derrida, quoted in Morton, *Dark Ecology*, 80.

105. James and Seshagiri, "Metamodernism," 88.

106. James and Seshagiri, "Metamodernism," 92.

107. M. Brown, "Periods and Resistances," 316.

108. Quoted in M. Brown, "Periods and Resistances," 311.

109. Hayot, *On Literary Worlds*, 154.

110. James and Seshagiri, "Metamodernism," 92.

111. Tung, "Baddest Modernism," 519.

112. In *Archaeologies of the Future*, Fredric Jameson claims that science fiction itself is a "modernist practice" because it attempts "to defamiliarize and restructure our experience of our own present" by providing us with contingent futures that follow from this present and "exposing how that present is deeply historical" (quoted in Tung, "Baddest Modernism," 519).

113. Charles Tung notes that Virginia Woolf and Olaf Stapledon enthusiastically read each other's works and corresponded in the late 1930s about Stapledon's *Star Maker* (1937) and Woolf's *The Years* (1937) (Tung, "Baddest Modernism," 528).

114. Quoted in Tung, "Baddest Modernism," 530–31.

115. McGurl, "Posthuman Comedy," 537. Similarly, Aaron Jaffe claims that one of the governing problems (if not *the* problem) for modernity is "the discovery of inhuman scale and the technical difficulties of optimizing proportional forms of access." Jaffe, "Introduction," 505.

116. McGurl, "Posthuman Comedy," 502.

117. Colebrook and Weinstein, "Preface," xix.

118. Colebrook and Weinstein, "Preface," xviii. See also Beckman, "Posthumanism and Narrativity," in which Beckman argues that the "post" of "posthuman" is not a "new step in evolution" of the human but rather a concept that can in turn be used to rethink the conceptual foundations of the human (51). See also Braidotti, "Posthuman Critical Theory," for an overview of anti-humanist and anti-anthropocentric posthumanist theory from the 1970s to the present. This is a lineage that includes thinkers from Michel Foucault, Jacques Derrida, Judith Butler, and Gilles Deleuze and Félix Guattari to Edward Said, Gayatri Spivak, Donna Haraway, Giorgio Agamben, Cary Wolfe, and Timothy Morton, along with Braidotti herself. Braidotti, following Agamben, notes that "the 'human' is not a neutral term, but rather one that indexes access to privileges and entitlements through processes of 'humanization' ('normalization') that are driven by and enforce power relations" (15).

In *What Is Posthumanism?* Cary Wolfe similarly argues that in contrast to a "humanist posthumanism" and a "posthumanist humanism," a "posthumanist posthumanism" has to do with "understanding . . . the very redefinition of what humanistic knowledge is" (126). A 2020 special issue of *parallax* on "Animal Borderlands" attends to the historicization of the categories of "animal" and "human." In Dominic O'Key's words from the introduction, "By destabilizing the supposedly fixed, supposedly inevitable categories of species difference—that is, by fracturing the opposing identities of humanity and animality—animal studies disrupts the polar opposites of the human and the animal, and, in doing so, conceives of human/animal difference as existing within shared borderlands, borderlands that are under constant negotiation, navigation and contestation" ("Animal Borderlands," 352).

119. In addition to Wynter's "Toward the Human/after Man" and Yusoff's *A Billion Black Anthropocenes*, see also, for example, Gumbs, *Spill*, *M Archive* (to be discussed in chapter 6); Jackson, *Becoming Human*; King, *The Black Shoals*; Leong, "The Mattering of Black Lives"; Sharpe, *In the Wake*; da Silva, *Unpayable Debt*.

120. Colebrook and Weinstein, "Introduction," 6; Haraway, *Staying with the Trouble*, 32.

121. The *Oxford English Dictionary* defines "inhuman" as "not pertaining to or in accordance with what is human, in form, nature, intelligence, etc.; not of the ordinary human type" (def. 2). See also Colebrook and Weinstein, "Introduction," where they define the "inhuman" as "a departure from the posthuman": "Sidestepping the *post*- and the inevitable reuptake of the human vestiges that the post- is meant to surpass, *inhuman* orients us to all that is not human, not just that which comes after the human" (5).

122. North, *The Yield*, xviii. Kafka's protagonists, North continues, "decline to use power, dismiss the *Ubermensch*, and would rather stay under the thumb of a master than become one" (xviii).

123. I am inspired here by the critical methodologies of Katherine McKittrick and Tiffany Lethabo King, the latter of whom theorizes British imperialist mapmaking projects in the eighteenth century as means of "exercis[ing] domination over irrational and sensual beings such as Black and Indigenous others existing at the margins of humanity" (King, *The Black Shoals*, 76). For King, a Black counter-geography emerges when we attend to the forms of what McKittrick calls the Black livingness that the

imperial map both makes present and tries to capture within a white cartographic imagination. King follows McKittrick's call to "betray the archive," to read maps and other imperialist archives for "what else happened," and to "shift . . . the analytical frame away from the suffering Black body and toward 'corelational texts, practices and narratives that emphasize black life'" (King, *The Black Shoals*, 78). I find the question of the "what else happened?" to be an extremely productive means toward greater attentiveness to what of our worlds and histories—what forms of relation, intimacy, and co-creation—both escapes and exceeds colonial, archival, and Western humanist gazes.

124. My understanding of Spinozist expression is indebted to (and inflected by) Deleuze's works *Spinoza: Practical Philosophy* and *Expressionism in Philosophy: Spinoza*.

125. Spinoza, *Ethics*, 75. Hereafter cited as *E* with accompanying Book (I–V) with Proposition (P), Demonstration (D), or Scholium (S).

126. Clevidence, *~~Flung~~ Throne*, 14.

127. Kafka, *Zürau Aphorisms*, 105.

128. Kafka, *Letters to Felice*, 447 (March 3, 1915).

129. In his analysis of Henri Bergson's *Laughter*, to be discussed in chapter 5, Mark McGurl writes, "Comedy can only be human, but it is also true that it arises only in the relation of humanity to something other" ("The Posthuman Comedy," 549).

130. McGurl, "Posthuman Comedy," 537.

131. DeLoughrey, *Allegories of the Anthropocene*, 4.

132. DeLoughrey, *Allegories of the Anthropocene*, 4.

133. James and Seshagiri, "Metamodernism," 88.

134. Gumbs, *Dub*, 198.

1. Growth

1. Heidegger, *The Fundamental Concepts of Metaphysics*, 307.

2. Heidegger, *The Fundamental Concepts of Metaphysics*, 307.

3. Heidegger, *The Fundamental Concepts of Metaphysics*, 308.

4. Heidegger, *The Fundamental Concepts of Metaphysics*, 308; italics in original.

5. Heidegger, *Ontology: The Facticity of Hermeneutics*, quoted in Elden, "Reading *Logos* as Speech," 285. Elden notes that Heidegger's insistence on defining *zoon logon echon* as languaging animal is meant to counter what Heidegger believed to be Roman antiquity's mistranslation of *logos* as *ratio* and "man" as *animale rationale*. It is language and the capacity for common discourse rather than "rationality" or "reason" that defines human being, according to Heidegger (285–86).

6. Kafka, "The Cares of a Family Man," 429.

7. In *Zoographies*, Matthew Calarco claims that Heidegger "fails to *think* the anthropocentric ground of metaphysics and the concept of subjectivity that flows from out of this ground" (33).

8. On the anthropocentrism of animal metaphor, Susan McHugh, for example, claims animal metaphor as "a strong defense for poetics in the service of anthropocentrism" and a trope that is "unable to bear animal agency" (*Animal Stories*, 7–8). In *Stalking the Subject: Modernism and the Animal*, Carrie Rohman similarly dismisses

metaphor as an inevitably anthropocentric tool of comparison. These claims are preceded by Deleuze and Guattari's assessment, in *Kafka: Toward a Minor Literature*, that "metaphor" carries with it a "whole anthropocentric entourage" that "threatens" to fix writing into a limiting "Oedipal" structure (36).

9. Nealon, *Plant Theory*, 50.

10. Deleuze, *Spinoza*, 44–45.

11. Taylor, *Beasts of Burden*, 57–58.

12. Low et al., "Cambridge Declaration," 2.

13. Low et al., "Cambridge Declaration," 2.

14. In relation to this idea, N. Katherine Hayles, in *How We Became Posthuman*, discusses how consciousness itself is a material product that is not solely a human possession and does not "guarantee the existence of the self" but rather is distributed among a variety of human, nonhuman, and cybernetic entities. Hayles writes:

> The construction of the posthuman is . . . deeply involved with boundary questions, particularly when the redrawing of boundaries changes the locus of self-hood. Shift the seat of identity from brain to cell, or from neocortex to brainstem, and the nature of the subject radically changes. . . . Conscious mind can be hijacked, cut off by mutinous cells, absorbed into an artificial consciousness, or back-propagated through flawed memory. . . . Whether consciousness is seen as a precious evolutionary achievement . . . or as an isolation room whose limits we are ready to outgrow, we can no longer simply *assume* that consciousness guarantees the existence of the self. In this sense, the posthuman subject is also a postconscious subject. (279–80)

15. As we will explore further in chapter 3, Mel Chen employs the term "animacy hierarchy" to describe this kind of ordering and valuing of being according to (socially and culturally determined) degrees of livingness or activity. Chen, *Animacies*, 2.

16. Serres, *The Five Senses*, 155.

17. For example, try not to cringe from the seemingly self-unaware classism of this sentence, from the same chapter, "Tables": "I have lived downwind of a fast food restaurant long enough to know how disgusting it is to be lacking in culture" (Serres, *The Five Senses*, 166).

18. Serres, *The Five Senses*, 163.

19. Serres, *The Five Senses*, 155.

20. Serres, *The Five Senses*, 161.

21. In Spinoza's original: "Deinde nemo scit, qua ratione quibusve mediis mens moveat corpus, neque quot motus gradus possit corpori tribuere, quantaque cum celeritate idem movere queat."

22. "Mens se ipsam non cognoscit nisi quatenus corporis affectionum ideas percipit" (*E* III, P23).

23. V. Woolf, "Modern Fiction," 190; my italics.

24. Nietzsche, "On Truth and Lie in an Extra-Moral Sense," 46.

25. Nietzsche, "On Truth and Lie in an Extra-Moral Sense," 44, 43.

26. Uexküll, *A Foray*, 135.

27. Deleuze, *Difference and Repetition*, 131.

28. Deleuze, *Difference and Repetition*, 130, 131.

29. Deleuze, *Difference and Repetition*, 131.

30. Deleuze, *Difference and Repetition*, 14.

31. Deleuze, *Difference and Repetition*, 136.

32. In Deleuze's original: "Qu'est-ce qu'une pensée qui ne fait de mal à personne, ni à celui qui pense, ni aux autres?"

33. Aristotle, *Rhetoric* III.10.2.

34. Aristotle, *Rhetoric* III.2.12.

35. Braidotti, *The Posthuman*, 70.

36. Spencer, *The Principles of Biology*, 205.

37. See also Michael Marder's *Plant-Thinking: A Philosophy of Vegetal Life*, which similarly argues, broadly, that plant life offers a deconstruction of the foundations of Western metaphysics, and Joseph Anderton, "Vegetating Life and the Spirit of Modernism in Kafka and Beckett."

38. In "The Animal That Therefore I Am," Jacques Derrida notes that in the Old Testament story of Abel and Cain, "Cain, the older brother, the agricultural worker, therefore the sedentary one, submits to having his offering of the fruits of the earth refused by God who prefers, as an oblation, the first-born cattle of Abel, the rancher" (410).

39. Quoted in Nealon, *Plant Theory*, 67, 66.

40. Nealon, *Plant Theory*, 68. In Hegel's terms, *schlecht Unendlichkeit* (also translated as "spurious infinity") and *wahre Unendlichkeit* ("good" or "true" infinity). See Wayne M. Martin's analysis of Hegelian infinity (in *Science of Logic* [1812, 1816]) in "In Defense of Bad Infinity."

41. Similarly, according to Elaine Miller, in *The Vegetative Soul*, Aristotle's "notion of the vegetative soul" in *De Anima* is "one component of the irrational part of the soul, namely that component responsible for simple growth and alteration" (187).

42. Hunter, *Essays and Observations*, 242.

43. As outlined in the introduction, the term "multispecies" here is meant to apply outside of the category of the living, in line with its contemporary ecocritical usage as applying to the abiotic and "nonliving" rather than in its strict Western scientific definition as dealing exclusively with what are considered living species. As van Dooren et al. write in their introduction to "Multispecies Studies," "This immersive approach [of multispecies studies] is now also increasingly being applied to forms of liveliness that many, but by no means all, of us would consider to be nonliving: from stones and weather systems to artificial intelligences and chemical species" (4).

44. See endnote 37.

45. Gaudreault, *From Plato to Lumière*, 72. Early cinema showings, particularly during the era of what Tom Gunning calls the "cinema of attractions" (the most prominent cinematic form until about 1907), often featured an onstage presence who directed the audience's attention to the film without commenting on, framing it, or otherwise explicitly mediating it. This monstrative as opposed to narrative function, Gaudreault argues, persisted well after the cinema of attractions era and has continually evolved into new forms across the development of cinema in the twentieth and twenty-first centuries. See Gunning, "The Cinema of Attractions," 64.

46. Ricoeur, "The Metaphorical Process," 145.

47. Wheelwright, *Metaphor and Reality*, 73.

48. Wheelwright, *Metaphor and Reality*, 78.

49. Quintilian, *Institutio Oratorio*, book VIII, chap. 6, 303; italics in original.

50. Nealon, *Plant Theory*, 68. While a proper analysis of the Hegelian infinite is far beyond the scope of this book, Wayne M. Martin's introductory overview of the good and bad/spurious infinite in Hegel's *Science of Logic* provides a helpful grounding. For Martin, the Hegelian bad infinite is an indefinite ongoingness, such as the movement of the mathematical number series toward an infinity, in excess of countability. Martin writes, "In this idea of 'going on and on and on' into an infinitely receding distance, we find one of the hallmarks of the bad infinite as Hegel understands it. To say that some thing is in-finite, *unendlich*, seems to mean that there is no getting to the end of it. One can go indefinitely far along the series—as long as one has the will and patience to go—yet one will never get to the end of the line" ("In Defense of Bad Infinity" 4). The infinity in question here, however, also only achieves its infinite nature by *negating* finitude rather than being unconditionally infinite or infinite in itself. This infinity, in other words, is only definable in opposition to the finite and so is no true infinity at all. When we try to define the infinity of the number series, we find that what numbers extend toward is not really infinity itself but rather just a further expanse of finitude: the number series extends to more finite numbers. In the *Science of Logic*, Hegel writes, "This infinity which is perpetually determined as the beyond of the infinite is to be described as the bad qualitative infinite" (136).

By contrast, the good or true infinite can be thought of as a self-determining infinity but one that is located *within* finite things, and in the self-relation of finite things rather than in a separate, universal sphere. In the *Science of Logic*, Hegel writes, "It is not in the sublating of finitude in general that infinity in general comes to be; the truth is rather that the finite is only this, through its own nature to become itself the infinite. The infinite is its *affirmative determination*, that which it truly is in itself" (138). In Hegelian philosopher Andrew Davis's gloss of (true) infinity's immanence in "Hegel's Idealism," "infinity *belongs* to the being of finite things themselves. Infinity is not a loss of individual being for the sake of absorption in abstract generality. . . . No finite thing is merely finite. It holds to its identity . . . and is thus related to its own finitude and is thus infinite" (183). The self-relationality of the "I," in this sense, is what is truly infinite: "The infinite is that activity through which I become what I am by transcending the limitation and opposition of what is not-me. I am infinite as a potentially limited being that overcomes limitation and relates to itself by means of infinite (circular) activities," such as "digestion, respiration, circulation, growth, decay, [and] thinking" (183). Similarly, Martin gives the Möbius strip as an example of good infinite self-relation: "A Möbius strip is twisted band, easily constructed from a strip of paper by closing it into a circular form after introducing a single twist. Why should we think of it as infinite? Well, like Hegel's circle it has no ends. But it is also mathematically distinctive in being a closed figure that has no inside and no outside. Hence in several respects it seems fitting to describe it as 'end-less.' The crucial thing, however, is that it does not go on and on and on forever. It is a totality; the whole thing is here present on my writing table. So we need not think of in-finitude (the absence of ends) in terms of the schema of trailing off endlessly into the distance" (5). Rather than infinity being defined as the negation of finitude, true infinity is more appropriately thought as the self-relational activity of a finite thing "negating its own negation" and thus affirming the infinity within itself. In Hegel's words, "It is the very nature of the finite thing to transcend itself, to negate its own negation and to become infinite" (*Science of Logic*, 138).

51. Hegel, *Aesthetics*, 404.

52. Davidson, "What Metaphors Mean," 46.

53. Davidson, "What Metaphors Mean," 46–47.

54. Davidson, "What Metaphors Mean," 37.

55. Davidson, "What Metaphors Mean," 47

56. Ricoeur, *The Rule of Metaphor*, 293.

57. Ricoeur, *The Rule of Metaphor*, 293.

58. Kafka, *The Metamorphosis*, trans. Muir and Muir, 91. See also Kata Gellen's analysis of this scene in chapter 1 of *Kafka and Noise*.

59. Kafka, *The Metamorphosis*, trans. Muir and Muir, 98; translation modified.

60. After Gregor's death, his body becomes completely legible to the Samsa family and fully containable within the terms of referential description: "As a matter of fact, Gregor's body was completely flat and dry; this was obvious now for the first time, really, since the body was no longer raised up by his little legs and nothing else distracted the eye." Kafka, *The Metamorphosis*, trans. Corngold, 40.

61. Kafka, *Die Verwandlung*, 54; my translation.

62. Kafka, *The Metamorphosis*, trans. Corngold, 40.

63. Kafka, *The Metamorphosis*, trans. Corngold, 40.

64. See Chris Danta, *Animal Fables after Darwin*, for a detailed analysis of metaphors of upright posture as a means of distinguishing humans from animals, particularly in the modernist fable tradition.

65. Kafka, *The Metamorphosis*, trans. Corngold, 91.

66. Derrida, "White Mythology," trans. Bass, 214. In *The Philosophy of Rhetoric* (1936), Ivor A. Richards makes a similar point about the function of metaphor exceeding the role of exemplum of some preexisting reality. Richards argues that "language," and metaphor particularly, does not just "hand over sensations bodily" or merely represent phenomena that we have already experienced but itself produces the reality of experience and produces "the meaning which the image and the original perception lack" (150, 151). Richards is arguing against eighteenth-century perspectives on metaphor, such as Lord Kames's, which held that the goal of a metaphor was to set the "principal subject" of the metaphor (i.e., the tenor) in "a strong light" through the use of the "figure" (the vehicle) used to describe it (99). This and similar arguments about metaphor at the time, Richards argues, "show . . . the eighteenth-century assumptions that figures are a mere embellishment or added beauty and that the plain meaning, the tenor, is what alone really matters and is something that, 'regardless of the figures,' might be gathered by the patient reader" (99–100). In contrast, Richards argues that the vehicle is never just an "embellishment" that helps us arrive at a given meaning (tenor) but that meaning is produced out of the interaction between vehicle and tenor: "The vehicle is not normally a mere embellishment of a tenor which is otherwise unchanged by it but that vehicle and tenor in co-operation give a meaning of more varied powers than can be ascribed to either" (100).

A major difference between Richards's and Derrida's accounts, however, lies in Richards's belief in the possibility of attaining a more thorough knowledge of how metaphors work and how they condition our subjectivity, our mental processes, with the hope that if the "causes and modes of the misinterpretation of words" are located, this can help us to "throw light upon and suggest a remedial discipline for more grievous disorders; that, as the small and local errors in our everyday misunderstandings with language are models in miniature of the greater errors which disturb the development of our personalities, their study may also show us more about how these large scale disasters may be avoided" (Richards, *The Philosophy of Rhetoric*, 136–37). Derrida, on the other hand, sees no such possibility of attaining what Richards, following Aristotle, calls "the command of metaphor" since language is resolutely beyond the command of human understanding, and the multifarious production of meanings out of the metaphors we employ is not something we can consciously or meaningfully direct.

67. Ricoeur, *The Rule of Metaphor*, 231.

68. Nealon, *Plant Theory*, 50.

69. Canguilheim, quoted in Derrida, "White Mythology," trans. Alan Bass, 261. See also George Lakoff and Mark Johnson, *Metaphors We Live By*, for further elaboration on this theme.

70. Blumenberg, *Shipwreck with Spectator*, 81, 83.

71. Blumenberg, *Shipwreck with Spectator*, 83.

72. Blumenberg, *Shipwreck with Spectator*, 83.

73. Blumenberg, *Shipwreck with Spectator*, 82.

74. Aristotle, *Rhetoric* III.2.12 (trans. Kirby).

75. Aristotle, *Poetics*, 1458a1: 26–27.

2. Sediments

1. Dolar, *A Voice and Nothing More*, 141.

2. Deleuze, *The Logic of Sense*, 79.

3. The sixth edition of the *Origin* was considerably revised in response to public criticism and published in 1872. Unless otherwise noted, all citations of the *Origin* are from the first edition of 1859.

4. Kant, *Critique of Judgment*, 109. See also Denise Gigante, "Too Much Life: Romantic Monstrosity in *Frankenstein*," for a further analysis of the interaction between biological and aesthetic theories of the monstrous in the late eighteenth and early nineteenth centuries.

5. Hunter, *Essays and Observations*, 239. Palmira Fontas de Costa in "The Meaning of Monstrosities in Charles Darwin's Understanding of the Origin of Species" notes that although Hunter's volume was only published posthumously in 1861, "Darwin's notebooks on the transmutation of species show that he knew [Hunter's] work through Richard Owen (1804–1892)" (77).

6. Hunter, *Essays and Observations*, 246.

7. Darwin actually excised this passage from the second edition of the work (but added it back to subsequent editions) after a number of his detractors cited it as evidence of the absurdity of the proposition that existing species have evolved from remote ancestors instead of being individually created.

8. J. Woolf, *The Mysteries of Lewis Carroll*, 191.

9. W. Smith, "Dodgson's Scientific Interests," 8.

10. Cohen, *Lewis Carroll*, 350; Stern, *Lewis Carroll's Library*, 17.

11. Dodgson's diary entries from 1853 to 1863 were unfortunately destroyed after his death, but other documentary evidence points unmistakably to his presence at the debate. Morton Cohen claims that Dodgson "served on the reception committee for the men of science from foreign countries and distant parts of the UK," making it extremely unlikely that he did not attend it (350). More directly, Jenny Woolf, in *The Mystery of Lewis Carroll*, records that Carroll "paid a considerable sum to attend the seminal debate on evolution between Thomas Huxley and Bishop Wilberforce" on June 30, 1860 (191). Finally, Roger Taylor and Edward Wakeling, in *Lewis Carroll: Photographer*, confirm Dodgson's attendance at the debate by pointing out that he took several signed and dated photographs of important figures who were present at the debate, including Wilberforce and Huxley, on June 30, 1860 and the previous and subsequent days of the weeklong event at Oxford Natural History Museum (41).

12. Freud coined the term "Darwinian trauma" in his 1917 essay "A Difficulty in the Path of Psycho-Analysis" (140).

13. These phrases come from an anti-vivisection letter written in 1875 to the editor of the *Pall Mall Gazette*, titled "Vivisection as a Sign of the Times," which blames "man's" assumed superiority over "the animal worlds" for economic, social, and gender inequality: "The world has seen and tired of the worship of Nature, of Reason, of Humanity; for this nineteenth century has been reserved the development of the most refined religion of all—the worship of Self. For that indeed, is the upshot of it all. The enslavement of his weaker brethren—'the labour of those who do not enjoy, for the enjoyment of those who do not labour'—the degradation of woman—the torture of the animal worlds—these are the steps of the ladder by which man is ascending to higher civilization" (Collingwood, *The Life and Letters of Lewis Carroll*, 170).

14. Kant, *Critique of Pure Reason*, 124.

15. Kant, *Critique of Pure Reason*, 124.

16. Nonsense generally is a well-worn concept in studies of Lewis Carroll's work, most of which stem from Elizabeth Sewell's classic theoretical account, *The Field of Nonsense* (1952). According to Sewell, what we term "nonsense" is simply "sense" deprived of an established framework of meaning: nonsense is "a carefully limited world, controlled and directed by reason, a construction subject to its own laws," but its particular order does not correspond with the order of "reality" (4–5). Another key text in this regard is Deleuze's *The Logic of Sense* (*Logique du sens*, 1969), which picks up on the sense-nonsense dichotomy from a slightly different angle. For Deleuze, as for Sewell, nonsense is simply what "says its own sense," or "produces" its own sense, without recourse to another frame of reference, or a higher authorizing category that would grant it meaning (in other words, a name cannot refer to its own class of being, such as "species," to attain meaning or "sense" but must refer to a "genus," the "genus" to a "family," and so on; a name that attempted to generate meaning without referring to an authorizing category outside of its own class would only produce nonsense) (79, 80). Sense also forms a continuum with nonsense, the latter being pure utterance prior to its fixing into propositional content. In a departure from Sewell, however, Deleuze theorizes "sense" as the arbitrary principle, or "doxa," by which all propositions and things are granted or denied significance ("sense" then also stands as an analogue to consciousness). In a more conservative vein, Francis Huxley's study of

Carroll, *The Raven and the Writing Desk* (1976), argues that nonsense is situated on the "limit" between what can be thought and what is "impossible" to think (10). The ultimate "aim" of nonsense is to work through the "self-contradictions" of language "to convert what lies on the other side of the reasonable limit" into understanding. For Huxley, unlike Deleuze, nonsense is "convert[ible]" into sense through the work of consciousness (10). Jean-Jacques Lecercle's *Philosophy of Nonsense: The Intuitions of Victorian Nonsense Literature* (1994) presents a more historically minded argument than either Sewell's, Deleuze's, or Huxley's. Lecercle reads nonsense as a "conservative-revolutionary genre" that is both "respectful of authority" in terms of its adherence to grammatical rules and structures and "joyously" subversive of these rules the same time (2). Nonsense thus both "supports the myth of an informative and communicative language and deeply subverts it" (3). For Lecercle, then, the emergence of nonsense as a literary and cultural genre can be traced to the rise of the Victorian school, as "the institution that develops the need for meaning and a reflexive attitude towards language, and channels them in socially acceptable ways" (4).

My account attempts to integrate the post- or inhumanist theoretical strain of Deleuze's analysis with the historicizing drive of Lecercle's. I read nonsense in Sewell's and Deleuze's terms as "sense" without established frameworks of meaning or reference and as that which resists integration with the determining force of consciousness. At the same time, I am interested in exploring the *Alice* books beyond their function within Deleuze's (anti-)philosophical project and, with Lecercle, looking at how Carroll's writing functions in relation to its historical milieu (even as I try to avoid the somewhat strict determinism of his account, which looks at nonsense primarily as a "by-product" of the educational system). As such, we attempt here a link between nonsense, as an inconceptual mode of expression, and the questions of evolutionary theory, anthropocentrism, fictions of development, and subjective identity that loomed so large in the author's mind and throughout the 1860s and 1870s generally.

17. Deleuze, *Logic of Sense*, 5.

18. Deleuze, *Logic of Sense*, 3.

19. Rackin, *Alice's Adventures*, 93.

20. Carroll, "'Alice' on the Stage," 294.

21. These word origins are given by Carroll in his letter of February 3, 1888 to the editors of the magazine *Jabberwock* (quoted in Cohen, *Lewis Carroll*, 443).

22. Carroll's illustrations are part of the *Alice's Adventures Under Ground* manuscript and do not appear in the published versions of the *Alice* books, which are illustrated by John Tenniel.

23. Carroll, "'Alice' on the Stage," 294.

24. The construction of species and varieties as categories also results from an inescapable anthropocentric bias on the part of scientists and observers, as Darwin notes: "I have been struck with the fact, that if any animal or plant in a state of nature be highly useful to man, or from any cause closely attract his attention, varieties of it will almost universally be found recorded. These varieties, moreover, will be often ranked by some authors as species" (125–26).

25. Wallace, "Introduction," 26.

26. My analysis thus departs from Gillian Beer's pathbreaking work *Darwin's Plots: Evolutionary Narrative in Darwin, George Eliot, and Nineteenth-Century Fiction* in which,

she argues analogy and metaphor are Darwin's primary methods for both expressing the "multifariousness" of a world no longer governed by the central animating "Idea" of theological design, and for finding a form in which to encapsulate this multifariousness (73–74). In Beer's analysis, analogy is based on "discovering structures common to diverse forms" and proceeds in search of a "complete resolution" that will irrevocably unite the two forms as structural parallels to "reveal actual identity" between them (74). Similarly, metaphor functions like hypothesis: metaphors are fictive speculations about a state of affairs that can only be proven true or false, or meaningful or meaningless, in futurity. As such, metaphors, like analogies, take on a fundamentally "narrative" function: their primary role is to breach formerly unexplored terrain as part of a "continuous truth-discovering process" (85). Metaphors are capable of producing new meaning, but this tendency toward "divergence" is checked by the ultimate stabilizing function of metaphor, and what Beer supposes is Darwin's desire to create "truth" in his writing, defined as a "connection between representation and actuality" (86). "Ideally," writes Beer, "Darwin would like to restore that which is metaphorically invoked to complete congruity with the material order, so that imagination is verified physically" (95).

Darwin's Plots assumes that establishing "a stable order" is the aim of Darwinian thought, that both analogy and metaphor work in the service of cementing a common identity among various forms of life, and that the *Origin*, with all of its emphasis on the unknowability and incomprehensibility of life and the natural world, is meant to serve as a "representation" of empirical reality. In Beer's words, Darwin "needs" "to establish ways in which language may be authenticated by natural order, so that his own discourse and argumentation may be 'naturalised,' and so moved beyond dispute" (49). What I am arguing, to the contrary, is that Darwin's *Origin* emphasizes the improvisatory unfolding of both beings and the language by which we understand being, for which traditional concepts and the logics of telos have proven inadequate to express.

I hesitate to affirm Beer's faith in scientific discourse—and the nineteenth-century novelistic form that she claims follows in its wake—as an "authoritative . . . substitute" for the "god-like omnipotence and omniscience open to the theistic narrator" that seeks to convert hypothesis into "truth" (149, 150). In my view, to model metaphor and the novel after scientific hypothesis as a mode of speculatively accessing a calculable notion of truth misses the concept-indifferent flux at the center of Darwin's (anti-)methodology / metaphorology, or his disestablishing of origin. Darwin denaturalizes both the metaphysical joining of words to beings and the status of scientific discourse as a representational model of the world and, in so doing, undermines the older models of truth, omniscience, and nature to which Beer's argument is still beholden.

27. Darwin, *The Descent of Man*, 154.

28. Darwin, *The Descent of Man*, 158, 173.

29. Carroll, "'Alice' on the Stage," 296.

30. Heidegger, *The Fundamental Concepts*, 321, 324.

31. Heidegger, *The Fundamental Concepts*, 321.

32. Nancy, *Listening*, 10.

33. Dolar, *A Voice and Nothing More*, 141.

34. *Oxford English Dictionary*: "Mock turtle."

35. Also confounded here is the hierarchical separation between the Gryphon's ancient mythological origin and the Mock Turtle's playful, pun-made origin: neither is privileged over the other and two monstrous forms coalesce into each other in Carroll's illustration, as I discuss below.

36. Attridge, "Unpacking the Portmanteau," 141.

37. Jacques Lacan's term for this unification of two incommensurate logics—language as signifier and language as sound—is *lalangue*. *Lalangue*, as Dolar explains in *A Voice and Nothing More*, holds these two logics in relation even as they remain differentiated from one another (143–44).

38. Culler, "The Call of the Phoneme," 22.

39. See footnote 21 above.

40. *Oxford English Dictionary*: "Jabber."

41. For a version of this argument, see Adriana Cavarero, *For More than One Voice: Toward a Philosophy of Vocal Expression* (2005). The definition of voice as an exclusively human possession goes back at least as far as Aristotle, as Mladen Dolar notes in *A Voice and Nothing More*. In *De Anima*, Aristotle claims the voice exclusively for ensouled beings: "Not every sound . . . made by an animal is voice . . . ; what produces the impact must have soul in it and must be accompanied by an act of imagination, for voice is a sound *with a meaning*" (quoted in Dolar, *A Voice and Nothing More*, 23). According to Michel Chion in *Sound: An Acoulogical Treatise*, there is a fluid continuum between sound and voice (rather than a human/nonhuman distinction), and the definition depends on which sounds or types of sound a certain community of listeners takes or refuses as significant. In Chion's succinct phrase, "Every sound to which one listens to for a long time becomes a voice" (51).

Additionally, in *The Fact of Resonance: Modernist Acoustics and Narrative Form*, Julie Beth Napolin makes the fascinating point that the idea of "determinacy" itself is linked to the possession of voice/*Stimme*: "The archaic meaning of determinacy (*Bestimmtheit*) indicates a voice, *Stimme*. To declare *das stimmt* is to say 'it is true' or 'it is so.' But beneath this reality is another: 'It has been voiced.' The meaning of declaration is to say 'it is.' A declarative voice is the voice in which determination speaks. To say 'it is' is to assume determinative power; it is also to say of some other quantity 'it is not'" (94).

42. This perhaps equally iconic image, and index of all biological and metaphorical production in the *Origin*, must have taken root in Carroll's consciousness at some point during his extensive studies of Darwinism.

43. Kant, *A Critique of Pure Reason*, 124.

44. Carroll, *Letters*, 1:589; italics in the original.

3. Dust

1. Rose, "Multispecies Knots," 136. Also see the introduction to this book, 8–10.

2. Foucault, *The Order of Things*, 18. *Convenientia* is the first of Foucault's four central modalities of resemblance, the others being *aemulatio*, analogy, and sympathy/antipathy. See the "Prose of the World" chapter and the "Four Similitudes" section of *The Order of Things*.

3. Foucault, *The Order of Things*, 19.

4. Breton, *Manifestos of Surrealism*, 20.

5. Breton, *Manifestos of Surrealism*, 37; italics in the original.

6. Breton, *Manifestos of Surrealism*, 36.

7. Flynn, *James Joyce*, 177.

8. From Breton's first *Manifesto of Surrealism*: "In my wake I raise up monsters that are lying in wait; they are not yet too ill-disposed toward me, and I am not lost, since I fear them. Here are 'the elephants with the heads of women and the flying lions' which used to make [Philippe] Soupault and me tremble in our boots to meet" (*Manifestos of Surrealism,* 40).

9. For a more detailed discussion of the stage directions in "Circe," see Katie Wales, "'Bloom Passes through Several Walls': The Stage Directions in 'Circe.'" The proto-surrealist character of Flaubert's *The Temptation* has been explored in works such as Victor H. Brombert's *Novels of Flaubert: A Study of Themes and Techniques* (1966). For a detailed account of Joyce's engagement with both surrealism and surrealist precursors (such as Flaubert and Aragon) in "Circe," see Scarlett Baron, *"Strandentwining Cable."*

10. Flynn, "'Circe' and Surrealism," 122.

11. Flynn, *James Joyce*, 176. In Apollinaire's original: "Il est juste que le dramaturge se serve / De tous les mirages qu'il a à sa disposition / Comme faisait Morgane sur le Mont-Gibel / Il est juste qu'il fasse parler les foules les objets inanimés / S'il lui plait / Et qu'il ne tienne pas plus compte du temps / Que de l'espace."

12. Baron, *"Strandentwining Cable,"* 146.

13. Foucault, introduction to *The Temptation*, xxix–xxx.

14. Bersani, "Flaubert," 42.

15. Bersani, "Flaubert," 42; italics in the original.

16. V. Woolf, *Diary Vol. III*. 203 (November 7, 1928).

17. For an interesting reading of blended vegetal-human-animal poetics and the "plastic figurations of metamorphosis" in 1930s surrealisms, see Joyce Cheng, "Mask, Mimicry, Metamorphosis: Rogier Caillois, Walter Benjamin and Surrealism in the 1930s."

18. From the closing chapter of the *Origin*: "It is interesting to contemplate a tangled bank, clothed with many plants of many kinds, with birds singing on the bushes, with various insects flitting about, and with worms crawling through the damp earth, and to reflect that these elaborately constructed forms, so different from each other, and dependent upon each other in so complex a manner, have all been produced by laws acting around us" (*OS* 397–98).

19. Acts of the Apostles 10:14–15.

20. As Foucault claims in his introduction, "it may be . . . that these creatures of unnatural issue escaped from the book, from the gaps between the open pages or the blank spaces between the letters" (*The Temptation of Saint Anthony*, xxx–xxxi).

21. This disordering, as we are later told in a proleptic utterance of Anthony's disciple Hilarion in his channeling of Lucifer, is an implicit result of the Copernican and Newtonian revolutions.

22. Night, in Milton's *Paradise Lost* (which we also explore further in chapter 5), is similarly given as another milieu of creation: "the wide womb of uncreated Night / Devoid of sense and motion" as Belial terms it in conversation with Satan. Milton, *Paradise Lost*, Book II: 150–51.

23. See Jutta-Annette Bruhn, "The Magic Lantern Slide after *The Temptation of Saint Anthony* by Jacques Callot" and Jean Seznec, "Flaubert and the Graphic Arts."

24. See the "Formation, Deformation, Reformation" chapter of Geoffrey Harpham's classic account, *On the Grotesque: Strategies of Contradiction in Art and Literature*. The word "grotesque" additionally has its origin in the thought of the underground: grotesque comes from the *grotto*, or underground cave. When the buried ruins of Nero's Domus Aurea were rediscovered and unearthed in 1480, as Harpham describes "grotesque" (*grottesche* or "grotto-esque") was the name given to the style of art of Nero's frescoes. These were decorated with human-animal and vegetal-human hybrid figures that themselves were even "derived from designs and ways of thinking much older than Rome" (27). Like the origins of species and of words, the grotesque's origins are a blasphemous ungrounding of humanistic sense. In perhaps the most explicit coarticulation of evolutionary and grotesque inhumanisms, H. G. Wells characterized his early sci-fi novel *The Island of Dr. Moreau* (1896) as a "theological grotesque" and a "flaming caricature" of humanity. (See Gordon Beauchamp, "*The Island of Dr. Moreau* as Theological Grotesque.") In Chris Danta's analysis in *Animal Fables after Darwin*, the theological grotesque "invert[s] the *scala natura* or divine order of things" by confounding "the divine, the spiritual, and the rational" with "the animal, the physical, and the irrational" (13–14).

25. Bakhtin, *Rabelais and His World*, 317.

26. Again, we will return to this motif and its Miltonic resonances in chapter 5.

27. Bakhtin, *Rabelais and His World*, 318.

28. Bakhtin, *Rabelais and His World*, 320.

29. Seznec, "Flaubert and the Graphic Arts," 187.

30. Redon, quoted in Seznec, "Flaubert and the Graphic Arts," 187.

31. Foucault, *The Order of Things*, 18.

32. We will return to the world makings of the forest and their implications for modernist and metamodernist planetarites in chapter 6.

33. In Act V, Lucifer assumes the form of Anthony's disciple, Hilarion, to, in Flaubert's words, "poison the mind of Anthony with hatred of the fathers of the church" (3).

34. The Anthonean imagination, as Leo Bersani argues, creates another reality that encompasses within itself "what we had assumed to be the real," and using it as "just another metaphorical resource in the development of similarities" ("Flaubert," 42).

35. Derrida, "White Mythology" (trans. Moore), 53.

36. See, for example, Blake Leland, "An Abode of Bliss: Plumtree's Potted Meat and the Allegory of the Theologians." In *Dubliners*, "simony" is the word Joyce uses to mark the collapse between economic and religious discourses.

37. The marketing of soap in Ireland and elsewhere, as Hye Ryoung Kil compellingly argues, was also bound up with late Victorian–era discourses of racial and social purity that accompanied British imperial expansion. As Kil explains in "Soap Advertisements in *Ulysses*," "with the rapid development of its manufacturing technology after the mid-century, soap assumed an important role in establishing personal and social boundaries for the Victorian middle class. Previously available only to the upper class, Anne McClintock notes, soap enabled members of the middle class to remove 'dirt,' the 'visible evidence of manual labor,' and appear clean and pure, thereby differentiating themselves from the dirty lower class as well as the corrupted aristocracy" (417). Further, "soap advertising, which depended on the distinction between clean and unclean, thus delivered a message that was more political than cultural: the

clean were the decent, civilized people who were equated with white Anglo-Saxons, while the unclean were the indecent savages personified by non-white colonials requiring cleansing and discipline" (417).

38. Joyce, *Selected Letters*, 272; italics in the original.

39. See, for example, Carver, "Moly: Bloom's Preservative; Correspondence and Function in *Ulysses*"; Schechner, *Joyce in Nighttown*; Sankar "Potatosoap"; Osteen, *The Economy of* Ulysses; Moshenberg, "The Capital Couple."

40. Joyce, *Selected Letters*, 272; Homer, *Odyssey* X: 339–41.

41. From Derrida's "Plato's Pharmacy":

> This *pharmakon*, this "medicine," this philter, which acts as both remedy and poison, already introduces itself into the body of the discourse with all its ambivalence. This charm, this spellbinding virtue, this power of fascination, can be—alternatively or simultaneously—beneficent or maleficent. The *pharmakon* would be a *substance*—with all that that word can connote in terms of matter with occult virtues, cryptic depths refusing to submit their ambivalence to analysis, already paving the way for alchemy—if we didn't have eventually to come to recognize it as antisubstance itself: that which resists any philosopheme, indefinitely exceeding its bounds as nonidentity, nonessence, nonsubstance; granting philosophy by that very fact the inexhaustible adversity of what funds it and the infinite abyss of what founds it. (70)

42. Joyce, *Selected Letters*, 272 (September 29, 1920).

43. Hemarajata, "Revisiting the Great Imitator," n.p.

44. Breton, *Manifestos of Surrealism*, 37.

45. R. Brown, "Everything in 'Circe,'" 233.

46. Cixous, "At Circe's," 387.

47. Quoted in Peake, *James Joyce*, 144.

48. Quoted in Fordham, "'Circe,'" 514.

49. Acts of the Apostles 10:14–15.

50. For further exploration of the question of animal and human rights in *Ulysses*, see Katherine Ebury, "The Beast and the Sovereign in 'Circe.'"

51. Rabelais, *Gargantua and Pantagruel*, 27.

52. Quoted in Nealon, *Plant Theory*, 66.

53. Bakhtin, *Rabelais and His World*, 21. See also the "Bloomsbody" chapter of Bernard McElroy, *Fiction of the Modern Grotesque*, for a further discussion of "Circe" in relation to theories of the grotesque.

54. The pig's foot, additionally, carries with it an overdetermined surrealist context, as act 1, scene 2 of Guillaume Apollinaire's *Les mamelles de Tirésias* (1920) features Thérèse (who later becomes the eponymous Tiresias) responding to her husband's crude sexualized request for "bacon" with "Cook your own feet like pigs' feet stew!" (2).

55. Gifford, Ulysses *Annotated*, 455.

56. See Carol J. Adams classic work, *The Sexual Politics of Meat: A Feminist-Vegetarian Critical Theory*, for a detailed account of this theme.

57. Channeling Apollinaire's Thérèse, the stage directions themselves seem to be responding to Bloom's animalizing gaze by telling him, in so many words, to "cook your own feet like pigs' feet stew!"

58. Heidegger, *The Fundamental Concepts of Metaphysics*, 196.

59. For Stephen, symbolic fatherhood is also tied in with the law-giving institutions of the Catholic Church and the British imperialist government. See, for example, Sheldon Brivic, *Joyce through Lacan and Žižek*; Vincent Cheng, *Joyce, Race, and Empire*; Andrew Gibson, *Joyce's Revenge*; Vicki Mahaffey, *Reauthorizing Joyce*; and Joseph Valente, *James Joyce and the Problem of Justice*.

60. Rabaté, *James Joyce, Authorized Reader*, 67.

61. Breton, *Manifestos of Surrealism*, 20; Harpham, *On the Grotesque*, 17.

62. As Catherine Flynn writes concerning the phantasmagoric nature of the objects/commodities in "Circe," "The objects of 'Circe' as listed in the stage directions of its dramatic script seem to have a fixed materiality, yet the constant transformations of the episode nonetheless render them as gestures, momentary performances that defy established ways of making sense." In Flynn's analysis, objects and aggregations of objects in "Circe" "stimulate a renewed and liberated perception of the world" and often lead to "a giddy, uncontrollable abundance." Flynn, "'Circe' and Surrealism," 124.

63. My thinking on this topic emerges from the extensive body of work in feminist science studies, Indigenous studies, and critical theory focused on the question of the ecofeminist mattering and ethics, toward which I can only gesture here. See, for example, Barad, *Meeting the Universe Halfway*; Bennett, *Vibrant Matter*; Chen, *Animacies*; Haraway, *Staying with the Trouble*; Liboiron, *Pollution Is Colonialism*; Murphy, "Alterlife"; Povinelli, *Geontologies* and *Between Gaia and Ground*; TallBear, "Caretaking Relations, Not American Dreaming"; Todd, "An Indigenous Feminist's Take"; Tsing, *The Mushroom at the End of the World*.

64. Chen, *Animacies*, 2, 13.

65. Chen, *Animacies*, 34. See also Zakiyyah Iman Jackson, *Becoming Human: Matter and Meaning in an Antiblack World*, for a deeper discussion of the racialization of animality and the intertwined discourses of anti-Black racism and speciesism.

66. Taylor, "Beasts of Burden," 195.

67. Taylor, *Beasts of Burden*, 10.

68. For a compelling elaboration on the transgressive dimensions of Bella/o, see the above-cited Cixous, "At Circe's, or the Self-Opener."

69. In *Animal Fables after Darwin*, Chris Danta compellingly claims that the modernist literary fable tradition is largely a means of reorienting human with earthly and animalistic being, through a turning toward ground. "Rather than lifting the human up out of the realm of biology," Danta writes, "fables cast the human down by casting the human as an animal" (2–3).

70. Cixous, "At Circe's," 394.

71. See Deleuze, *Masochism: Coldness and Cruelty*.

72. Sacher-Masoch, *Venus in Furs* (appendix II: "Two Contracts of von Sacher-Masoch"), 278.

73. Sacher-Masoch, *Venus in Furs*, 195.

74. Borrowing from Don Gifford's annotations to *Ulysses*, David Rando, in "The Cat's Meow: *Ulysses*, Animals, and the Veterinary Gaze," notes that the name Rudolph "stems from Old German 'fame' (*hrothi*) and 'wolf' (*vulf*)," which might make Rudy "the most important animal in *Ulysses*" (538). Rando reads Rudy as an animal body, and more specifically as a wolf, and looks at Bloom's encounter with Rudy as a "diagnostic or clinical scene" in which the silent Rudy "appears to Bloom and to readers as a body that must be interpreted" (538).

75. Joyce, *Selected Letters*, 272; italics in the original.

76. These shavings, earlier called "woodshavings," are likely from the "toppling masonry" dislodged by Stephen when he lifts his ashplant into the chandelier of the brothel in "Circe": "He lifts his ashplant high with both hands and smashes the chandelier. Time's livid final flame leaps and, in the following darkness, ruin of all space, shattered glass and toppling masonry" (*U* 15.4243–45).

77. Aristotle, *Poetics*, 1458a1: 26–27.

78. Aristotle, *Rhetoric* III.2.12.

79. Adorno, *Philosophy of New Music*, 37.

80. Kafka, *The Aphorisms*, 105.

4. Soils

1. In Kafka's original: "Die Demut gibt jedem, auch dem einsam Verzweifelnden, das stärkste Verhältnis zum Mitmenschen, und zwar sofort, allerdings nur bei völliger und dauernder Demut. Sie kann das deshalb, weil sie die wahre Gebetsprache ist, gleichzeitig Anbetung und festeste Verbindung. Das Verhältnis zum Mitmenschen ist das Verhältnis des Gebetes, das Verhältnis zu sich das Verhältnis des Strebens; aus dem Gebet wird die Kraft für das Streben geholt."

2. Benjamin, "Franz Kafka," 802.

3. "Imagine," too, is a visual term: from *imago* / "image." It is almost impossible, at least within English, to express thought about the future through sensory modalities other than the visible.

4. We will revisit the visual, and sky-oriented, dimension of *theōria* in chapter 6, particularly in reference to John Durham Peters's analysis of this term in *The Marvelous Clouds*.

5. Kafka, *The Aphorisms*, 58.

6. North, *The Yield*, 29.

7. North, *The Yield*, 52.

8. North, *The Yield*, 54.

9. North, *The Yield*, 56. The right to assign claims over being, North argues in *The Castle*, has been "wrenched away from the ontological and given over to the political" (59).

10. Quoted in Saint-Amour, "Introduction," 443. For more on weak theory in modernism and in related contexts, see the special issue of *Modernism/modernity* on "Weak Theory" (September 2018).

11. Saint-Amour, "Introduction," 455–56.

12. Saint-Amour, "Introduction," 456.

13. North, *The Yield*, xviii. Kafka's protagonists, North continues, "decline to use power, dismiss the *Ubermensch*, and would rather stay under the thumb of a master than become one" (xviii).

14. Kafka, *Diary*, 34, 35. Original German version at www.kafkaproject.org and www.gutenberg.spiegel.de / buch / tagebucher-1910-1923-162. Translation modified where indicated.

15. Kafka, *Diary*, 35; translation modified.

16. Kafka, *Diary*, 35; translation modified. In the original, the last sentence reads, "Das letztere wird durch den obigen Anfang wieder nicht widerlegt."

17. Kafka, *Diary*, 34; translation modified.

18. Sokel, "Kafka's Poetics," 40.

19. Kafka, *Diary*, 35.

20. For example, in *Kafka's Zoopoetics*, Naama Harel writes, "An animal narrative, then, is significantly different from a human narrative, which progresses chronologically and linearly. Presenting the story in the present tense, depicting habitual, routine activities—in itself a departure from the literary norm—is well suited to a nonhuman animal character, whose life consists of repetitive routines" (126). See also Coetzee, "Time, Tense and Aspect."

21. Sussman, "The Calculable," 1145.

22. Corngold, *Franz Kafka*, 284.

23. The uninterpretability of Kafka's works has become a critical commonplace, perhaps beginning with Walter Benjamin's 1934 essay on Kafka in which he claims that Kafka "took all conceivable precautions against the interpretation of his writings" (804). Accounts from Theodor Adorno, Albert Camus, Heinz Politzer, Gerhard Neumann, Roland Barthes, Deleuze and Guattari, Giorgio Agamben, and Judith Butler all discuss various dimensions of the impossibility of interpreting Kafka. See Naama Harel, *Kafka's Zoopoetics*, 6–10, for a good summary of these and other accounts.

24. The traditional definition of instinct is "an innate propensity in organized beings . . . manifesting itself in acts which appear to be rational but are performed without conscious design or intentional adaptation of means to ends" (*Oxford English Dictionary*). In the *Origin*, Darwin refuses to provide a strict definition of instinct but claims that it is linked to behaviors, or "habits," that work for the preservation of a particular species, and thus are naturally selected to be passed on to future generations of that species. Instincts themselves are not "specially endowed or created" but are "small consequences of one general law, leading to the advancement of all organic beings, namely, multiply, vary, let the strongest live and weakest die" (*OS* 247). Yet "none of these characters of instinct are universal" since "a little dose . . . of judgment or reason, often comes into play, even in animals very low in the scale of nature" (225). While conceiving of instinct as a set of innate behaviors, Darwin refuses a totalizing definition of what is and is not instinct, and further, rejects the claim that animal behavior is purely driven by instinct, allowing for the intertwinement of instinctual drives and higher-order cognitive functions in animal behavior. See also Brian Massumi's analysis of instinct in "The Supernormal Animal" and *What Animals Teach Us about Politics*.

25. Geier, *Kafka's Nonhuman Form*, 39.

26. Driscoll, *Towards a Poetics of Animality*, 171.

27. Kafka, *The Aphorisms*, 71.

28. Sussman, "The All-Embracing Metaphor," 104.

29. Marx, *Economic and Philosophic Manuscripts*, 106.

30. Marx, *Economic and Philosophic Manuscripts*, 107.

31. Marx, *Economic and Philosophic Manuscripts*, 107.

32. A term popularized by composer and early sound theorist Pierre Schaeffer in 1966 and originating from the word *akousmatikoi*, or the disciples of Pythagoras who were only allowed to listen to him lecture while he was hidden behind a curtain, "acousmatic" sound refers to "a noise that one hears without seeing what causes it." Schaeffer, "Acousmatics," 77.

33. See Dolar, "The Burrow of Sound" and *A Voice and Nothing More*; Kane, *Sound Unseen*; Gellen, *Kafka and Noise*.

34. In Chion's analysis, the powers of the *acousmêtre* are "to be everywhere, to see all, to know all, and to have complete power. In other words, ubiquity, panopticism, omniscience, omnipotence." Chion, *The Voice in Cinema*, 24.

35. In *A Voice and Nothing More*, Dolar writes, "Ultimately, *there is no such thing as disacousmatization*. The source of the voice cannot be seen, it stems from an undisclosed and structurally concealed interior, it cannot possibly match what we see. . . . Every emission of the voice is by its very essence *ventriloquism*. . . . The fact that we see the aperture does not demystify the voice; on the contrary, it enhances the enigma" (70; italics in the original).

36. Dolar, "Burrow of Sound," 120.

37. In homage to Adorno's influential maxim on Kafka's fiction: "Each sentence says 'interpret me,' and none will permit it" ("Notes on Kafka," 246).

38. North, *The Yield*, 10.

39. See *Zürau Aphorisms* 69 and 109.

40. In Kafka's original, the last phrase is written as "Aber alles blieb unverändert, das" (no period). Most renditions complete the phrase by excising the floating *das* at the end, and the Muirs' translation adds a period to the end to give an even more pronounced sense of closure. See Schuster, "Enjoy Your Security."

41. Kafka, *The Aphorisms*, 108.

42. Kafka, *Letters to Felice*, 447. In Kafka's original: "Hättest Du Dich nicht auf den Boden unter das Getier gelegt, hättest Du auch den Himmel mit den Sternen nicht sehn können und wärest nicht erlöst worden. Du hättest vielleicht die Angst des Aufrechtstehns gar nicht überlebt."

43. In *Kafka's Zoopoetics*, Naama Harel notes, "The Yiddish term *Luftmenschen* (which literally means 'air people'), referring to impractical contemplative people with no definite income, has been suggested as the source of this neologism" (112).

44. Kafka, *Letters to Felice*, 447 (March 3, 1915).

45. Aristotle, *Rhetoric* III.2.12.

46. Nancy, "Listening to Philosophy, Silence, and Self," n.p.

47. Earlier in the story, the dog-philosopher explicitly critiques the Enlightenment idea of "universal progress" through the advance of knowledge: "People often praise the universal progress made by the dog community throughout the ages, and probably mean by that more particularly the progress in knowledge. Certainly knowledge is progressing, its advance is irresistible, it actually progresses at an accelerating speed, always faster, but what is there to praise in that? It is as if one were to praise someone because with the years he grows older, and in consequence comes nearer and nearer to death with increasing speed" (299).

48. Winkelman, "Kafka's 'Forschungen eines Hundes,'" 207.

49. Norris, "Kafka's Hybrids," 25. See also Kata Gellen's *Kafka and Noise* for a similar interpretation of "Investigations," as well as Naama Harel's *Kafka's Zoopoetics*.

50. Norris, "Kafka's Hybrids," 24.

51. Aristotle, *Poetics*, 1458a1: 26–27.

52. Gellen, *Kafka and Noise*, 172.

53. As Gellen writes in "Noises Off: Cinematic Sound in Kafka's 'Burrow,'" "['Investigations'], unlike 'The Burrow,' is a farce. . . . All the mysteries of dogdom can even-

tually be resolved by the reader, a member of the human race who eventually figures out that humans are responsible for producing all the seemingly uncaused and inexplicable effects experienced by the dogs. . . . This means, in turn, that topics that they consider to be worthy of intensive study are mundane and obvious from the human perspective" (122).

54. On the question of instinct, see endnote 24 above.

55. Kafka, *The Aphorisms*, 69.

56. Kafka, *The Aphorisms*, 24. In Kafka's original: "Das Glück begreifen, daß der Boden, auf dem du stehst, nicht größer sein kann, als die zwei Füße ihn bedecken."

57. Kafka, *The Aphorisms*, 25.

58. Milton, *Paradise Lost*, Book IV: 657–58; 661–67.

5. Sky

1. V. Woolf, *The Diary of Virginia Woolf, Vol. V, 1936–41*, 162. For more on the idea of *Between the Acts* marking the end of civilization, see, for example, Michele Pridmore-Brown, "1939–40: Of Virginia Woolf, Gramophones, and Fascism"; Alex Zwerdling, *Virginia Woolf and the Real World*; Gillian Beer, "*Between the Acts*: Resisting the End," in *Virginia Woolf: The Common Ground*; and Paul Saint-Amour, *Tense Future: Modernism, Total War, Encyclopedic Form*.

2. DeLoughrey, *Allegories of the Anthropocene*, 4.

3. V. Woolf, "The Value of Laughter," 58.

4. V. Woolf, "The Value of Laughter," 58. Comedy is also gendered, for Woolf, "of the sex of the graces and the muses." When confronted with "masculine" tragedy's "spirit of solemnity," comedy flees toward "the bosoms of her sisters" and toward the making of a queer sociality through "irresistible laughter" (58).

5. V. Woolf, "The Value of Laughter," 58.

6. McGurl, "The Posthuman Comedy," 549.

7. Gillian Beer, *Virginia Woolf: The Common Ground*.

8. Bergson, *Laughter*, 4–5.

9. McCole, *Walter Benjamin and the Antinomies of Tradition*, 136. Benjamin's *The Origin of German Tragic Drama* (1928), quoted in DeLoughrey, *Allegories of the Anthropocene*, 57.

10. Bakhtin, *Rabelais and His World*, 12.

11. Bakhtin, importantly, critiques Bergson's emphasis on what he calls the "negative," mocking or satirical aspects of the comic rooted in detachment and distance: "Let us stress once more that for the Renaissance (as for the antique sources described above) the characteristic trait of laughter was precisely the recognition of its positive, regenerating, creative meaning. This clearly distinguishes it from the later theories of the philosophy of laughter, including Bergson's conception, which bring out mostly its negative functions" (71).

12. In "The Comedy of Nature: Darwinian Feminism in Virginia Woolf's *Between the Acts*," Sam See similarly argues that the conventions of tragedy in *Between the Acts* are subsumed into the novel's larger comic vision: "Woolf's comedy of nature employs the tragic structure of decline as a good: as a submission and return to a Darwinian understanding of nature as infinitely heterogeneous and transformative, one

wherein 'species are not immutable' but exist in 'infinitely varied diversities'" (643). See also Judy Little, *Comedy and the Woman Writer: Woolf, Spark, and Feminism*, and to be discussed shortly, Melba Cuddy-Keane, "The Politics of Comic Modes in Virginia Woolf's *Between the Acts*."

13. McGurl, "The Posthuman Comedy," 539.

14. Kafka, "Cares of a Family Man," 428 translation modified.

15. See endnote 41 in chapter 2 of this book, on Julie Beth Napolin's analysis of voice as determinacy. Odradek's indeterminacy, as *Unbestimmter*, might then also be thought at his non-voicedness, which of course does not preclude (and perhaps even intensifies) the expressive noisiness of Odradek's laughter, speech, and other sonances.

16. Cuddy-Keane, "The Politics of Comic Modes," 275. See also Christopher Ames, "Carnivalesque Comedy in *Between the Acts*."

17. Cuddy-Keane, "The Politics of Comic Modes," 275, 280. Citing classical historian Jane Harrison (whose works Woolf may very well have read), Cuddy-Keane claims that "while the chorus [in Greek drama] appears in both tragedy and comedy, it is the communal genre of comedy that provides the chorus with its true identity. Following Aristotle, Harrison traces the origins of tragedy to the *leaders* of the dithyramb, the spring dance, but the chorus, she argues, derived from the earlier stage of purely communal dance" (275).

18. V. Woolf, *Diary, Vol. V*, 178. "L." refers to Leonard Woolf.

19. For compelling articulations of Woolf's theorization of animality in *Between the Acts*, see, for example, Peter Adkins, *The Modernist Anthropocene*; Christina Alt, *Virginia Woolf and the Study of Nature*; Derek Ryan, *Virginia Woolf and the Materiality of Theory: Sex, Animal, Life*; Vicki Tromanhauser, "Animal Life and Human Sacrifice in Virginia Woolf's *Between the Acts*."

20. V. Woolf, "The Reader," 429. Both "Anon" and "The Reader" were potential chapters for a larger work to be titled *Reading at Random*.

21. See Steve Ellis, *British Writers and the Approach of World War II*, 214.

22. V. Woolf, "Anon," 382.

23. Westling, "Virginia Woolf and the Flesh of the World," 860. See also Stefanie Heine, "Forces of Unworking in Virginia Woolf's 'Time Passes.'"

24. Milton, *Paradise Lost*, Book II: 895, 894.

25. Milton, *Paradise Lost*, Book II: 891–94.

26. We can thus also think Woolf's and La Trobe's aesthetic forays along the lines of what Ben de Bruyn has called the "multispecies soundscape," which is conceived as a gathering place for nonhuman sounds outside of human noise and humanistic determinations of significance. In *The Novel and the Multispecies Soundscape*, de Bruyn argues that while "human activities are causing animal voices to disappear" in the twentieth and twenty-first centuries,

> they are also adding increasing volumes of noise to the environment, outcompeting the signals of other species in a sonic contest that explains, scientists argue, why the vocalizations of animals like bats, whales, tamarins, and nightingales become louder in circumstances where the presence of human noises would otherwise mask their attempts at communication. . . . A similar effect obtains in the cultural realm . . . as the omni-presence of human noise in contemporary society gives rise to a louder natural soundscape in the novels of

environmentally sensitive writers too, who are amplifying these fragile voices at a time when the available bandwidth has narrowed considerably. (5)

27. Milton, *Paradise Lost*, Book II: 150–51.

28. In Mary Adams's analysis, Milton's "Night . . . is feminine and matter," in line with the traditional view of "Old Night" as, in Walter Clyde Curry's words, "a feminine divinity who is the receptacle of paternal causes, transmitting into all posterior things the generative power of the gods." Chaos and eldest Night, according to John Rumrich, are the dark feminine counterpart of God's creative essence in which "God is the confused matter of chaos even as he is the creative virtue of light" and "chaos is God's womb, essential to his deity." M. Adams, "Fallen Wombs," 170 (Curry quoted in same place); Rumrich, "Milton's God," 1043.

29. *Atopos* is a central motif in Timotheus Vermuelen and Robin van den Akker's theorization of contemporary metamodernism, to be discussed further in chapter 6.

30. V. Woolf, *Diary, Vol. III*, 139 (June 18, 1927). Spacing and punctuation in the original.

31. V. Woolf, *Diary, Vol. III*, 203 (November 7, 1928).

32. V. Woolf, *Flush*, 39, 38.

33. V. Woolf, *Flush*, 130.

34. V. Woolf, *Flush*, 130.

35. Kendall-Morwick, "Mongrel Fiction," 520. See also Jeanne Dubino, "The Bispecies Environment, Coevolution, and *Flush*" and Derek Ryan, "'Was It Flush, or Was It Pan?': Virginia Woolf, Ethel Smyth, and Canine Biography."

36. Maureen Chun, in "Between Sensation and Sign: The Secret Language of *The Waves*," discusses a similar linguistic dynamic at work in *The Waves* in which Woolf attempts, in the novel's soliloquies and interludes, to "unsettle" the "symbolic function of language" by employing words and images that bear with them traces of their sensuous origins, rather than serving merely as representational indicators of some other reality (54). For more on the question of nonhuman or inhuman phenomenologies, see, for example, Ben de Bruyn, *The Novel and the Multispecies Soundscape*; Rachel Mundy, *Animal Musicalities*; Jacob Smith, *Eco-Sonic Media*.

37. Wells's *Outline of History* (1920) attempts to provide what the author calls "a plain history of life and mankind" from the beginnings of life on earth to all of human history, ending with the Treaty of Versailles in 1920 (the year of publication for the first edition). Peter Adkins and Vicki Tromanhauser both note that the work that Swithin is reading, titled *Outline of History*, is likely an amalgamation of Wells's *Outline of History* and George Macaulay Trevelyan's *History of England* (1926). See Adkins, *The Modernist Anthropocene*; Tromanhauser, "Animal Life and Human Sacrifice"; as well as Gillian Beer, *Virginia Woolf: The Common Reader*.

38. V. Woolf, "White's Selborne," 190. Like White, Mrs. Swithin in *Between the Acts* is also obsessed with tracking and analyzing the migration patterns of the swallow.

39. V. Woolf, "White's Selborne," 191.

40. Morton, *Ecology without Nature*, 8.

41. Morton, *Ecology without Nature*, 8.

42. White, *The Natural History*, 252.

43. Thordarson and Self, "Atmospheric and Environmental Effects," 7.

44. Thordarson and Self, "Atmospheric and Environmental Effects," 1.

45. Thordarson and Self, "Atmospheric and Environmental Effects," 13, 14.

46. White, *The Natural History*, 253.

47. Milton, *Paradise Lost*, Book I: 594–99. See also Tobias Menley's *Climate and the Making of Worlds: Toward a Geohistorical Poetics*, a wide-ranging account of planetary and climate history that proceeds from Milton's *Paradise Lost*.

48. In "A Defence of Poetry," Percy Shelley famously writes of poets as "the mirrors of the gigantic shadows which futurity casts upon the present" (535).

49. White, *The Natural History*, 255. The entirety of this fascinating paragraph—the work's penultimate—reads, "Though the clouds at South Lambeth, near London, were at that juncture thin and light, and no storm was in sight, nor within hearing, yet the air was strongly electric; for the bells of an electric machine at that place rang repeatedly, and fierce sparks were discharged" (255).

50. White, *The Natural History*, 214; V. Woolf, "White's Selborne," 191.

51. In her account of sound and sound technology in "Virginia Woolf, Sound Technologies, and the New Aurality," Melba Cuddy-Keane offers a helpful grounding in the difficulties of encountering sound through the visual medium of writing, particularly in Woolf's experimentation with the sonicity of written language:

> Besides the problem of mediating sound through a visually oriented discourse, there is the fundamental problem of mediating sound through language at all—the inevitable translation of sound into a conceptual category that takes place in the process of verbalization. This second concern would perhaps seem less troubling for an analysis of the representation of sound in literature, where sound has already been transcribed in linguistic form. But there is a significant difference between the linguistic representation of a sound and the linguistic conceptualization of it (sonicity as opposed to semantics). (70)

See also Tim Armstrong, "Player Piano: Poetry and Sonic Modernity," and Ben de Bruyn, *The Novel and the Multispecies Soundscape*.

52. Among the extensive recent theoretical and philosophical body of thought on noise, especially significant are Douglas Kahn's analysis, in *Noise, Water, Meat*, of noise as that which interferes with semantic meaning, in both auditory and visual contexts; Michel Chion's relativity argument in *Sound: An Acoulogical Treatise* that "the assessment of noise as noise and of music as music is . . . a matter of cultural and individual context" (57); and Paul Valéry's claim that "a sound makes into a semi-presence the whole system of sounds—and that is what primitively distinguishes sound from noise" (quoted in Nancy, *Listening*, 10). See also Jacques Attali, *Noise: The Political Economy of Music* (1977) and Luigi Russolo's 1913 Futurist manifesto, *The Art of Noises*.

53. In "1939–40: Of Virginia Woolf, Gramophones, and Fascism," Michelle Pridmore-Brown analyzes the more insidious dimension of the universality offered by the phonograph, found in the capacity of phonographic expression to transform "individuals into a herd that can be controlled by a charismatic leader." Alternatively, the novel can offer different modalities of listening that "short-circuit the herd impulse" by "exploiting the noise or static inherent in communications technology" (408). For more on the role of the gramophone in *Between the Acts*, see also Bonnie Kime Scott, "The Subversive Mechanics of Gramophone" and Pamela L. Caughie, "Virginia Woolf: Radio, Gramophone, and Broadcasting."

54. S. Butler, "Dogs and Phonographs," 156. See also S. Butler, *The Ancient Phonograph*.

55. Quoted in S. Butler, "Dogs and Phonographs," 156. For more on the colonialist dimensions of phonographic inscriptive and listening practices in the late nineteenth and early twentieth centuries, see Roshanak Kheshti's *Modernity's Ear: Listening to Race and Gender in World Music.*

56. S. Butler, "Dogs and Phonographs," 156. "In other words," Butler continues, "the phonograph itself models the very kind of promiscuous listening it makes available to subsequent auditors" (156).

57. Kittler, *Gramophone, Film, Typewriter*, 32–33. Concerning comparisons between the phonograph and the human brain, Kittler writes, "The principal difference between the brain and the phonograph is that the metal disk of Edison's still rather primitive machine remains deaf to itself; there is no transition from movement to consciousness" (32). Theories of the brain as a "conscious phonograph," primarily from Jean-Marie Guyau, ultimately betray the impersonality of phonographic listening by which it attains both its possibility and its power:

> But consciousness, the quality that Guyau ascribes to the brain in order to celebrate the latter as an infinitely perfected phonograph, would result in an infinitely inferior one. Rather than hearing the random acoustic events forcing their way into the bell-mouth in all their real-time entropy, Guyau's conscious phonograph would attempt to understand and thus corrupt them. Once again, alleged identities or meaning or even functions of consciousness would come into play. Phonographs do not think, therefore they are possible. (33)

58. Quintilian, *Institutio Oratorio*, Book 8, chap. 6, 303.

59. Derrida, *White Mythology*, trans. Moore, 41.

60. Kahn, *Noise, Water, Meat*, 26.

61. Kahn, *Noise, Water, Meat*, 22.

62. Kahn, *Noise, Water, Meat*, 243, 242. On the strange confluence between Cage and Pollock's artistic methods in the 1950s, Kahn writes, "The floodgates cracked open to a concurrence of sound, performance, and water and fluidity among the arts, a dissolution of disciplinary constraints and discrete bounds of the object, and an increased osmosis between art and life" (260). See also Mowitt and Tazudeen, "Introduction to Ecological Soundings," 127.

63. Louvel, *"Ecrire la couleur,"* 243.

64. Similarly, Pridmore-Brown notes in "1939–40: Of Virginia Woolf, Gramophones, and Fascism" that La Trobe's use of words might be more about the transmission of affect rather than the generation of meaning. Caughie writes, "In an early draft of the novel, [Woolf] directly speculated about La Trobe's use of words not to convey meaning but to evoke emotion (and ultimately self-recognition). 'Supported by simple chords,' she wrote, La Trobe aims 'to sink words into the layer beneath the hats, trousers, skirts and other tokens of class, rank and calling,' to address the invisible currents of that collective mind. In this sense, the gramophone functions like a 'harmonium' (a word Woolf considered using in the draft) that permeates and consolidates cultural material" (413).

65. See chapter 1 of this book, 53–54.

66. V. Woolf, "The Mark," 88, 83.

67. V. Woolf, "The Mark," 84. See also Aaron Jaffe's insightful analysis of "The Mark on the Wall" in his introduction to the special issue of *Modernism/modernity*, "Modernist Inhumanisms."

68. Summarizing Giles's proximity to early twentieth-century fascist attitudes, Alex Zwerdling writes in *Virginia Woolf and the Real World*,

> Giles's behavior is, despite his patriotic feeling for England, very close to the Fascist threat he fears. His aggressive masculinity and Nordic looks remind one of the Master Race; William Dodge sees him as "the muscular, the hirsute, the virile" (127). He is filled with hatred and contempt: for "old fogies who sat and looked at views over coffee and cream" (66); for the homosexual William, whom he thinks of as "a toady; a lickspittle" (75); even for his wife. He is a good indigenous example of an ethos Virginia Woolf had seen in the first days of Italian Fascism and described in *A Room of One's Own*: "I began to envisage an age to come of pure, of self-assertive virility, such as . . . the rulers of Italy have already brought into being" (*AROO*, 154). In *Three Guineas* [1937] she made it clear that this was no longer a vision of the future but of the present. (308)

69. See p. 9 of this book on Chaucer's usage of "monster."

70. Similarly, Sarah Cole, in *At the Violet Hour: Modernism and Violence in England and Ireland*, offers an interesting reading of this scene as the creation of a kind of abstract art, suggesting that art itself is intimately bound up with the violent conditions out of which it emerges. The act of symbolization itself, in other words, is a mode of violence.

71. Alt, *Virginia Woolf and the Study of Nature*, 165.

72. Cole also notes that the trope of "action . . . had by the middle of the 1920s become a political fetish, with special resonance for fascist and protofascist rhetoric, hence carrying troubling affiliations for the pacifist Woolf" (*At the Violet Hour*, 201). For more accounts of Giles's association with fascism, see also Zwerdling, *Virginia Woolf and the Real World*, 307; DiBattista, *Virginia Woolf's Major Novels: The Fables of Anon*, 205; and Wood, *Virginia Woolf's Late Cultural Criticism*, 129.

73. Blumenberg, *Shipwreck with Spectator*, 82.

74. V. Woolf, "Modern Fiction," 190

75. V. Woolf, *The Waves*, 4, 140.

76. V. Woolf, *The Waves*, 194.

77. Gilbert White, too, remarks on Selborne's swallows flying to Africa and back during the winters but only "from hedge to hedge" "the summer long" (*Selborne*, 33). "Hedge-hogs abound in my gardens and fields," he later recounts, emphasizing the "little round holes they dig" and the "deep and warm hybernaculam [they make] with leaves and moss, in which they conceal themselves" (67, 68).

78. See chapter 2 of this book, 67–69.

79. In a letter of November 4, 1767, White writes,

> As to the short-winged soft-billed birds, which come trooping in such numbers in the spring, I am at a loss even what to suspect about them. I watched them narrowly this year, and saw them abound till about Michaelmas, when they appeared no longer. Subsist they cannot openly among us, and yet elude

> the eyes of the inquisitive: and, as to their hiding, no man pretends to have found any of them in a torpid state in the winter. But with regard to their migration, what difficulties attend that supposition! that such feeble bad fliers (who the summer long never flit but from hedge to hedge) should be able to traverse vast seas and continents in order to enjoy milder seasons amidst the regions of Africa! (*Selborne*, 33)

80. The idea of a threshold of audition, and of perception itself, emerges from the philosophy of Gottfried Wilhelm Leibniz, who in the preface to the *New Essays* of 1704 suggests the existence of "tiny perceptions" (*petites perceptions*) that persist beneath the surface of what the human mind, because of its attunement toward objects that demand its attention, normally perceives. Leibniz writes,

> At every moment there is an infinity of perceptions in us, but without apperception and without reflection—that is, changes in the soul itself, which we do not consciously perceive, because these impressions are either too small or too numerous, or too homogenous, in the sense that they have nothing sufficiently distinct in themselves; but combined with others, they do have their effect and make themselves felt in the assemblage, at least confusedly. It is in this way that custom makes us ignore the motion of a mill or of a waterfall, after we have lived nearby for some time. It is not that this motion ceases to strike our organs and that there is nothing corresponding to the soul, . . . but that the impressions in the soul and body, lacking the appeal of novelty, are not sufficiently strong to attract out attention and memory, which are applied only to more demanding objects. (295)

For an extended discussion on the role of Leibniz's *petites perceptions* in Woolf's fiction, see Maureen Chun, "Between Sensation and Sign: The Secret Language of *The Waves*." See also "The Ear" chapter of Chion's *Sound: An Acoulogical Treatise* for further analysis of the differences between visual and sonic thresholds of perception and the difficulties of objectively measuring the "intensity" of perceived sounds (24).

81. Kristeva, *Revolution in Poetic Language*, 26, 29, 27.

82. Chion, *Sound*, 12–13.

83. Kristeva, *Revolution in Poetic Language*, 29.

84. Uexküll, *A Foray*, 94, 148.

85. Cuddy-Keane, "The Politics of Comic Modes in Virginia Woolf's *Between the Acts*," 281.

86. V. Woolf, "The Value of Laughter," 58.

87. Ryan, "'The Reality of Becoming,'" 549.

88. V. Woolf, *The Waves*, 4.

89. Cavarero, *For More than One Voice*. See footnote 42 in chapter 2 of this book.

90. Pettman, *Sonic Intimacy*, 79.

91. In the words of Kafka's *Zürau Aphorism* 24, as discussed in the previous chapter: "Grasp the good fortune that the ground on which you stand cannot be any bigger than the two feet covering it" (24). In the original: "Das Glück begreifen, daß der Boden, auf dem du stehst, nicht größer sein kann, als die zwei Füße ihn bedecken."

92. See the introduction of this book, 8.

93. Woolf is borrowing these lines from an unknown poet quoted in E. K. Chambers and F. Sidgwick's collection, *Early English Lyrics*, poem 33, quoted in Brenda Silver, "'Anon' and 'The Reader': Woolf's Last Essays," 402.

94. Silver, in her notes to "Anon," connects the "green scar" more explicitly to *Between the Acts* and to the motif of the furrowing of the earth. Silver references an earlier draft of "Anon" in which Woolf writes, "For when we see the great Elizabethan house—Wilton say, or Penshurst—we see the stranded relic, that has emerged from its surroundings. Could we see it as it said that the airman sees a village <field> from the air with the scars of other villages <old roads> and other houses on it. we should see the great house connected with the village" (405–6).

In Silver's gloss:

> The reference to the airman's perspective is reminiscent of Mr. Oliver's comment in *Between the Acts* that "From an aeroplane . . . you could see, plainly marked, the scars made by the Britons; by the Romans; the plough, when they ploughed the hill to grow wheat in the Napoleonic wars." These scars are evoked again at the end of the novel when we are told that "The pilgrims had bruised a lane on the grass." The antiquaries who tell us of the survival of the old roads include Trevelyan, who wrote of ancient trackways "like the tract along the south edge of the North Downs, long afterwards known and used as the 'Pilgrims Way' to Canterbury, and still at places available to the pedestrian as it was four thousand and more years ago," and of the Roman roads that in some stretches "are reserved for the Briton or Saxon who still fares on foot; and are . . . to be traced as green lanes." (406)

95. Adkins, *The Modernist Anthropocene*, 178.

96. V. Woolf, *A Passionate Apprentice*, 203.

97. Rilke, *Duino Elegies*, 59. In Rilke's original: "Erde, ist nicht dies, was du willst: unsichtbar / in uns erstehn?"

98. V. Woolf, *Diary, Vol. V*, 276–77.

99. "L. & V." refer to two elm trees outside of Woolf's Monks House residence (where she wrote this entry), which she named "Leonard" and "Virginia." Woolf's ashes are now buried in the ground beneath "V."

100. Louvel, "Ecrire la couleur," 243.

101. Barad, "Nature's Queer Performativity," 126, 129.

102. Barad, *Meeting the Universe Halfway*, 185.

103. Derrida, *Voice and Phenomenon*, 67, 68.

6. Forest, Ocean, Planet (Green & Blue)

1. Milton, *Paradise Lost*, Book II: 624–26. These lines form the epigraph to *~~Flung~~ Throne.*

2. Clevidence, *~~Flung~~ Throne*, 29.

3. Clevidence, *~~Flung~~ Throne*, 29.

4. Clevidence, *~~Flung~~ Throne*, 62.

5. Clevidence, *~~Flung~~ Throne*, 105.

6. Clevidence, *~~Flung~~ Throne*, 8.

7. Clevidence, *~~Flung~~ Throne*, 14.

8. Clevidence, *~~Flung~~ Throne*, 120.

9. Ahmed, *Queer Phenomenology*, 78.

10. Dembroff, "Escaping the Natural Attitude," 19–20.

11. Byrne, "Are Women Adult Human Females?" 3.

12. Hird, "Naturally Queer." See also, for example, Robin Dembroff, "Beyond Binary: Genderqueer as Critical Gender Kind" and "Escaping the Natural Attitude about Gender"; Greta LaFleur, *The Natural History of Sexuality in Early America*; and Catriona Mortimer-Sandilands, "Fear of a Queer Plant?"

13. See, for example, R. H. F. Hunter and T. Greve, "Intersexuality in Pigs: Clinical, Physiological, and Practical Considerations"; Katrina Karzakis, *Fixing Sex: Intersex, Medical Authority, and Lived Experience*; R. A. McFeely, W. C. Hare, and J. D. Biggers, "Chromosome Studies in 14 Cases of Intersex in Domestic Mammals"; Daniel Vaiman and Eric Pailhoux, "Mammalian Sex Reversal and Intersexuality: Deciphering the Sex-Determination Cascade."

14. Hird, "Naturally Queer," 85. See also Anna Tsing on the queerness of fungal reproduction in *The Mushroom at the End of the World: On the Possibility of Life in Capitalist Ruins*. Tsing writes, "Fungi propagate through many mechanisms, and sexual reproduction through the mating of germinated spores is just one. A good deal of fungal propagation is clonal. . . . Fungi also propagate through asexual spores. . . . For some species, sexual reproduction is absent or rare" (237).

15. Tsing, *The Mushroom at the End of the World*, 85.

16. James and Seshagiri, "Metamodernism," 88.

17. James and Seshagiri, "Metamodernism," 88.

18. Vermeulen and van den Akker, "Notes on Metamodernism."

19. Brunton, "Whose (Meta)modernism?" 74.

20. Brunton, "Whose (Meta)modernism?" 69, 73.

21. Gumbs, *M Archive*, 23.

22. Gumbs, *M Archive*, 36.

23. Gumbs, *M Archive*, 36, 37. Of the mud mothers, Gumbs writes, "and their roots grew deeper and their branches reached for sky"; "their breathing grew the abyss" (36, 37).

24. Yusoff, *A Billion Black Anthropocenes*, xiii.

25. Gumbs, *M Archive*, xi.

26. Gumbs, *M Archive*, 200.

27. Gumbs, *M Archive*, 6, 7.

28. Gumbs, *M Archive*, 36.

29. In *Becoming Human: Matter and Meaning in an Antiblack World*, Jackson compellingly argues that "Eurocentric humanism needs blackness as a prop to erect whiteness: to define its own limits and to designate humanity as an achievement as well as to give form to the category of 'the animal'" (4).

30. As Brunton writes in "Whose (Meta)modernism?" "Rather than fixing desire in metamodern aesthetics as a desire for a home in the future (utopia), we might do well to locate those strands of metamodernism that express a desire for a home in the present, rooted in a reactivation of the unrealized past" (73).

31. Gumbs, *Dub*, 8.

32. Gumbs, *Dub*, 17, 65.

33. Shaughnessy, *The Octopus Museum*, 50.

34. Shaughnessy, *The Octopus Museum*, 50.

35. Shaughnessy, *The Octopus Museum*, 57; italics in the original.

36. Shaughnessy, *The Octopus Museum*, 62, 63.

37. In Derrida's account of signification, as discussed in the introduction, every contingent mark needs a preexisting "inscribable surface" as a system of meaning or sense within a linear order of time by which the mark counts as (or is re-marked [*re-marqué*] as) a significant expression and not just an aleatory scribble or noise.

38. Waldman, *Manatee/Humanity*, 20.

39. Waldman, *Manatee/Humanity*, 74.

40. Waldman, *Manatee/Humanity*, 21.

41. Waldman, *Manatee/Humanity*, 69. Shaughnessy, *The Octopus Museum*, 49.

42. Waldman, *Manatee/Humanity*, 74.

43. In this vein, Elisabeth Mann Borgese writes in *The Oceanic Circle*, "The ocean is a medium different from the earth. . . . It forces us to think differently. The medium itself, where everything flows and everything is interconnected, forces us to 'unfocus,' to shed our old concepts and paradigms, to 'refocus' on a new paradigm" (quoted in DeLoughrey, "Submarine Futures of the Anthropocene," 37).

44. DeLoughrey, "Submarine Futures of the Anthropocene," 62.

45. DeLoughrey, "Submarine Futures of the Anthropocene," 66, 64; italics in the original.

46. Shaughnessy, *The Octopus Museum*, 49.

47. Gumbs, *Dub*, 65.

48. Gumbs, *Dub*, 72–73.

49. Milton, *Paradise Lost*, Book II: 891–92, 911, 916. In John Rumrich's influential reading of Book II of *Paradise Lost*, Miltonic chaos "indicates . . . a material potency that is the precondition of creation. . . . Chaos thus represents an indeterminate material principle whose complex disorder persists dynamically in any order." Rumrich, "Milton's God," 1041.

50. Columbus, "Journal of the First Voyage of Columbus," 154.

51. The term "conquistador humanism" is borrowed from Tiffany Lethabo King's *The Black Shoals*. See endnote 77 of the introduction of this book.

52. Anonymous, *Arcana of Science and Art*, 167.

53. Anonymous, *Arcana of Science and Art*, 166–67; italics in the original

54. Pratt, *Imperial Eyes*, 15.

55. Pratt, *Imperial Eyes*, 9.

56. Pratt, *Imperial Eyes*, 180.

57. Howard, "Swim Your Ground," 321.

58. Howard, "Swim Your Ground," 322; Ingersoll, *Waves of Knowing*. See also Paul Gilroy, *The Black Atlantic*; Hester Blum, "The Prospect of Oceanic Studies"; Daniel Brayton, *Shakespeare's Ocean*; Elisabeth Mann Borgese, *The Oceanic Circle*; Elizabeth DeLoughrey, "Submarine Futures of the Anthropocene"; and Stefan Helmreich, *Alien Ocean*.

59. DeLoughrey, "Submarine Futures of the Anthropocene," 33.

60. DeLoughrey, "Submarine Futures of the Anthropocene," 34.

61. Ingersoll, *Waves of Knowing*, 33.

62. Clevidence, ~~*Flung*~~ *Throne*, 29. Additionally, in "Forest Forms and Ethical Life," Eduardo Kohn argues that "a forest could be formally defined as any arrangement of minds that comes to manifest—as an emergent property—a higher-order mind." And further, "A 'forest,' then, is a general but precise concept, and the sylvan world to which it gestures exceeds the tropics as well as those ecosystems made up primarily of woody creatures" (409). Both forests and oceans are and are not themselves, each saying the excess of its manifestations, and endlessly receptive to new forms and new minds to manifest with.

63. In Flaubert's original, "Des polypiers, qui ont l'air de sycomores, portent des bras sur leurs branches."

64. In a review of Gladding's book, entomologist Dezene Huber explains regarding the "communication" of bark beetles:

> Once [bark beetles] attack a tree, the parents and then the hatched brood etch their marks on the wood of the tree as they mine through the tissues just under the bark. The under-the-bark patterns of bark beetle galleries are about as diverse as the number of bark beetle species. In general terms, one or both of the parents excavate a main gallery, the female lays eggs on the walls of that tunnel, and then the hatched larvae bore outward from the parental gallery. This means that an interested observer can strip back the bark of a tree—whether attacked recently or in the rather distant past—and can measure the length of the parental gallery and the length and number of the larval galleries. These measurements and counts give information on the number of eggs laid, the number of larvae hatched, and the success of the larvae among other things. ("Book Review," n.p.)

65. Gladding, *Translations from Bark Beetle*, 7.

66. Waldman, *Manatee/Humanity*, 74.

67. Kohn, *How Forests Think*, 72.

68. Kohn, *How Forests Think*, 8, 9.

69. Wheeler, "The Biosemitic Turn," quoted in Moe, "Towards a Poetics of Gaia," 453. See also Jesper Hoffmeyer, *Biosemiotics: An Examination into the Signs of Life and the Life of Signs*.

70. Peters, *The Marvelous Clouds*, 380.

71. Peters, *The Marvelous Clouds*, 6.

72. Peters, *The Marvelous Clouds*, 69.

73. Jody Gladding and Jen Bervin, "Three Dimensions: An Interview" (February 1, 2017), quoted in Moe, "Towards a Poetics of Gaia," 460.

74. Gellen, "Noises Off," 122.

75. Gladding, *Translations from Bark Beetle*, 67.

76. Gladding, *Translations from Bark Beetle*, 46; bolding in the original.

77. Katz, "Small Pests, Big Problems."

78. Katz, "Small Pests, Big Problems."

79. Milton, *Paradise Lost*, Book II: 624–25.

80. Nakayasu, *The Ants*, 67, 76.

81. Huxley, "Philosophical Ants," 179.

82. Nakayasu, *The Ants*, 39.

83. Peters, *The Marvelous Clouds*, 255.

84. Nakayasu, *The Ants*, 86.

85. Nakayasu, *The Ants*, 39.

86. Hovanec, *Animal Subjects*, 146.

87. Huxley, "Philosophical Ants," 178.

88. Huxley, "Philosophical Ants," 180, 182.

89. Huxley, "Philosophical Ants," 183.

90. "Myrmecine," it should be noted, is not technically an adjectival form for the entire family "Ant," even as it has taken on this connotation. Myrmecine, rather, names a particular subfamily of ants, "Myrmicinae."

91. Quoted in Huxley, "The Societies of Ants," 819.

92. Huxley, "The Societies of Ants," 819.

93. Julian Huxley espoused eugenicist views throughout his career, serving as vice president (1937–44) and president (1959–62) of the British Eugenics Society (after having joined in 1931) along with cosigning the *Eugenics Manifesto* of 1939. However, beginning in the early 1930s, he spent considerable efforts to oppose his Social Darwinist beliefs in human progress through an improvement in the gene pool to the thanatopolitics and "pseudoscience" of the European fascist and National Socialist movements. See, for example, Garland E. Allen, "Julian Huxley and the Eugenical View of Human Evolution"; Pauline Mazumdar, *Eugenics, Human Genetics, and Human Failings*; and Paul Weindling, "Julian Huxley and the Continuity of Eugenics in Twentieth-century Britain."

94. Huxley, "The Societies of Ants," 820.

95. Nakayasu, *The Ants*, 30.

96. Cacciari quoted in Agamben, *Potentialities*, 172.

97. Derrida, "Before the Law," 191.

98. Kafka, *The Complete Stories*, 4.

99. Nakayasu, *The Ants*, 4.

100. Huxley, "The Societies of Ants," 820.

101. Will Dunham, "How Many Ants Are Crawling on Earth? Scientists Say 20 Quadrillion," Reuters, September 22, 2022, https://www.reuters.com/lifestyle/science/ants-go-marching-one-by-one-20-quadrillion-them-2022-09-20.

102. James, *Modernist Futures*, 2.

103. James, *Modernist Futures*, 30.

104. Vermuelen and van den Akker, "Notes on Metamodernism."

105. Vermuelen and van den Akker, "Notes on Metamodernism."

106. For more, see Peter Adkins's strident account of the relations between modernist aesthetics and climate change in *The Modernist Anthropocene*.

107. *M Archive*, as noted above, is "written in collaboration with the survivors, the far-into-the-future witnesses to the realities we are making possible or impossible with our present apocalypse" (xi).

108. Vermeulen and van den Akker, "Periodising the 2000s," 2.

109. Milton, *Paradise Lost*, Book II: 891–95.

110. Milton, *Paradise Lost*, Book II: 150–51.

111. Milton, *Paradise Lost*, Book II: 911–14. As John Rumrich writes in "Milton's God and the Matter of Chaos," "The ontological deficiency of chaos indicates instead a material potency that is the precondition of creation" (1043).

112. Gumbs, *M Archive*, 6.

113. Gumbs, *M Archive*, 6, 7.

114. In *Paradise Lost* as well, Chaos and eldest Night are the dark feminine counterpart of God's creative essence in which, according to John Rumrich, "God is the confused matter of chaos even as he is the creative virtue of light" and "chaos is God's womb, essential to his deity." Rumrich, "Milton's God and the Matter of Chaos," 1043.

115. Gumbs, *Dub*, 198.

116. Woolf, *Passionate Apprentice*, 203.

117. Clevidence, *~~Flung~~ Throne*, 23.

118. *Nemesis* is also the name of the Greek goddess of retribution, proportion, and justice.

119. Milton, *Paradise Lost*, Book IV: 657–58.

120. Milton, *Paradise Lost*, Book IV: 661–67.

121. Milton, *Paradise Lost*, Book IV: 656, 664.

122. Deleuze, "Pure Immanence," 26.

123. Deleuze, "Immanence: A Life," 26.

124. Eliot, *The Waste Land* V: 385–88.

125. In his foundational essay, "Apostrophe," Jonathan Culler writes of the vocative of poetic apostrophe: "We might then posit a third level of reading where the vocative of apostrophe is a device which the poetic voice uses to establish with an object a relationship which helps constitute him. The object is treated as a subject, an *I* which implies a certain type of *you* in its turn. One who successfully invokes nature is one to whom nature might, in its turn, speak" (63).

Bibliography

Adams, Carol J. *The Sexual Politics of Meat: A Feminist-Vegetarian Critical Theory*. New York: Continuum, 2010.

Adams, Mary. "Fallen Wombs: The Origins of Death in Miltonic Sexuality." *Milton Studies* 29 (1992): 165–79.

Adkins, Peter. *The Modernist Anthropocene: Nonhuman Life and Planetfary Change in James Joyce, Virginia Woolf, and Djuna Barnes.* Edinburgh: Edinburgh University Press, 2022.

Adorno, Theodor W. "Notes on Kafka." In *Prisms*, translated by Samuel Weber and Shierry Weber, 243–271 Cambridge, MA: MIT Press, 1983.

——. *Philosophy of New Music*. Translated by Robert Hullot-Kenner. Minneapolis: University of Minnesota Press, 2006.

Ahmed, Sara. *Queer Phenomenology: Orientations, Objects, Others*. Durham, NC: Duke University Press, 2006.

Alaimo, Stacy. *Bodily Natures: Science, Environment, and the Material Self.* Bloomington: Indiana University Press, 2010.

Allen, Garland E. "Julian Huxley and the Eugenical View of Human Evolution." In *Julian Huxley: Biologist and Statesman of Science*, ed. C. Kenneth Waters and Albert Van Helden, 193–222. Houston: Rice University Press, 1992.

Alt, Christina. *Virginia Woolf and the Study of Nature.* Cambridge: Cambridge University Press, 2010.

Amatya, Alok, and Ashley Dawson. "Literature in an Age of Extraction: An Introduction." *Modern Fiction Studies* 66, no. 1 (2020): 1–19.

Ames, Christopher. "Carnivalesque Comedy in *Between the Acts*." *Twentieth-Century Literature* 44, no. 4 (1998): 394–408.

Anderson, Mark M. "Sliding Down the Evolutionary Ladder? Aesthetic Autonomy in *The Metamorphosis*." In *The Metamorphosis: A Norton Critical Edition*, edited by Stanley Corngold, 154–71. New York: Norton, 1996.

Anderson, Nicole. "Pre- and Posthuman Animals: The Limits and Possibilities of Animal-Human Relations." In *Posthumous Life: Theorizing beyond the Posthuman*, edited by Jami Weinstein and Claire Colebrook, 17–42. New York: Columbia University Press, 2017.

Anderton, Joseph. "Vegetating Life and the Spirit of Modernism in Kafka and Beckett." *Modernism/modernity* 26, no. 4 (2019): 805–27.

Anker, Elizabeth. "Where Was Moses When the Candle Went Out? Infinity, Prophecy, and Ethics in Spinoza and 'Ithaca.'" *James Joyce Quarterly* 44, no. 4 (2007): 661–77.

Apollinaire, Guillaume. *Les mamelles de Tirésias; L'enchanteur pouirrissant; Couleur du temps.* Paris: Poesie / Gallimard, 1972.

Aristotle. *Poetics.* Translated by Ingram Bywater. In *The Complete Works of Aristotle, Vol II.* Princeton: Princeton University Press, 1984.

——. *Rhetoric.* Translated by J. H. Freese. *Aristotle in 23 Volumes, Vol. 22.* London: William Heinemann Ltd., 1926.

Armstrong, Tim. "Player Piano: Poetry and Sonic Modernity." *Modernism/modernity* 14, no. 1 (2007): 1–19.

Attali, Jacques. *Noise: The Political Economy of Music.* Manchester: Manchester University Press, 1977.

Attridge, Derek. "Unpacking the Portmanteau, or Who's Afraid of *Finnegans Wake*?" In *On Puns: The Foundation of Letters*, ed. Jonathan Culler, 140–55. Oxford: Basil Blackwell, 1988.

Bakhtin, Mikhail. *Rabelais and His World.* Translated by Helene Iswolsky. Bloomington: Indiana University Press, 1984.

Banfield, Ann. *The Phantom Table: Woolf, Fry, Russell and the Epistemology of Modernism.* Cambridge: Cambridge University Press, 2000.

Barad, Karen. *Meeting the Universe Halfway: Quantum Physics and the Entanglement of Matter and Meaning.* Durham, NC: Duke University Press, 2007.

——. "Nature's Queer Performativity." *Qui Parle* 19, no. 2 (2011): 121–58.

Baron, Scarlett. *"Strandentwining Cable": Joyce, Flaubert, and Intertextuality.* Oxford: Oxford University Press, 2012.

Baucom, Ian. "History 4°: Postcolonial Method and Anthropocene Time." *Cambridge Journal of Postcolonial Inquiry* 1, no. 1 (2014): 123–42.

Baudelaire, Charles. *The Painter of Modern Life.* Translated by Jonathan Mayne. London: Phaidon, 1995.

Bauer, Esther K. "The Power of the Look: Franz Kafka's 'The Cares of a Family Man.'" In *Kafka's Creatures: Animals, Hybrids, and Other Fantastic Beings*, ed. Marc Lucht and Donna Yarri, 157–74. Plymouth, UK: Lexington Books, 2010.

Beauchamp, Gorman. "*The Island of Dr. Moreau* as Theological Grotesque." *Papers on Language and Literature* 15, no. 4 (1979): 408–17.

Beckman, Frida. "Posthumanism and Narrativity: Beginning Again with Arendt, Derrida, and Deleuze." In *Posthumous Life: Theorizing beyond the Posthuman*, edited by Claire Colebrook and Jami Weinstein, 43–64. New York: Columbia University Press, 2017.

Beer, Gillian. "*Between the Acts*: Resisting the End," in *Virginia Woolf: The Common Ground.* Ann Arbor: University of Michigan Press, 1996.

——. *Darwin's Plots: Evolutionary Narrative in Darwin, George Eliot, and Nineteenth-Century Fiction.* Cambridge: Cambridge University Press, 2000.

Benjamin, Walter. "Franz Kafka: On the Tenth Anniversary of His Death." In *Selected Writings: Volume 2: 1927–1934.* Edited by Michael Jennings, Howard Eiland, and Gary Smith. Translated by Rodney Livingstone, 794–856. Cambridge, MA: Harvard University Press, 1999.

Bennett, Jane. *Vibrant Matter: A Political Ecology of Things.* Durham, NC: Duke University Press, 2010.

Bergson, Henry. *Laughter: An Essay on the Meaning of the Comic.* Translated by Cloudesley Brereton and Fred Rothwell. New York: Macmillan, 1914.

Bersani, Leo. "Flaubert: The Politics of Mystical Realism." *Massachusetts Review* 11, no. 1 (1970): 35–50.

Blattner, William. "Ontology, the A Priori, and the Primacy of Practice: An Aporia in Heidegger's Early Philosophy." In *Transcendental Heidegger*, edited by Steven Galt Crowell and Jeff Malpas, 10–27. Redwood City, CA: Stanford University Press.

Blum, Hester. "The Prospect of Oceanic Studies." *PMLA* 125, no. 3 (2010): 670–77.

Blumenberg, Hans. *Shipwreck with Spectator: Paradigm of a Metaphor for Existence*. Translated by Steven Rendall. Cambridge, MA: MIT Press, 1997.

Borgese, Elisabeth Mann. *The Oceanic Circle: Governing the Seas as a Global Resource*. Tokyo: United Nations University Press, 1998.

Bown, Nicola. "'Entangled Banks': Robert Browning, Richard Gadd, and the Darwinian Grotesque." In *Victorian Culture and the Idea of the Grotesque*, edited by Colin Todd, Paul Barlow, and David Amigoni, 119–39. Aldershot, UK: Ashgate, 1999.

Braidotti, Rosi. "Posthuman Critical Theory." *Journal of Posthuman Studies* 1, no. 1 (2017): 9–25.

——. *The Posthuman*. Cambridge: Polity, 2013.

Brayton, Dan. *Shakespeare's Ocean: An Ecocritical Exploration*. Charlottesville: University of Virginia Press, 2012.

Breton, André. *Manifesto of Surrealism*. In *Manifestos of Surrealism*. Translated by Richard Seaver and Helen R. Lane, 1–47. Ann Arbor: University of Michigan Press, 1972.

Brivic, Sheldon. *Joyce through Lacan and Žižek: Explorations*. New York: Palgrave Macmillan, 2008.

Brobjer, Thomas. *Nietzsche's Philosophical Context: An Intellectual Biography*. Urbana: University of Illinois Press, 2008.

Brogan, Jacqueline Vaught. *Stevens and Simile: A Theory of Language*. Princeton: Princeton University Press, 1986.

Brombert, Victor H. *Novels of Flaubert: A Study of Themes and Techniques*. Princeton: Princeton University Press, 2016.

Broswimmer, Franz. *Ecocide: A Short History of the Mass Extinction of Species*. London: Pluto, 2002.

Brown, Andrew. "'Un assez vague Spinozisme': Flaubert and Spinoza." *Modern Language Review* 91, no. 4 (1996): 848–65.

Brown, Jayna. *Black Utopias: Speculative Life and the Music of Other Worlds*. Durham, NC: Duke University Press, 2021.

Brown, Marshall. "Periods and Resistances." *Modern Language Quarterly* 62, no. 4 (2001): 309–16.

Brown, Richard. "Everything in 'Circe.'" In *Reading Joyce's "Circe,"* edited by Andrew Gibson, 222–40. Amsterdam: Rodopi, 1994.

Bruhn, Jutta-Annette. "The Magic Lantern Slide after *The Temptation of Saint Anthony* by Jacques Callot." *Journal of Glass Studies* 38 (1996): 268–76.

Brunton. James. "Whose (Meta)modernism? Metamodernism, Race, and the Politics of Failure." *Journal of Modern Literature* 41, no. 3 (2018): 60–76.

Budgen, Frank. *James Joyce and the Making of* Ulysses. London: Grayson and Grayson, 1934.

Butler, Judith. "Kafka and the Poetics of Non-arrival." Lecture presented at European Graduate School, October 2, 2011. Video recording, 1:25:24. https://egs.edu/lecture/judith-butler-kafka-and-the-poetics-of-non-arrival-2011/.

Butler, Shane. "Dogs and Phonographs." *Parallax* 26, no. 2 (2020): 151–62.

——. *The Ancient Phonograph*. New York: Zone Books, 2005.

Byrne, Alex. "Are Women Adult Human Females?" *Philosophical Studies* 177, no. 12 (2020): 3783–803.

Calarco, Matthew. *Zoographies: The Question of the Animal from Heidegger to Derrida*. New York: Columbia University Press, 2008.

Carroll, Lewis. Alice's Adventures in Wonderland *and* Through the Looking-Glass. London: Penguin, 1998.

——. *The Letters of Lewis Carroll, Volumes I and II*. Edited by Morton Cohen. New York: Oxford University Press, 1979.

Carver, Craig. "Moly: Bloom's Preservative; Correspondence and Function in *Ulysses*." *James Joyce Quarterly* 12, no. 4 (1975): 414–22.

Caughie, Pamela L. "Virginia Woolf: Radio, Gramophone, and Broadcasting." In *The Edinburgh Companion to Virginia Woolf and the Arts*, edited by Maggie Humm, 332–47. Edinburgh: Edinburgh University Press, 2010.

Cavarero, Adriana. *For More than One Voice: Toward a Philosophy of Vocal Expression*. Translated by Paul Kottman. Stanford: Stanford University Press, 2005.

Ceballos, Gerardo, Paul R. Ehrlich, Anthony D. Barnosky, Andrés García, Robert M. Pringle, and Todd M. Palmer. "Accelerated Modern Human–Induced Species Losses: Entering the Sixth Mass Extinction." *Science Advances* 1, no. 5 (2015): e1400253.

Cervenak, Sarah Jane. *Black Gathering: Art, Ecology, Ungiven Life*. Durham, NC: Duke University Press, 2021.

Chakrabarty, Dipesh. "The Climate of History: Four Theses." *Critical Inquiry* 35 (2009): 197–222.

Chen, Mel Y. *Animacies: Biopolitics, Racial Mattering, and Queer Affect*. Durham, NC: Duke University Press, 2012.

Cheng, Joyce. "Mask, Mimicry, Metamorphosis: Rogier Caillois, Walter Benjamin and Surrealism in the 1930s." *Modernism/modernity* 16, no. 1 (2009): 61–86.

Cheng, Vincent. *Joyce, Race, and Empire*. Cambridge: Cambridge University Press, 1995.

Chion, Michel. *Sound: An Acoulogical Treatise*. Translated by James A. Steintrager. Durham, NC: Duke University Press, 2015.

——. *The Voice in Cinema*. Translated by Claudia Gorbman. New York: Columbia University Press, 2019.

Choi, Charles Q. "Early Whales Had Legs." *Live Science*, September 11, 2008. https://www.livescience.com/7564-early-whales- legs.html.

Chow, Juliana. *Nineteenth-Century American Literature and the Discourse of Natural History*. Cambridge: Cambridge University Press, 2022.

Chun, Maureen. "Between Sensation and Sign: The Secret Language of *The Waves*." *Journal of Modern Literature* 36, no. 1 (2012): 53–70.

Cicero. *On the Orator (De Oratorio), Book III*. Translated by Harris Rackman. Cambridge, MA: Harvard University Press, 1942.

Cixous, Hélène. "At Circe's, or the Self-Opener." *Boundary 2* 3, no. 2 (1975): 387–97.

Clevidence, Cody-Rose. *Beast Feast*. Boise: Ahsahta Press, 2014.

——. *~~Flung~~ Throne*. Boise: Ahsahta Press, 2018.

Coetzee, John M. "Time, Tense and Aspect in Kafka's 'The Burrow.'" *MLN* 96, no. 3 (1981): 556–79.

Cohen, Morton N. *Lewis Carroll: A Biography*. New York: Knopf, 1995.

Cole, Sarah. *At the Violet Hour: Modernism and Violence in England and Ireland*. Oxford: Oxford University Press, 2012.

Colebrook, Claire, and Jami Weinstein, "Introduction: Critical Life Studies and the Problems of Inhuman Rites and Posthumous Life." In *Posthumous Life: Theorizing beyond the Posthuman*, edited by Claire Colebrook and Jami Weinstein, 1–14. New York: Columbia University Press, 2017.

——. "Preface: Postscript on the Posthuman," in *Posthumous Life*, op. cit.

Collingwood, Stuart. *The Life and Letters of Lewis Carroll*. London: T. Fisher Unwin, 1898.

Columbus, Christopher. "Journal of the First Voyage of Columbus." In *Journal of Christopher Columbus*, edited and translated by Clements R. Markham, 15–193. London: Hakluyt Society, 1893.

Cooper, Melinda. *Life as Surplus: Biotechnology and Capitalism in the Neoliberal Era*. Seattle: University of Washington Press, 2008.

Corngold, Stanley. *Franz Kafka: The Necessity of Form*. Ithaca: Cornell University Press, 1988.

——. "*The Metamorphosis*: Metamorphosis of the Metaphor." In *The Metamorphosis: A Norton Critical Edition*, edited by Stanley Corngold, 79–106. New York: Norton, 1996.

Crist, Eileen. "On the Poverty of Our Nomenclature." In *Anthropocene or Capitalocene: Nature, History, and the Crisis of Capitalism*, edited by Jason Moore, 14–33. Oakland, CA: PM Press, 2016.

Crutzen, Paul, and Eugene Stoermer. "The Anthropocene." *International Geosphere-Biosphere Programme Newsletter* 41 (2000): 17–18.

Cuddy-Keane, Melba. "The Politics of Comic Modes in Virginia Woolf's *Between the Acts*." *PMLA* 105, no. 2 (1990): 273–85.

——. "Virginia Woolf, Sound Technologies, and the New Aurality." In *Virginia Woolf in the Age of Mechanical Reproduction: Music, Cinema, Photography, and Popular Culture*, edited by Pamela L. Caughie, 69–96. New York: Garland, 2000.

Culler, Jonathan. "Apostrophe." *Diacritics* 7, no. 4 (1977): 59–69.

——. "The Call of the Phoneme: Introduction." In *On Puns: The Foundation of Letters*, edited by Jonathan Culler, 1–16. Oxford: Basil Blackwell, 1988.

Culler, Jonathan, ed. *On Puns: The Foundation of Letters*. Oxford: Basil Blackwell, 1988.

da Costa, Palmira Fontas. "The Meaning of Monstrosities in Charles Darwin's Understanding of the Origin of Species." In *Darwin, Evolution, Evolutionisms*, edited by Ana Leonor Pereira, Joao Rui Pita, and Pedro Ricardo Fonseca, 75–83. Coimbra: Coimbra University Press, 2011.

Danta, Chris. *Animal Fables after Darwin*. Cambridge: Cambridge University Press, 2018.

Darwin, Charles. *The Descent of Man, and Selection in Relation to Sex*. London: Penguin, 2004.
——. *The Origin of Species*. Peterborough, ON: Broadview, 2003.
da Silva, Denise Ferreira. *Unpayable Debt*. London: Sternberg Press, 2022.
Davidson, Donald. "What Metaphors Mean." *Critical Inquiry* 5, no. 1 (1978): 31–47.
Davis, Heather, and Zoe Todd. "On the Importance of a Date, or, Decolonizing the Anthropocene." *ACME: An International Journal for Critical Geographies* 16, no. 4 (2017): 761–80.
Dawson, Ashley. *Extinction: A Radical History*. New York: OR Books, 2016.
de Bruyn, Ben. *The Novel and the Multispecies Soundscape*. New York: Palgrave-Macmillan, 2020.
Deleuze, Gilles. "Coldness and Cruelty," in *Masochism*. Translated by Jean McNeil, 9-142. New York: Zone Books, 1989.
——. *Difference and Repetition*. Translated by Paul Patton. New York: Columbia University Press, 1994.
——. *Expressionism in Philosophy: Spinoza*. Translated by Martin Joughin. New York: Zone Books, 1992.
——. "Immanence: A Life," in *Pure Immanence: Essays on A Life*. Translated by Anne Boyman, 25–33. Brooklyn, NY: Zone Books, 2001.
——. *Spinoza: Practical Philosophy*. Translated by Robert Hurley. San Francisco: City Lights Books, 1988.
——. *The Logic of Sense*. Translated by Mark Lester. New York: Columbia University Press, 2004.
Deleuze, Gilles, and Felix Guattari. *A Thousand Plateaus: Capitalism and Schizophrenia*. Translated by Brian Massumi. Minneapolis: University of Minnesota Press, 1987.
——. *Kafka: Toward a Minor Literature*. Translated by Dana Polan. Minneapolis: University of Minnesota Press, 1986.
DeLoughrey, Elizabeth. *Allegories of the Anthropocene*. Durham, NC: Duke University Press, 2019.
——. "Submarine Futures of the Anthropocene." *Comparative Literature* 69, no. 1 (2017): 32–44.
Dembroff, Robin. "Beyond Binary: Genderqueer as Critical Gender Kind." *Philosophers' Imprint* 20, no. 9 (2020): 1–23.
——. "Escaping the Natural Attitude about Gender." *Philosophical Studies* 178 (2021): 983–1003.
Derrida, Jacques. "And Say the Animal Responded?" In *Zoontologies: The Question of the Animal*, edited by Cary Wolfe, 121–46. Minneapolis: University of Minnesota Press, 2003.
——. "Before the Law." In *Acts of Literature*. Edited by Derek Attridge. Translated by Avital Ronell and Christine Roulston, 181–220. New York: Routledge, 1992.
——. *Glas*. Translated by John P. Leavey. Lincoln: University of Nebraska Press, 1985.
——. "Plato's Pharmacy." In *Dissemination*, translated by Barbara Johnson, 61–171. London: Athlone, 1981.
——. "Structure, Sign, and Play in the Discourse of the Human Sciences." In *Writing and Difference*, translated by Alan Bass, 351–70. London: Routledge, 2001.

———. "The Animal That Therefore I Am (More to Follow)." Translated by David Wills. *Critical Inquiry* 28, no. 2 (2002): 369–418.

———. *Voice and Phenomenon: Introduction to the Problem of the Sign in Husserl's Philosophy*. Translated by Leonard Lawlor. Evanston, IL: Northwestern University Press, 2011.

———. "White Mythology: Metaphor in the Text of Philosophy." In *Margins of Philosophy*. Translated by Alan Bass, 207–71. Chicago: University of Chicago Press, 1982.

———. "White Mythology: Metaphor in the Text of Philosophy." Translated by Tim Moore. *New Literary History* 6, no. 1 (1974): 5–74.

DiBattista, Maria. *Virginia Woolf's Major Novels: The Fables of Anon*. New Haven: Yale University Press, 1980.

Dolar, Mladen. *A Voice and Nothing More*. Boston: MIT Press, 2006.

———. "The Burrow of Sound." *Differences: A Journal of Feminist Cultural Studies* 22, no. 2–3 (2011): 112–39.

Doyle, Laura, and Laura Winkiel. *Geomodernisms: Race, Modernism, Modernity*. Bloomington: Indiana University Press, 2005.

Driscoll, Kári. *Towards a Poetics of Animality: Hofmannstahl, Rilke, Pirandello, Kafka*. PhD diss., Columbia University, 2014.

Dubino, Jeanne. "The Bispecies Environment, Coevolution, and Flush." In *Virginia Woolf: Twenty-First Century Perspectives*, edited by Jeanne Dubino, Gill Lowe, Vara Neverow, and Kathryn Simpson, 131–47. Edinburgh: Edinburgh University Press, 2015.

Ebury, Katherine. "The Beast and the Sovereign in 'Circe': Human and Animal Rights in Joyce Studies." *Textual Practice* 36, no. 2 (2022): 223–41.

Edwards, Karen L., Derek Ryan, and Jane Spencer, eds. *Reading Literary Animals: Medieval to Modern*. New York: Routledge, 2020.

Elden, Stuart. "Reading *Logos* as Speech: Heidegger, Aristotle, and Rhetorical Politics." *Philosophy & Rhetoric* 38, no. 4 (2005): 281–301.

Eliot, Thomas Stearns. *The Waste Land*. In *The Complete Poems and Plays, 1909–1950*. New York: Harcourt, 1971.

Ellis, Steve. *British Writers and the Approach of World War II*. Cambridge: Cambridge University Press, 2015.

Empson, William. *Some Versions of Pastoral*. New York: New Directions, 1974.

Emrich, Wilhelm. *Franz Kafka: A Critical Study of His Writings*. New York: Frederick Ungar, 1968.

Faris, Wendy. "Bloomsbury's Beasts." *Yearbook of English Studies* 37, no. 1 (2007): 107–25.

Ferdinand, Malcom. *Decolonial Ecology: Thinking from the Caribbean World*. Translated by Anthony Paul Smith. Cambridge: Polity, 2022.

Flaubert, Gustave. *La tentation de Saint Antoine*. Paris: Garnier-Flammarion, 1967.

———. *The Temptation of Saint Anthony*. Translated by Lafcadio Hearn. New York: Modern Library, 2001.

Fleishman, Ian Thomas. "The Rustle of the Anthropocene: Kafka's Odradek as Ecocritical Icon." *Germanic Review: Literature, Culture, Theory* 92, no. 1 (2017): 40–62.

Flynn, Catherine. "'Circe' and Surrealism: Joyce and the Avant-Garde." *Journal of Modern Literature* 34, no. 2 (2011): 121–38.

——. *James Joyce and the Matter of Paris*. Cambridge: Cambridge University Press, 2019.

Fordham, Finn. "'Circe' and the Genesis of Multiple Personality." *James Joyce Quarterly* 45, no. 3–4 (2008): 507–520.

Foucault, Michel. Introduction to *The Temptation of Saint Anthony*, by Gustave Flaubert, xxiii–xliv. Translated by Donald F. Bouchard and Sherry Simon. New York: Modern Library, 2001.

——. *The Birth of the Clinic: An Archaeology of Medical Perception*. Translated by Alan Sheridan. New York: Vintage, 1994.

——. *The Order of Things. An Archaeology of the Human Sciences*. Translated by Alan Sheridan. New York: Vintage, 1994.

Freud, Sigmund. "A Difficulty in the Path of Psycho-Analysis." In *The Standard Edition of the Complete Psychological Works, Vol. 17*, edited by James Strachey, 135–44. London: Hogarth, 1953–56.

Friedman, Susan Stanford. *Planetary Modernisms: Provocations on Modernity across Time*. New York: Columbia University Press, 2015.

Froula, Christine. *Virginia Woolf and the Bloomsbury Avant-Garde: War, Civilization, Modernity*. New York: Columbia University Press, 2005.

Garcia, Edgar. *Signs of the Americas: A Poetics of Pictography, Hieroglyphics, and Khipu*. Chicago: University of Chicago Press, 2020.

Gaudreault, Andre. *From Plato to Lumière: Narration and Monstration in Literature and Cinema*. Translated by Timothy Barnard. Toronto: University of Toronto Press, 2009.

Geier, Ted. *Kafka's Nonhuman Form: Troubling the Boundaries of the Kafkaesque*. London: Palgrave Macmillan, 2016.

Gellen, Kata. *Kafka and Noise: The Discovery of Cinematic Sound in Literary Modernism*. Evanston, IL: Northwestern University Press, 2019.

——. "Noises Off: Cinematic Sound in Kafka's 'The Burrow.'" In *Mediamorphosis: Kafka and the Moving Image*, edited by Shai Biderman and Ido Lewit, 111–29. New York: Columbia University Press, 2016.

Gibson, Andrew. *Joyce's Revenge: History, Politics, and Aesthetics in* Ulysses. Oxford: Oxford University Press, 2002.

Gifford, Don. Ulysses *Annotated*. Berkeley: University of California Press, 1989.

Gigante, Denise. "Too Much Life: Romantic Monstrosity in *Frankenstein*." *Common Reader: A Journal of the Essay*, no. 8 (2018). https://commonreader.wustl.edu/c/too-much-life/.

Gilroy, Paul. *The Black Atlantic: Modernity and Double Consciousness*. London: Verso, 1993.

Giraud, Eva Haifa. *What Comes after Entanglement? Activism, Anthropocentrism, and an Ethics of Exclusion*. Durham, NC: Duke University Press, 2019.

Gladding, Jody. *Translations from Bark Beetle*. Minneapolis: Milkweed, 2014.

Glissant, Édouard. *The Ripening*. Translated by Frances Frenaye. New York: George Braziller, 1959.

Goldman, Jane. "Ce chien est à moi: Virginia Woolf and the Signifying Dog." *Woolf Studies Annual* 13 (2007): 49–86.

——. "The Dog That Therefore Woolf Follows." In *Virginia Woolf and the Natural World: Selected Papers from the Twentieth Annual International Conference on*

Virginia Woolf, edited by Kristin Czarnecki and Carrie Rohman, 125–32. Clemson, SC: Clemson University Digital Press, 2011.

Goodman, Paul. *Kafka's Prayer*. New York: Vanguard Press, 1947.

Gross, Ruth. "Kafka's Short Fiction." In *The Cambridge Companion to Kafka*, edited by Julian Preece, 80–94. Cambridge: Cambridge University Press, 2002.

Grusin, Richard. Introduction to *The Nonhuman Turn*, edited by Richard Grusin, vii–xxix. Minneapolis: University of Minnesota Press, 2015.

Gumbs, Alexis Pauline. *Dub: Finding Ceremony*. Durham, NC: Duke University Press, 2020.

——. *M Archive: After the End of the World*. Durham, NC: Duke University Press, 2018.

——. *Spill: Scenes of Black Feminist Fugitivity*. Durham: Duke University Press, 2016.

Gunning, Tom. "The Cinema of Attraction: Early Cinema, Its Spectator, and the Avant-Garde." *Wide Angle* 8, no. 3–4 (1986): 63–70.

Haraway, Donna. *Staying with the Trouble: Making Kin in the Chthulucene*. Durham, NC: Duke University Press, 2016.

——. "Symbiogenesis, Sympoiesis, and Art Science Activisms for Staying with the Trouble." In *Arts of Living on a Damaged Planet: Ghosts and Monsters of the Anthropocene*, edited by Anna Tsing, Heather Swanson, Elaine Gan, and Nils Bubandt, M25–50. Minneapolis: University of Minnesota Press, 2017.

——. *When Species Meet*. Minneapolis: University of Minnesota Press, 2008.

Harel, Naama. *Kafka's Zoopoetics: Beyond the Human-Animal Barrier*. Ann Arbor: University of Michigan Press, 2020.

Harpham, Geoffrey. *On the Grotesque: Strategies of Art and Contradiction in Art and Literature*. Aurora, CO: Davies Group, 2006.

Hayles, N. Katherine. *How We Became Posthuman: Virtual Bodies in Cybernetics, Literature, and Informatics*. Chicago: University of Chicago Press, 1999.

Hayman, Eleanor. "Future Rivers of the Anthropocene or Whose Anthropocene Is It? Decolonising the Anthropocene!" *Decolonization: Indigeneity, Education, and Society* 6, no. 2 (2018): 77–92.

Hayman, Ronald. *K: A Biography of Kafka*. London: Phoenix Press, 2001.

Hayot, Eric. *On Literary Worlds*. Oxford: Oxford University Press, 2012.

Hegel, G. W. F. *Aesthetics: Lectures on Fine Art, Volume 1*. Translated by Thomas Malcolm Knox. Oxford, Oxford University Press, 1975.

——. *Science of Logic*. Translated by Arnold Vincent Miller. Amherst, NY: Prometheus Books, 1998.

Heidegger, Martin. *The Fundamental Concepts of Metaphysics: World, Finitude, Solitude*. Translated by William MacNeill and Nicholas Walker. Bloomington: Indiana University Press, 1995.

Heine, Stefanie. "Forces of Unworking in Virginia Woolf's 'Time Passes.'" *Textual Cultures* 12, no. 1 (2019): 120–36.

Helmreich, Stefan. *Alien Ocean: Anthropological Voyages in Microbial Seas*. Berkeley: University of California Press, 2009.

Helms, Eleanor. "The Difficult Task of Being Real: Odradek, the Kittenlamb, and the Historical Individual." In *Kafka's Creatures: Animals, Hybrids, and Other Fantastic Beings*, edited by Marc Lucht and Donna Yarri, 81–100. Lanham, MD: Lexington Books, 2010.

Hemarajata, Peera. "Revisiting the Great Imitator: The Origin and History of Syphilis." *American Society for Microbiology*, June 17, 2019. https://asm.org/Articles/2019/June/Revisiting-the-Great-Imitator,-Part-I-The-Origin-a.

Herwig, Malte. "The Unwitting Muse: Jakob von Uexküll's Theory of Umwelt and Twentieth- Century Literature." *Semiotica* 134, no. 1 (2001): 553–93.

Hillel, Daniel. "Terra-aqua vitae: The Role of Soil and Water in Supporting Civilization." *Nutrient Cycling in Agroecosystems* 61 (2005): 231–35.

Hillmann, Heinz. "Das Sorgenkind Odradek." *Zeitschrift für deutsche Philologie* 86, no. 2 (1967): 197–210.

Hird, Myra J. "Naturally Queer." *Feminist Theory* 5, no. 1 (2004): 85–89.

Hoffmeyer, Jesper. *Biosemiotics: An Examination into the Signs of Life and the Life of Signs*. Scranton, PA: University of Scranton Press, 2009.

Homer. *Odyssey*. Translated by Robert Fagles. New York: Penguin, 1996.

Hovanec, Caroline. *Animal Subjects: Literature, Zoology, and British Modernism*. Cambridge: Cambridge University Press, 2018.

Howard, Jonathan. "Swim Your Ground: Towards a Black and Blue Humanities." *Atlantic Studies* 20, no. 2 (2023): 308–30.

Huber, Dezene. "Book Review: *Translations from Bark Beetle* by Jody Gladding." Retrieved 2014. http://blogs.ubc.ca/huber/2014/10/24/book-review-translations-for-bark-beetle-by-jody-gladding/.

Hunter, John. *Essays and Observations on Natural History, Anatomy, Physiology, Psychology, and Geology, Vol. I*. London: John Van Voorst, Paternoster Row, 1861.

Hunter, Ronald Henry, and Treve Gorben. "Intersexuality in Pigs: Clinical, Psychological, and Practical Considerations." *Acta Veterinaria Scandinavica* 37 (1996): 1–12.

Huxley, Francis. *The Raven and the Writing Desk*. New York: Harper and Row, 1976.

Huxley, Julian. *Ants*. New York: AMS Press, 1930.

——. *Essays of a Biologist*. New York: Knopf, 1923.

Ingersoll, Karin Amimoto. *Waves of Knowing: A Seascape Epistemology*. Durham, NC: Duke University Press, 2016.

Jackson, Zakiyyah Iman. *Becoming Human: Matter and Meaning in an Antiblack World*. New York: New York University Press, 2020.

Jaffe, Aaron. "Introduction: Who's Afraid of the Inhuman Woolf?" *Modernism/modernity* 23, no. 3 (2016): 491–513.

James, David. *Modernist Futures: Innovation and Inheritance in the Contemporary Novel*. Cambridge: Cambridge University Press, 2012.

James, David, and Urmila Seshagiri. "Metamodernism: Narratives of Continuity and Revolution." *PMLA* 129, no. 1 (2014): 87–100.

Jameson, Fredric. *A Singular Modernity: Essay on the Ontology of the Present*. London: Verso, 2002.

Jensen, J. Vernon, "Return to the Wilberforce-Huxley Debate." *British Journal for the History of Science* 21, no. 2 (1988): 161–79.

Joyce, James. *Collected Letters, Volume I*. Edited by Richard Ellmann. New York: Viking, 1966.

——. *Ulysses*. Edited by Hans Walter Gabler. New York: Vintage, 1984.

Kafka, Franz. *The Blue Octavo Notebooks*. Translated Ernst Kaiser and Eithne Wilkins. Cambridge, MA: Exact Change, 1991.

——. "The Burrow," in *The Complete Stories*. Translated by Willa and Edwin Muir. New York: Schocken, 1971.
——. "The Cares of a Family Man," in *The Complete Stories*. Translated by Edwin and Willa Muir. New York: Schocken, 1971.
——. *Diary: Vol. II, 1914–1923*. Translated by Martin Harry Greenberg. West Haven, CT: Pendulum Press, 1987.
——. *Die Verwandlung*. Teddington, UK: Echo Library, 2008.
——. "Investigations of a Dog," in *The Complete Stories*. Translated by Willa and Edwin Muir. New York: Schocken, 1971.
——. "Josephine the Singer, or the Mouse Folk," in *The Complete Stories*. Translated by Willa and Edwin Muir. New York: Schocken, 1971.
——. *Letters to Felice*. Translated by Elizabeth Duckworth and James Stern. New York: Penguin, 2016.
——. *The Metamorphosis, in The Complete Stories*. Translated by Willa and Edwin Muir. New York: Schocken, 1971.
——. *The Metamorphosis*. Translated by Stanley Corngold. New York: Norton, 1996.
——. "A Report to an Academy," in *The Complete Stories*. Translated by Willa and Edwin Muir. New York: Schocken, 1971.
——. *The Zürau Aphorisms,* in *The Aphorisms*. Translated by Willa Muir, Edwin Muir, and Michael Hoffman. New York: Schocken, 2015.
Kahn, Douglas. *Noise, Water, Meat: A History of Sound in the Arts*. Cambridge, MA: MIT Press, 1999.
Kane, Brian. *Sound Unseen: Acousmatic Sound in Theory and Practice*. Oxford: Oxford University Press, 2014.
Kant, Immanuel. *Critique of Judgment*. Translated by Werner S. Pluhar. Indianapolis: Hackett, 1987.
——. *Critique of Pure Reason*. Translated by Marcus Weigelt. London: Penguin, 2007.
Karera, Axelle. "Blackness and the Pitfalls of Anthropocene Ethics." *Critical Philosophy of Race* 7, no. 1 (2019): 32–56.
Karzakis, Katrina. *Fixing Sex: Intersex, Medical Authority, and Lived Experience*. Durham, NC: Duke University Press, 2008.
Katz, Cheryl. "Small Pests, Big Problems: The Global Spread of Bark Beetles." *Yale Environment 360*, September 21, 2017. https://e360.yale.edu/features/small-pests-big-problems-the-global-spread-of-bark-beetles.
Kendall-Morwick, Karalyn. "Mongrel Fiction: Canine Bildung and the Feminist Critique of Anthropocentrism in Woolf's *Flush*." *Modern Fiction Studies* 60, no. 3 (Fall 2014): 506–26.
Kheshti, Roshanak. *Modernity's Ear: Listening to Race and Gender in World Music*. New York: New York University Press, 2015.
Kil, Hye Ryoung. "Soap Advertisements in *Ulysses*: The Brooke's Monkey Brand Ad and the Capital Couple." *James Joyce Quarterly* 47, no. 3 (2010): 417–26.
King, Tiffany Lethabo. *The Black Shoals: Offshore Formations of Black and Native Studies*. Durham, NC: Duke University Press, 2019.
Kirby, John T. "Aristotle on Metaphor." *American Journal of Philology* 118, no. 4 (1997): 517–54.
Kirksey, Eben, ed. *The Multispecies Salon*. Durham, NC: Duke University Press, 2014.

Kittler, Friedrich. *Gramophone, Film, Typewriter.* Translated by Geoffrey Winthrop-Young and Michael Wutz. Stanford: Stanford University Press, 1999.
Kohn, Eduardo. "Forest Forms and Ethical Life." *Environmental Humanities* 14, no. 2 (2022): 401–18.
——. *How Forests Think: Towards an Anthropology Beyond the Human*. Berkeley: University of California Press, 2013.
Kristeva, Julia. *Powers of Horror: An Essay on Abjection*. Translated by Leon S. Roudiez. New York: Columbia University Press, 1982.
——. *Revolution in Poetic Language*. Translated by Margaret Waller. New York: Columbia University Press: 1984.
LaFleur, Greta. *The Natural History of Sexuality in Early America*. Baltimore: Johns Hopkins University Press, 2020.
Lakoff, George, and Mark Johnson. *Metaphors We Live By*. Chicago: University of Chicago Press, 2003.
Latour, Bruno. *Politics of Nature: How to Bring the Sciences into Democracy*. Translated by Catherine Porter. Cambridge, MA: Harvard University Press, 2004.
——. *We Have Never Been Modern*. Cambridge: MA: Harvard University Press, 1991.
Lecercle, Jean-Jacques. *Philosophy of Nonsense: The Intuitions of Victorian Nonsense Literature*. London: Routledge, 1994.
Leibniz, Gottfried Wilhelm. *Philosophical Essays*. Translated by Rogier Ariew and Daniel Garber. Indianapolis: Hackett, 1989.
Leland, Blake. "An Abode of Bliss: Plumtree's Potted Meat and the Allegory of the Theologians." *James Joyce Quarterly* 52, no. 1 (2014): 37–53.
Leong, Diana. "The Mattering of Black Lives: Octavia Butler's Hyperempathy and the Promise of the New Materialisms." *Catalyst: Feminism, Theory, Technology* 2, no. 2 (2016): 1–35.
Leopold, Aldo. *A Sand County Almanac and Sketches Here and There*. New York: Oxford University Press, 1968.
Liboiron, Max. *Pollution Is Colonialism*. Durham, NC: Duke University Press, 2021.
Liddell, Henry George, and Robert Scott. *A Greek-English Lexicon*. Edited by Henry Stuart Jones. Oxford: Clarendon, 1940. https://www.perseus.tufts.edu/hopper/.
Lippit, Akira Mizuta. *Electric Animal: Towards a Rhetoric of Wildlife.* Minneapolis: University of Minnesota Press, 2000.
Little, Judy. *Comedy and the Woman Writer: Woolf, Spark, and Feminism*. Lincoln, NE: University of Nebraska Press, 1983.
Louvel, Liliane. "Ecrire la couleur: Un défi poétique histoires de bleu et de vert." *Interfaces* 33 (2012): 243–55.
Low, Philip. *The Cambridge Declaration on Consciousness.* Edited by Jaak Panksepp, Diana Reiss, David Edelman, Bruno Van Swinderen, Philip Low, and Christof Koch. Francis Crick Memorial Conference on Consciousness in Human and non-Human Animals, Churchill College, University of Cambridge. July 7, 2012.
Lyotard, Jean-François. *The Inhuman: Reflections on Time*. Translated by Geoffrey Bennington and Rachel Bowlby. Stanford: Stanford University Press, 1991.
Mahaffey, Vicki. *Reauthorizing Joyce*. New York: Cambridge University Press, 1988.

Marder, Michael. *Plant-Thinking: A Philosophy of Vegetal Life*. New York: Columbia University Press, 2013.

Martin, Wayne. "In Defense of Bad Infinity: A Fichtean Response to Hegel's *Differenzschrift*." *Hegel Bulletin* 28, no. 1–2: 168–87.

Marx, Karl. *Economic and Philosophic Manuscripts of 1844*. Translated by Martin Milligan. New York: Dover, 2007.

Massumi, Brian. "The Supernormal Animal." In *The Nonhuman Turn*, edited by Richard Grusin, 1–18. Minneapolis: University of Minnesota Press, 2015.

——. *What Animals Teach Us about Politics*. Durham, NC: Duke University Press, 2014.

Mazumdar, Pauline. *Eugenics, Human Genetics, and Human Failings*. New York: Routledge, 1992.

McCarthy, Jeffrey M. *Green Modernism: Nature and the English Novel, 1900 to 1930*. New York: Palgrave-Macmillan, 2015.

McCole, John. *Walter Benjamin and the Antinomies of Tradition*. Ithaca: Cornell University Press, 1993.

McElroy, Bernard. *Fiction of the Modern Grotesque*. London: Palgrave-Macmillan, 1989.

McFeely R. A., W. C. Hare, and J. D. Biggers. "Chromosome Studies in 14 Cases on Intersex in Domestic Mammals." *Cytogenetics* 6, no. 3 (1967): 242–53.

McGurl, Mark. "The Posthuman Comedy." *Critical Inquiry* 38, no. 3 (2012): 533–53.

McHugh, Susan. *Animal Stories: Narrating across Species Lines*. Minneapolis: University of Minnesota Press, 2011.

Menley, Tobias. *Climate and the Making of Worlds: Toward a Geohistorical Poetics*. Chicago: University of Chicago Press, 2021.

Miller, Elaine. *The Vegetative Soul: From Philosophy of Nature to Subjectivity in the Feminine*. Albany: State University of New York Press, 2002.

Milton, John. *Paradise Lost*. Edited by John Leonard. New York: Penguin, 2000.

Moe, Aaron. "Towards a Poetics of Gaia: Biosemiotics and Jody Gladding's *Translations from Bark Beetle*." *ISLE: Interdisciplinary Studies in Literature and Environment* 25, no. 3 (2018): 450–67.

Moore, Jason. "The Rise of Cheap Nature." In *Anthropocene or Capitalocene: Nature, History, and the Crisis of Capitalism*, edited by Jason Moore, 78–115. Oakland, CA: PM Press, 2016.

Mortimer-Sandilands, Catriona. "Fear of a Queer Plant?" *Gay and Lesbian Quarterly* 23, no. 3 (2017): 419–29.

Mortimer-Sandilands, Catriona, and Bruce Erickson, eds. *Queer Ecologies: Sex, Nature, Politics, Desire*. Bloomington: Indiana University Press, 2010.

Morton, Timothy. *Dark Ecology: For a Logic of Future Coexistence*. New York: Columbia University Press, 2016.

——. *Ecology without Nature*. Cambridge, MA: Harvard University Press, 2007.

Moshenberg, Daniel. "The Capital Couple: Speculating on *Ulysses*." *James Joyce Quarterly* 25, no. 3 (1988): 333–47.

Moten, Fred. *In the Break: The Aesthetics of the Black Radical Tradition*. Minneapolis: University of Minnesota Press, 2003.

Mundy, Rachel. *Animal Musicalities: Birds, Beasts, and Evolutionary Listening*. Middletown, CT: Wesleyan University Press, 2018.

Murphy, Michelle. "Alterlife and Decolonial Chemical Relations." *Cultural Anthropology* 32, no. 4 (2017): 494–503.

Nakayasu, Sawako. *The Ants*. Los Angeles: Les Figues Press, 2014.

Nancy, Jean-Luc. *Listening*. Translated by Charlotte Mandell. New York: Fordham University Press, 2007.

——. "Listening to Philosophy, Silence, and Self." Interview with *Four by Three Magazine*, October 30, 2016. https://www.christine-jakobson.squarespace.com/issue/silence/listening-to-jean-luc-nancy.

Napolin, Julie Beth. *The Fact of Resonance: Modernist Acoustics and Narrative Form*. New York: Fordham University Press, 2020.

Nealon, Jeffrey T. *Plant Theory: Biopower and Vegetable Life*. Stanford: Stanford University Press, 2016.

Neverow, Vera. "The Woolf, the Horse, and the Fox: Recurrent Motifs in *Jacob's Room* and *Orlando*." In *Virginia Woolf and the Natural World: Selected Papers from the Twentieth Annual International Conference on Virginia Woolf*, edited by Kristin Czarnecki and Carrie Rohman, 116–24. Clemson, SC: Clemson University Digital Press, 2011.

Nietzsche, Friedrich. "On Truth and Lie in an Extra-Moral Sense." In *The Portable Nietzsche*, translated by Walter Kaufmann, 42–46. New York: Viking, 1968.

——. "On Truth and Lying in a Non-moral Sense." Translated by Ronald Speirs. In *The Norton Anthology of Theory and Criticism*, edited by Vincent Leitch, 764–74. New York: Norton, 2001.

——. *The Will to Power*. Translated by Walter Kaufmann and R. J. Hollingdale. New York: Vintage, 1968.

Nixon, Rob. *Slow Violence and the Environmentalism of the Poor.* Cambridge, MA: Harvard University Press, 2011.

Norris, Margot. *Beasts of the Modern Imagination: Darwin, Nietzsche, Ernst, Kafka, and Lawrence*. Baltimore: Johns Hopkins University Press, 1985.

——. "Kafka's Hybrids: Thinking Animals and Mirrored Humans." In *Kafka's Creatures: Animals, Hybrids, and Other Fantastic Beings*, edited Marc Lucht and Donna Yarri, 17–32. Lanham, MD: Lexington Books, 2010.

North, Paul. *The Yield: Kafka's Atheological Reformation*. Stanford: Stanford University Press, 2015.

O'Key, Dominic. "Animal Borderlands: An Introduction." *Parallax* 25, no. 4 (2019): 351–57.

Osteen, Mark. *The Economy of* Ulysses*: Making Both Ends Meet*. Syracuse, NY: Syracuse University Press, 1995.

Pawel, Ernst. *The Nightmare of Reason: A Life of Franz Kafka*. New York: Farrar, Strauss, and Giroux, 1992.

Peake, Charles. *James Joyce, the Citizen and Artist*. Stanford: Stanford University Press, 1977.

Peters, John Durham. *The Marvelous Clouds: Towards a Philosophy of Elemental Media*. Chicago: University of Chicago Press, 2015.

Pettman, Dominic. *Sonic Intimacy: Voice, Species, Technics*. Stanford: Stanford University Press, 2017.

Povinelli, Elizabeth A. *Between Gaia and Ground: Four Axioms of Existence and the Ancestral Catastrophe of Late Liberalism*. Durham, NC: Duke University Press, 2021.

——. *Geontologies: A Requiem to Late Liberalism*. Durham, NC: Duke University Press, 2016.

Pridmore-Brown, Michele. "1939–40: Of Virginia Woolf, Gramophones, and Fascism." *PMLA* 113, no. 3 (1998): 408–21.

Pryor, Sean. "Inhuman Words: Philology, Modernism, Poetry." *Modernism/modernity* 23, no. 3 (2016): 491–513.

Quintilian. *Institutio Oratorio*. Translated by Donald A. Russell. Cambridge, MA: Harvard University Press, 2002.

Rabaté, Jean-Michel. *James Joyce, Authorized Reader*. Baltimore: Johns Hopkins University Press, 1991.

Rabelais, François. *Gargantua and Pantagruel*. Translated by Michael Andrew Screech. New York: Penguin, 2006.

Rackin, Donald. *Alice's Adventures in Wonderland and Through the Looking-Glass: Nonsense, Sense, and Meaning*. New York: Twayne, 1991.

Raleigh, John Henry. "Bloom as a Modern Epic Hero." *Critical Inquiry* 3, no. 3 (1977): 583–98.

Rando, David. "The Cat's Meow: *Ulysses*, Animals, and the Veterinary Gaze." *James Joyce Quarterly* 46, no. 3–4 (2009): 529–43.

Redfearn, Graham. "We Are Approaching the Trumpocene, a New Epoch Where Climate Change Is Just a Big Scary Conspiracy." *The Guardian*, October 21, 2016. https://www.theguardian.com/environment/planet-oz/2016/oct/21/we-are-approaching-the-trumpocene-a-new-epoch-where-climate-change-is-just-a-big-scary-conspiracy.

Reynaga, Tahia Thaddeus. "Agents of the Forgotten: Animals as the Vehicles of Shame in Kafka." In *Kafka's Creatures: Animals, Hybrids, and Other Fantastic Beings*, edited by Marc Lucht and Donna Yarri, 67–80. Lanham, MD: Lexington Books, 2010.

Richards, Ivor Armstrong. *The Philosophy of Rhetoric*. New York: Oxford University Press, 1936.

Richardson, John. "Nietzsche contra Darwin." *Philosophy and Phenomenological Research* 65, no. 3 (2002): 537–75.

Ricoeur, Paul. "The Metaphorical Process as Cognition, Imagination, and Feeling." *Critical Inquiry* 5, no. 1 (1978): 143–59.

Rilke, Theodor Maria von. Duino Elegies *and* The Sonnets to Orpheus. Translated by Stephen Mitchell. New York: Vintage, 2009.

Rohman, Carrie. *Stalking the Subject: Modernism and the Animal*. New York: Columbia University Press, 2008.

Rose, Deborah Bird. "Multispecies Knots of Ethical Time." *Environmental Philosophy* 9, no. 1 (2012): 127–40.

Rubenstein, Michael, and Justin Neuman, eds. *Modernism and Its Environments*. London: Bloomsbury Academic, 2020.

Rumrich, John. "Milton's God and the Matter of Chaos." *PMLA* 110, no. 5 (1995): 1035–46.

Russolo, Luigi. *The Art of Noises*. Translated by Robert Filliou. N.p.: Ubu Classics, 2004. http://www.artype.de/Sammlung/pdf/russolo_noise.pdf.

Ryan, Derek. "Posthumanist Interludes: Ecology and Ethology in *The Waves*." In *Virginia Woolf: Twenty-First Century Perspectives*, edited by Jeanne Dubino, Gill

Lowe, Vara Neverow, and Kathryn Simpson, 148–66. Edinburgh: Edinburgh University Press, 2015.

——. "'The Reality of Becoming': Deleuze, Woolf, and the Territory of Cows." *Deleuze Studies* 7, no. 4 (2013): 537–61.

——. *Virginia Woolf and the Materiality of Theory: Sex, Animal, Life.* Edinburgh: Edinburgh University Press, 2013.

——. "'Was it Flush, or Was it Pan?': Virginia Woolf, Ethel Smyth, and Canine Biography." In *Reading Literary Animals*, edited by Karen L. Edwards, Derek Ryan, and Jane Spencer, 264–81. New York: Routledge, 2019.

Saint-Amour, Paul. "Introduction: Weak Theory, Weak Modernism." *Modernism/modernity* 25, no. 3 (2018): 437–59.

——. *Tense Future: Modernism, Total War, Encyclopedic Form.* Oxford: Oxford University Press, 2015.

Sankar, Nandini Ramesh. "'Potatosoap': Moly in James Joyce's *Ulysses*." *ANQ: A Quarterly Journal of Short Articles, Notes, and Reviews* (2022). https:doi.org/10.1080/0895769X.2022.2059752.

Schaeffer, Pierre. "Acousmatics." In *Audio Culture: Readings in Modern Music*, edited by Christoph Cox and Daniel Warner, 95–102. New York: Bloomsbury, 2017.

——. *Treatise on Musical Objects.* Translated by Christine North and John Dack. Oakland: University of California Press, 2017.

Schechner, Mark. *Joyce in Nighttown: A Psychoanalytic Inquiry into Ulysses*. Berkeley: University of California Press, 1974.

Schuster, Aaron. "Enjoy Your Security: On Kafka's 'The Burrow.'" *e-flux*, no. 113 (November 2020). https://www.e-flux.com/journal/113/359952/enjoy-your-security-on-kafka-s-the-burrow/.

Scott, Bonnie Kime. 'The Subversive Mechanics of Gramophone.' In *Virginia Woolf in the Age of Mechanical Reproduction*, edited by Pamela L. Caughie, 97–114. New York: Garland, 2000.

See, Sam. "The Comedy of Nature: Darwinian Feminism in Virginia Woolf's *Between the Acts*." *Modernism/modernity* 17, no. 3 (2010): 639–67.

Serres, Michel. *The Five Senses: A Philosophy of Mingled Bodies*. Translated by Margaret Sankey and Peter Cowley. London: Bloomsbury, 2016.

Sewell, Elizabeth. *The Field of Nonsense.* London: Chatto and Windus, 1952.

Seznec, Jean. "Flaubert and the Graphic Arts." *Journal of the Warburg and Courtauld Institutes* 8 (1945): 175–90.

Sharpe, Christina. *In the Wake: On Blackness and Being*. Durham, NC: Duke University Press, 2016.

Shaughnessy, Brenda. *The Octopus Museum: Poems*. New York: Knopf Doubleday, 2021.

Shechner, Mark. *Joyce in Nighttown: A Psychoanalytic Inquiry into* Ulysses. Berkeley: University of California Press, 1974.

Shelley, Percy. "A Defence of Poetry." In *Shelley's Poetry and Prose*, edited by Donald H. Reiman and Neil Fraistat, 478–508. New York: Norton, 2002.

Shukin, Nicole. *Animal Capital: Rendering Life in Biopolitical Times*. Minneapolis: University of Minnesota Press, 2009.

Silver, Brenda. "'Anon' and 'The Reader': Woolf's Last Essays." *Twentieth Century Literature* 25, no. 3/4 (1979): 356–441.

Simpson, Kathryn. "'Lappin and Lapinova': A Woolf in Hare's Clothing?" In *Virginia Woolf and the Natural World: Selected Papers from the Twentieth Annual International Conference on Virginia Woolf*, edited by Kristin Czarnecki and Carrie Rohman, 151–56. Clemson, SC: Clemson University Digital Press, 2011.

Smith, Jacob. *Eco-Sonic Media*. Oakland: University of California Press, 2015.

Smith, W. John. "Dodgson's Scientific Interests with Reference to the Books in His Library." *Jabberwocky: The Journal of the Lewis Carroll Society* 14, no. 1 (1984/85): 3–9.

Smythe, Karen. "Virginia Woolf's Elegiac Enterprise." *Novel: A Forum on Fiction* 26, no. 1 (1992): 64–79.

Snyder, Verne P. "Kafka's 'Burrow': A Speculative Analysis." *Twentieth-Century Literature* 27, no. 2 (1981): 113–26.

Sokel, Walter. "Kafka's Poetics of the Inner Self." *Modern Australian Literature* 11, no. 3–4 (1978): 37–58.

——. *The Myth of Power and the Self: Essays on Franz Kafka*. Detroit: Wayne State University Press, 2002.

Spencer, Herbert. *The Principles of Biology*. London: Williams and Norgate, 1864.

Spinoza, Benedict de. *Ethics*. Translated by Edwin Curley. London: Penguin, 1996.

Stallman, Robert C. Divine Hospitality in the Pentateuch: A Metaphorical Perspective on God as Host. PhD diss., Westminster Theological Seminary, 1999.

Stern, Jeffrey, ed. *Lewis Carroll's Library: A Facsimile Edition of the Catalogue of the Auction Sale following C. L. Dodgson's Death in 1898*. Silver Spring, MD: Lewis Carroll Society of North America, 1981.

Stine, Peter. "Franz Kafka and Animals." *Contemporary Literature* 22, no. 1 (1981): 58–80.

Stroumsa, Guy G. "Myth as Enigma: Cultural Hermeneutics in Late Antiquity." In *Untying the Knot: On Riddles and Other Enigmatic Modes*, edited by Galit Hasan-Rokem and David Shulman, 271–83. New York: Oxford University Press, 1996.

Sultzbach, Kelly. *Ecocriticism in the Modernist Imagination*. Cambridge: Cambridge University Press, 2016.

Sunder Rajan, Kaushik. *Biocapital: The Constitution of Postgenomic Life*. Durham, NC: Duke University Press, 2006.

Sussman, Henry. "The All-Embracing Metaphor: Reflections on 'The Burrow.'" *Glyph: Johns Hopkins Textual Studies* 1 (1977): 99–131.

——. "The Calculable, the Incalculable, and the Rest: Kafka's Virtual Environment." *Modern Language Notes* 127, no. 5 (2012): 1144–70.

TallBear, Kim. "Caretaking Relations, Not American Dreaming." *Kalfou: A Journal of Comparative and Relational Ethnic Studies* 6, no. 1 (2019): 24–41.

Taylor, Roger, and Edward Wakeling. *Lewis Carroll: Photographer*. Princeton: Princeton University Press, 2002.

Taylor, Sunaura. *Beasts of Burden: Animal Disability and Liberation*. New York: New Press, 2017.

——. "Beasts of Burden: Disability Studies and Animal Rights." *Qui Parle* 19, no. 2 (2011): 191–222

Thordarson, Thorvaldur, and Stephen Self. "Atmospheric and Environmental Effects of the 1783–1784 Laki Eruption: A Review and Reassessment." *Journal of Geophysical Research* 108, D1 (2003): 1–27.

Timbs, John, ed. *Arcana of Science and Art: Or an Annual Register of Useful Inventions and Improvements.* London: John Limbird, 1831.

Todd, Zoe. "An Indigenous Feminist's Take on the Ontological Turn: 'Ontology' Is Just Another Word for Colonialism." *Journal of Historical Sociology* 29, no. 1 (2016): 4–22.

Tromanhauser, Vicki. "Animal Life and Human Sacrifice in Virginia Woolf's *Between the Acts*." *Woolf Studies Annual* 15 (2009): 67–90.

Tsing, Anna. *The Mushroom at the End of the World: On the Possibility of Life in Capitalist Ruins*. Princeton: Princeton University Press, 2015.

Tsing, Anna, Heather Swanson, Elaine Gain, and Nils Bubadnt, eds. *Arts of Living on a Damaged Planet: Ghosts and Monsters of the Anthropocene*. Minneapolis: University of Minnesota Press, 2017.

Tung, Charles M. "Baddest Modernism: The Scales and Lines of Inhuman Time." *Modernism/modernity* 23, no. 3 (2016): 515–38.

Vaiman, Daniel, and Eric Pailhoux. "Mammalian Sex Reversal and Intersexuality: Deciphering the Sex-Determination Cascade." *Trends in Genetics* 16, no. 11 (2000): 488–94.

Valente, Joseph. *James Joyce and the Problem of Justice: Negotiating Sexual and Colonial Difference*. Cambridge, Cambridge University Press, 1995.

——. *The Myth of Manliness in Irish Culture: 1880–1922*. Urbana: University of Illinois Press, 2011.

van Dooren, Thom, Eben Kirksey, and Ursula Münster, Introduction to Special Issue, *Multispecies Studies: Cultivating Arts of Attentiveness*, *Environmental Humanities* 8, no. 1 (2016): 1–23.

Vermeulen, Timotheus, and Robin van den Akker. "Notes on Metamodernism." *Journal of Aesthetics & Culture* 2, no. 1 (2017). https://doi.org/10.3402/jac.v2i0.5677.

——. "Periodising the 2000s." In *Metamodernism: History, Affect, and Depth after Postmodernism*, edited Robin van den Akker, Alison Gibbons, and Timotheus Vermeulen, 1–20. London: Rowman and Littlefield, 2017.

Vialles, Noelle. *Animal to Edible*. Translated by J. A. Underwood. Cambridge: Cambridge University Press, 1994.

von Sacher-Masoch, Leopold. *Venus in Furs*. Translated by Jean McNeil. In *Masochism*, 143–293. New York: Zone Books, 1991.

von Uexküll, Jakob. *A Foray into the Worlds of Animals and Humans: With a Theory of Meaning*. Translated by Joseph D. O'Neil. Minnesota: University of Minneapolis Press, 2010.

Waldman, Anne. *Manatee/Humanity*. London: Penguin, 2009.

Wales, Katie. "'Bloom Passes through Several Walls': The Stage Directions in 'Circe.'" In *Reading Joyce's "Circe,"* edited by Andrew Gibson, 241–76. Amsterdam: Rodopi, 1994.

Wallace, Jeff. "Introduction: Difficulty and Defamiliarisation." In *Charles Darwin's Origin of Species: New Interdisciplinary Essays*, edited by David Amigoni and Jeff Wallace, 1–46. Manchester: Manchester University Press, 1995.

Wegenbach, Klaus. *Kafka*. Translated by Ewald Osers. London: Haus Publishing, 2003.

Weil, Kari. *Thinking Animals: Why Animal Studies Now?* New York: Columbia University Press, 2012.

Weindling, Paul. "Julian Huxley and the Continuity of Eugenics in Twentieth-Century Britain." *Journal of Modern European History* 10, no. 4 (2012): 480–99.

Wells, Herbert George. *The Outline of History: Being a Plain History of Life and Mankind*. New York: Macmillan, 1921.

Westling, Louise. "Virginia Woolf and the Flesh of the World." *New Literary History* 30, no. 4 (1999): 855–75.

Wheeler, Wendy. "The Biosemitic Turn: Abduction, or, the Nature of Creative Reason in Nature and Culture." In *Ecocritical Theory: New European Approaches*, edited by Axel Goodbody and Kate Rigby, 270–82. Charlottesville: University of Virginia Press, 2011.

Wheelwright, Philip. *Metaphor and Reality*. Bloomington: Indiana University Press, 1962.

White, Gilbert. *The Natural History & Antiquities of Selborne in the County of Southampton*. London: Macmillan, 1900.

Wiegand, Hermann. "Franz Kafka's 'The Burrow' (*Der Bau*): An Analytical Essay." *PMLA* 87 (1972): 152–66.

Winkelman, John. "Kafka's 'Forschungen eines Hundes.'" *Monatshefte* 59, no. 3 (1967): 204–16.

Wolfe, Cary. *What Is Posthumanism?* Minneapolis: University of Minnesota Press, 2010.

Wood, Alice. *Virginia Woolf's Late Cultural Criticism: The Genesis of The Years, Three Guineas, and Between the Acts.* London: Bloomsbury Academic, 2013.

Woolf, Jenny. *The Mystery of Lewis Carroll*. New York: St. Martin's, 2010.

Woolf, Virginia. "Anon," in "Woolf's Last Essays," edited by Brenda Silver. *Twentieth Century Literature* 25, no. 3/4 (1979): 356-441.

——. *Between the Acts*. San Diego: Harcourt Brace, 1970.

——. *The Diary of Virginia Woolf, Vol. III, 1925-30*. Edited by Anne Olivier Bell. San Diego: Harvest, 1981.

——. *The Diary of Virginia Woolf, Vol. IV, 1931–35*. Edited by Anne Olivier Bell. San Diego: Harvest, 1982.

——. *The Diary of Virginia Woolf, Vol. V, 1936–41*. Edited by Anne Olivier Bell. San Diego: Harvest, 1984.

——. *Flush*. San Diego: Harcourt Brace, 1983.

——. "The Mark on the Wall." In *The Complete Shorter Fiction of Virginia Woolf*, edited by Susan Dick, 83–89. San Diego: Harvest, 1989.

——. "Modern Fiction." In *The Common Reader: Second Edition*, edited by Andrew McNeillie, 146–54. London: Hogarth Press, 1925.

——. *A Passionate Apprentice: The Early Journals 1897–1909*. Edited by Mitchell A. Leaska. San Diego: Harcourt Brace, 1990.

——. *To the Lighthouse*. Orlando: Harcourt Brace, 1989.

——. "The Value of Laughter." In *Collected Essays Vol. I*, edited by Andrew McNeillie, 58–60. London: Hogarth, 1967.

——. *The Waves*. San Diego: Harcourt Brace, 2006.

——. "White's Selborne." In *The Essays of Virginia Woolf, Volume Six (1933–41)*, edited by Stuart Clarke, 189–94. London: Hogarth, 2011.

Wynter, Sylvia. "Unsettling the Coloniality of Being/Power/Truth/Freedom: Towards the Human, After Man, Its Overrepresentation—An Argument." *CR: The New Centennial Review* 3, no. 3 (2003): 257–337.

Yeats, William Butler. "The Song of Wandering Aengus." In *The Collected Poems of W. B. Yeats*, edited by Richard J. Finneran, 59. New York: Simon and Schuster, 1996.

Yusoff, Kathryn. *A Billion Black Anthropocenes or None.* Minneapolis: University of Minnesota Press, 2019.

——. "Geologic Realism: On the Beach of Geologic Time." *Social Text* 37, no. 1 (2019): 1–26.

Zilcosky, John. "'Samsa war Reisender': Trains, Trauma, and the Unreadable Body." In *Kafka for the Twenty-First Century*, edited by Stanley Corngold and Ruth Gross, 179–206. Rochester, NY: Camden House, 2011.

Zwerdling, Alex. *Virginia Woolf and the Real World.* Berkeley: University of California Press, 1986.

INDEX